DESIGNING SOCIAL SYSTEMS IN A CHANGING WORLD

Contemporary Systems Thinking

Series Editor: Robert L. Flood
University of Hull
Hull, United Kingdom

DESIGNING SOCIAL SYSTEMS IN A CHANGING WORLD

Bela H. Banathy

International Systems Institute
Carmel, California and
Saybrook Graduate School
San Francisco, California

PLENUM PRESS • NEW YORK AND LONDON

Library of Congress Cataloging-in-Publication Data

Banathy, Bela H.
    Designing social systems in a changing world / Bela H. Banathy.
        p.   cm. -- (Contemporary systems thinking)
    Includes bibliographical references (p.) and index.
    ISBN 0-306-45251-0
    1. Social systems--Planning. 2. Social engineering.   I. Title.
    II. Series.
    HM73.B253 1996
    301--dc20   96-36781
                                                                CIP

ISBN 0-306-45251-0

© 1996 Plenum Press, New York
A Division of Plenum Publishing Corporation
233 Spring Street, New York, N. Y. 10013

Printed in the United States of America

To Eva

I must Create a System, or be enslav'd by another
    Man's;
I will not Reason and Compare; my business is to
    Create.

—William Blake, Jerusalem, pl. 10, 1.20

Preface

What is presented in this work is based on a lifetime involvement in learning and exploring; in research, development, and applications; helping others to learn; and living and working in many organizational settings. An unwavering belief in human betterment has guided me through the decades, as with many others I have joined in creating resources, opportunities, and programs by which people, groups, organizations, and communities are enabled to develop and fulfill their individual and collective potentials and become the best they can.

In the course of the last decade, however, I have become increasingly convinced that even if people fully develop their potential, they cannot give direction to their lives, they cannot forge their destiny, they cannot take charge of their future—unless they also develop competence to take part directly and authentically in the design of the systems in which they live and work, and reclaim their right to do so. This is what true empowerment is about.

This book is about enabling such empowerment. It offers resources and programs by which individuals, groups, and organizations can learn to create a common ground, collectively define values and qualities they seek to realize, envision ideal images of a desired future, and bring those images to life by engaging in the disciplined inquiry of social systems design.

I am also convinced that this kind of empowerment, when learned and exercised by families, groups, organizations, and social and societal systems of all kinds, is the only hope we have to give direction to our evolution, to create a democracy that truly represents the aspiration and will of people, and to create a society about which all of us can feel good.

The ideas, the propositions, and the learning arrangements presented in this work have been influenced by many. What I have learned about systems and design inquiry I have learned with many others, in many ways, over several decades. And I am grateful to them. Nevertheless, I wish to acknowledge the very special help of a few friends who reviewed the manuscript and offered advice and guidance. The help of B. Antal Banathy, Aleco Christakis, Paul Hood, Lynn Jenks, Tad Frantz, and Gordon Rowland is much treasured and appreciated. Following their review, the seven of us spent a week in an intensive conversation at the 1995 International Conference of Social Systems Design, exploring the implications of the work for the further development of resources and programs in design learning and design applications. I also wish to thank Jean Sims for her editorial help.

I am most grateful to Ken Derham, Managing Director of Plenum Publishing Company, for embracing the idea proposed by this work, and thank his most capable editorial staff, Jeff Gilbert and his colleagues, who have worked on and guided the production of the book.

Contents

1

Introduction

There is an increasing realization of the massive societal changes and transformations that are reflected in the new realities of the postindustrial information/knowledge era. These changes touch the lives of every person, family, community, and nation and define the future of humanity. However, we are entering the twenty-first century with organizations designed during the nineteenth. Improvement or restructuring of existing systems, based on the design of the industrial machine age, does not work any more. Only a radical and fundamental change of perspectives and purposes, and the redesign of our organizations and social systems, will satisfy the new realities and requirements of our era.

Questions arise. What is our role in these massive changes? Are we only spectators? Are we destined to be victims of these changes? Do we have to relegate decisions affecting our lives to those who represent us? Are we at the mercy of experts who design systems for us? Or is there a role for us in shaping our future and the future of the systems to which we belong? Is there a way for us to participate in giving direction to the evolution of our systems, our communities, and our society?

These are some of the questions that are the genesis of this book. These are the questions with which I have struggled and worked over the last four decades. I address these questions in this book in order to explore approaches by which we can individually and collectively contribute to creating a better future for ourselves, for our systems and communities, and for future generations. The core idea of this work is that the design of social systems is a future-creating, collective human activity. People in social systems engage in design in order to devise and implement systems based on their vision of what those systems should be. Or they may redesign their existing system in order to realize their changing aspirations and coevolve with the emerged realities and expectations of their environment.

In an age when the speed, intensity, and complexity of change increases constantly and exponentially, the ability to shape change—rather than being its victims or spectators—depends on our competence and willingness to guide the purposeful evolution of our systems, our communities, and our society. The

1

method by which we can guide change is systems design. Collective design capability empowers us to exercise truly participative democracy. It enables us to take part in decisions affecting our lives and guide the activities that can enrich the quality of our lives and add value to the systems in which we live.

Many people in our social systems are not yet aware of the power of systems design. Education in design and the professional practice of design are limited to a few technical professions. If we are serious about "empowering" so that we can take charge of our lives and our systems, then we have to create opportunities that teach us what design is as a human activity, how it works, and how it can be applied in the contexts of our lives and our systems. Life is a journey. Making use of the power of design enables us to give direction to this journey and shape our destiny.

The two-pronged purpose of this book can be best defined in view of the ideas presented above. First, the book aims at guiding the reader to understand what systems design is, how it works, why we need it. The second purpose is to develop an appreciation of the power we can gain by acquiring ever-unfolding insight and competence in systems design. The purposes are addressed by an intensive exploration and development of the various knowledge bases of social systems design, and by providing activities that enable the user of the book to construct his/her own meaning of systems design and apply it in functional contexts.

In developing the book, I worked from two major sources. One is my own research and development and teaching in the systems and design sciences and findings that have emerged from my design work in organizational, social, and educational settings over the span of four decades. The second source is an in-depth review of the knowledge base in systems and design inquiry and the exploration of various fields of knowledge that might have relevance to systems design.

The book is differentiated from other works in the field by its relevance to all fields of social systems. As a rule, books on systems design address architectural, engineering, industrial, environmental, and business applications. In the design literature, there is today a lack of attention to the fields of social work, health, and other helping services; to education and human development; to community and volunteer agencies; and to the field of public policy. The content of this book makes contributions to these fields by presenting knowledge about social systems and social systems design and introducing design approaches, models, and methods appropriate to these fields. Still, the work provides useful learning to all fields of human systems, including business and industry. Furthermore, the presentation is comprehensive, drawing upon a wide range of literature on systems design, and on fields of knowledge that are relevant to systems design.

The knowledge base introduced and interpreted includes the literature on (1) design research, theory, philosophy, and methodology, as well as applications; (2) systems science research and its applications; (3) planning and problem solving;

(4) decision and conclusion-oriented disciplined inquiry; (5) philosophy and theory, relevant to systems and design inquiry; and (6) organizational behavior, communication, societal evolution, creativity, ethics, and sociocultural issues.

1.1. The Outline of the Work

1.1.1. Part I: Understanding Design

An understanding of social systems design unfolds as we explore definitions of systems design, focus on its use in social systems, examine design as a disciplined inquiry, and review various approaches and methods to systems design.

1.1.1.1. Chapter 2: What Is Systems Design? Why Do We Need It?

Systems design is defined and highlighted in the context of social systems. It is differentiated from other types of pursuits, such as planning, improvement, and restructuring. Design is characterized as a decision-oriented disciplined inquiry. The nature and characteristics of design situations are described. The three cultures (the sciences, the humanities, and design) are juxtaposed and the societal need for the building of a design culture though education is advocated. Four orientations toward relating to change are discussed and the role of design is defined in a changing world of new realities.

1.1.1.2. Chapter 3: The Product and Process of Systems Design

The discussion begins by answering two questions: When should we initiate design and what is the outcome of design? Answers to these questions mark the beginning and end points of design inquiry. Between the two markers we have the process of design, which is elaborated by discussing research on design and reviewing the process models of several design scholars. In the main body of the chapter, the three main strategies of designing—transcending the existing system; envisioning an image of the future; and, based on the image, transforming the system by design—are elaborated.

1.1.1.3. Chapter 4: The Design Landscape

Following an overview of the historical landscape of design, the current design landscape is mapped at two levels. At the conceptual level, the categories of design are identified as the generic, the general, the field related, and the situation specific. At the operational level, the current evolution of design is highlighted and a rich picture is presented of the various design approaches, models, methods, and tools.

1.1.2.   *Part II: Adding Value to Systems Design and How Systems Design Adds Value to Society*

Part I explored and interpreted the current knowledge and experience bases of systems design. In Part II, following a discussion of systems and design thinking, fields of knowledge are explored that could add value to systems design. In conclusion, a vision of how design can add value to society is captured.

1.1.2.1.   Chapter 5: Design, a Multidimensional Inquiry

Design, as a human activity, has several dimensions. Design has its own ways of knowing and thinking. Systems and design thinking are explored as conceptual bases of social systems design. The wisdom of involving multiple perspectives in design is explored. Design is portrayed as creative experience. It always involves communication; it is ideal seeking and should be guided by an ethical stance.

1.1.2.2.   Chapter 6: Getting Ready for Design

Following a discussion on the imperative of building a designing community, the following questions are addressed: Who should be the designers? How can we develop organizational capacity and human capability to engage in design? How do we build the designing system? How do we match design methods with the type of system we design? How do we design the design program?

1.1.2.3.   Chapter 7: Evaluation and Value Adding

Pitfalls and underconceptualizations in systems design are considered and their consequences are contemplated. Design evaluation should guard against both pitfalls and underconceptualizations. Perspectives and ways and means of evaluating the processes and products of design are explored and the issue of what qualities to seek in design is discussed. In moving toward the idea of what value design adds to society, the notion of societal evolution guided by purposeful design is introduced and the challenge of developing a creating/designing society is highlighted.

1.1.2.4.   Chapter 8: The Journey Continues

In the last chapter, arrangements are discussed that would guide the synthesis of core ideas and organizing perspectives identified in the course of working the activities. Proposals are made for a continuing development of design knowledge and design application.

1.2. The Design of the Book

A distinction can be made between Part I and Part II. Part I presents material from the current knowledge base of social systems design. Much of Part II explores new fields that could add value to design. The last four sections of Part II contemplate values that design can add to society.

The chapters build on each other, leading to an ever more in-depth knowl-

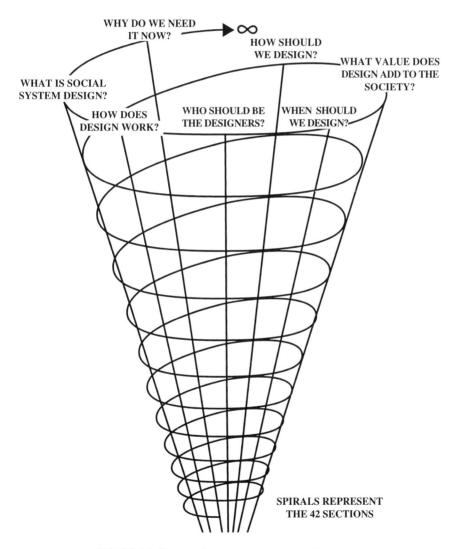

FIGURE 1.1. The overall organizational scheme of the work.

edge of design. The application activities provide opportunities for the learner to construct an understanding of design, formulate core ideas about design, and apply these to a system of interest.

In describing the work, I have repeatedly used the term "book." In fact, it is not a book in the conventional sense, not even a textbook. It is designed as a resource for design learning and design action.

Figure 1.1 depicts the overall organizational scheme of the work. The seven vertical lines of the figure represent the core issues addressed: What is design? Why do we need it? How does it work? How should we design? When should we design? Who should be the designers? What values does design add to the society? The knowledge base presented in the work is organized in response to these questions.

The spirals represent the succeeding sections that introduce the learning material. The learner moves up on the spiral as he or she works with the text and the activities. In the course of this move, the seven questions are revisited and the learner acquires ever greater depth and breadth of design knowledge and experience. This "revisiting" occurs in changed contexts and changed situations. What would be considered "redundancy" in a conventional book is here a purposeful feature of learning and applying.

1.3.   The Learning Strategy

The work is a resource for design learning and design applications. The activities offer the "personal learning space" of the learner. In this space the learner works with the text material, carries out the activities, and accomplishes three tasks.

Task #1. This task calls for the internalization of the ideas and the information presented in the text. The learner creates his or her own understanding and meaning of various aspects of design and transforms those into "personal knowledge." It is through this process that the learner cumulatively develops understanding of what design is, how it works, and how it can be used.

Task #2. Working with the text and the activities, the learner will formulate core ideas about design, developing organizing perspectives that can guide thinking about design and working with social systems design.

Task #3. This task is accomplished by applying the core ideas, the organizing perspectives, and the competence gained in engaging in design activities to systems of the learner's interest. Keep in mind that learning not applied to contexts that are real to the learner is learning not completed. As a guide to the use of the material I suggest that:

- The richer, the more extended, the more inclusive the sets of core ideas you generate are, the more powerful your understanding of social system design will be.

- The more integrated a specific set of core ideas is and the more integrated the sets of core ideas are, the higher the internal consistency is within a set and among the sets.
- The more in-depth and thorough the application of the core ideas in functional contexts selected by you is, the higher the design competence you attain.

I

Understanding Design

An understanding of design in general and social systems design in particular unfolds as you work with Chapters 2–4.

Chapter 2 presents a range of definitions of systems design and highlights design inquiry in the context of social systems. Design is characterized as a decision-oriented disciplined inquiry, and the need to develop a design culture is brought forth. Then various ways people relate to change are discussed and the crucial role of design in a rapidly changing world of new realities is elaborated.

In Chapter 3, we ask: When should we initiate design? And what is its outcome? It is between the answers to these questions that the core of this work—the process of design—resides. The design inquiry process is elaborated as we review research on design, explore various design strategies, and address the three main strategies of design: transcending, envisioning, and transforming.

Chapter 4 maps the design landscape. First, the conceptual level is explored: the general, the generic, the field related, and the situation specific. At the operational level a rich picture is presented of the various design approaches, methods, and tools, and an emerging computer-aided design technology is described.

2

What Is Design? Why Do We Need It?

Developing an understanding of systems design in the context of social systems is the task of the first four sections of this chapter. The last three sections discuss the rationale for acquiring competence in systems design. Section 2.1 contains definitions of design as quoted from design scholars and characterizes social systems. Section 2.2 clarifies the meaning of design by distinguishing it from planning, improvement, and problem solving. Section 2.3 describes design as disciplined inquiry and its relationship to other modes of inquiry is shown. Section 2.4 explores the characteristics of design situations. Section 2.5 defines "design culture" and differentiates it from the cultures of the sciences and humanities. Section 2.6 explores various modes of how people relate to change. Section 2.7 asks the questions: Why is it important to develop competence in systems design and, particularly, why is it important at this time?

2.1. Defining Design and Social Systems

In this section we explore a range of definitions of design, as formulated by various design scholars, and characterize social systems, which constitute the functional context of systems design. Activities will help the reader to formulate and synthesize core ideas about the design of social systems.

2.1.1. Definitions of Design

A selected set of definitions, offered below, reflects the thinking of design scholars of the last three decades and shows the variety of views they hold about design. The definitions convey the notion that design is practiced by many professions, in many different ways, and is applied in various contexts.

- Design is the initiation of change in man-made things (Jones, 1966).
- Design is the use of scientific principles, technical information, and imag-

ination in the definition of a system to perform specific functions with maximum economy and efficiency (Archer, 1966).

- A purposeful activity, design is directed toward the goal of fulfilling human needs (M. Asimow, 1962).
- Design simulates what we want to make before we make it, as many times as may be necessary to feel confident in the final result (Booker, 1964).
- Design is an imaginative jump from present facts to future possibilities (Page, 1966).
- The designer intends to change a segment of the universe. His motivation is consequential action, not understanding or explanation. . . . He designs whatever purpose he has in his mind and devises a scheme to accomplish this purpose (Rittel, 1973).
- A creative activity, design brings into being something new and useful that has not existed previously (Reswick, 1965).
- Design is the solution to the sum of the needs of a particular set of circumstances (Matchett, 1968).
- Design is a continuum of processes, an endless but moving chain of development, realization, and evaluation, directed toward purposeful creation (Van der Ryn, 1966).
- Design is primarily a thought process and communication process, transferring ideas into action by communication. It is a natural function, expressed in the many activities we engage in. For the teleologist, design means the conscious attempt to create a better world. For the antiteleologist design is the conscious part of action (Churchman, 1971).
- Design is concerned with how things ought to be. The designer devises a course of action aimed at changing existing situations into preferred ones (Simon, 1969).
- The entire activity from the stage of realization of a need to change to translating the image of the future system into reality is termed design (Mathur, 1978).
- Design is a new way of resolving basic human conflicts, critical for securing a safe passage to a desirable human future (Weisbord, 1992).
- Design is the investigation of contemplated and present systems to formulate, through the ideal systems concept, the most effective systems (Nadler, 1981).
- Design generates, organizes, and evaluates a large number of alternatives; keeping focused on the best possible or most ideal solution, rather than on collecting and analyzing data about the problem (Nadler and Hibino, 1990).
- Design is the core of purposeful and creative action of the active building of relations between man and his world (Jantsch, 1975).

- The act of designing is the prescription or model of the finished work in advance of its embodiment (Archer, 1984).
- Design aims to conceive the idea of a desired system and prepare a description of it (Banathy, 1979).
- Design is integral to all life and all human activity (Papanek, 1972).
- Design is the translation of information in the form of requirements, constraints, and experience into potential solutions which are considered by the designer to meet required performance characteristics (Luckman, 1984).
- Design consists primarily of six types of activity: intelligence, analysis, synthesis, choice, communication, and interpretation. The implementation of design is its concrete phase. The failure of any one of the six fundamental types will usually assure failure to implement (Warfield, 1990).
- Design is seen as a process of "variety reduction" with the very large number of potential solutions reduced by external constraints and by the designer's own cognitive structures (Darke, 1984).
- Design is initiated by using a very broad brush in sketching the first version. Then, details are gradually added. The process continues until a sufficiently detailed design is obtained that enables us to carry it out (Ackoff, 1981).
- Reengineering is the fundamental rethinking and radical redesign of processes to achieve dramatic improvement in measures of performance (Hammer and Champy, 1993).

The authors of these definitions portray a great variety of perspectives and represent many design professions, including architecture, environmental design, industrial design, organizational design, and social systems design.

A most intriguing statement made by Jones (1980) places design in a broader perspective. He says that design is a question of living, not a planning of life not yet lived. Jones thinks this idea seems nonsense if applied to designing by professionals, but, seen as part of a historic shift from product thinking to process thinking, isn't it just this planning that we have overlooked? Papanek (1972) shares this broad perspective in that he considers design integral to all human life and all human activity. He says that "any attempt to separate design, to make it a thing by itself, works counter to the inherent value of design as the primary matrix of life" (p. 23). In the same vein, Churchman (1971) sees design as communication among people enabling collective action and the transfer of the conception of the selected solution alternative into action.

Reflection on the definitions and the discussion above may lead to the formulation of ideas about what design is and why we need it. Reflection is an ongoing part of understanding and learning to use design. It enables the user of this book to integrate the ideas that emerge from the various definitions and to formulate organizing perspectives that may guide design activity.

Activity #1

Review the definitions and synthesize/construct your own definition of design as a purposeful human activity. Then, select out those definitions that are appropriate to social systems. Enter your findings in your workbook.

2.1.2. Social Systems

Peter Checkland (1981) presents a comprehensive characterization of social systems, which he calls human activity systems. He suggests ways we might think about them and how we might work with and change them. Human activity systems are very different from natural and engineered systems. Checkland writes,

> Natural and engineered systems cannot be other than what they are. Human activity systems, on the other hand, are manifested through the perceptions of human beings who are free to attribute a variety of meanings to what they perceive. There will never be a single (testable) account of human activity systems, only a set of possible accounts, all valid according to particular Weltanschauungen. (p. 14)

> Given man and his abilities, we have the huge range of human activity systems, from the one-man-with-a-hammer at one extreme to the international political systems at the other. (p. 21)

Checkland (1981) defines human activity systems as a collection of structured sets of activities that make up the system, coupled with a collection of activities concerned with processing information, making plans, performing, monitoring performance, etc. He says that a human activity system is a notional system that expresses some purposeful human activity that could be found in the real world. Such systems are intellectual constructs that we can use in a debate about possible changes in social systems.

Boulding (1980) proposes a hierarchy of nine levels of complexity and systems types. Social systems that operate at level seven are symbol-processing systems. People in social systems have shared models of reality. Social systems are also multi-echelon systems with a set of roles tied together by channels of communication. As a social organization, the system is a collection of individuals acting in concert. People in social systems are concerned with the meaning of messages, the nature and dimension of value systems, symbolization, and the complexity of human emotions.

Ackoff and Emery (1972) characterize human activity systems as purposeful systems whose members are purposeful individuals who intentionally and collectively formulate objectives. These systems are social organizations "in which the state of the part can be determined only by reference to the state of the system. The effect of change in one part or another is mediated by changes in the state of the whole" (p. 218).

According to Argyris and Schon (1978), a social group becomes an organization when it devises and uses procedures for "making decisions in the name of the collectivity, delegating to individuals the authority to act for the collectivity, and setting boundaries between the collectivity and the rest of the world" (p. 13).

Social systems are value-guided systems (Laszlo, 1972). Insofar as they are independent of biological need fulfillment and the reproductive needs of the species, social systems satisfy not body needs but values. But in what form they do so depends on the specific kind of values people happen to have.

Viewing human activity systems from an evolutionary perspective, Jantsch (1980) suggests that according to the dualistic paradigm, adaptation is a response to something that has evolved outside of the system. He notes, however, that with the emergence of the self-organizing paradigm a scientifically founded nondualistic view became possible. This view is process oriented and establishes that evolution is an integral aspect of self-organization. True self-organization incorporates self-transcendence, the creative reaching out of a system beyond its boundaries. Jantsch concludes that creation is the core of evolution; it is the joy of life, not just adaptation, and not just securing survival.

Reflections

In our journey toward understanding design, we started out with a review of a large set of definitions of design. You have reflected on those definitions and formulated your own. Because the purpose of this work is understanding the role and significance of design in the context of social systems, several definitions and characterizations of social systems were introduced.

Social systems were characterized as purposeful systems in which creative design can guide evolution and direct social and societal development. Therefore, design becomes the central activity in social systems, and competence in design becomes a capability of the highest value.

The next activity will enable you to work with the definitions of design and social systems presented in this section and synthesize the ideas in those definitions.

Activity #2

(1) Review the definitions of social systems introduced above, synthesize them, and formulate your own characterization of social systems. (2) With your own characterization of social systems as a reference, review the definition of

design you constructed in Activity #1 and formulate your redefinition of systems design. Enter your findings in your workbook.

2.2. Comparison between Design and Other Modes of Inquiry

The journey toward understanding design continues as we explore differences and similarities between systems design and other modes of inquiry. In this section, we distinguish "systematic" design from "systemic" design, science from design, art from design, and design from planning, problem solving, improvement, and restructuring.

2.2.1. Systematic versus Systemic

Several designers of an earlier generation viewed design as an orderly sequence of activities. This type of design is called "systematic" since it involves steps or phases in logical and linear arrangements. Jones (1980) noted that systematic design keeps logic and imagination, as well as problem and solution, apart by an effort of will and by external rather than internal means. This view of design is held by those who are engaged primarily in engineering design. Gregory (1963) says that a systematic approach highly constrains design if it is used in nonengineering contexts. Design that goes beyond engineering needs constant creative input, which requires flexibility and intuition.

Glegg (1971), who portrays a "systemic" view of design, points out three basic aspects of such a view: (1) specialized techniques by which one approaches a particular design problem situation, such as design methods and tools; (2) generalized rules that are not confined to a single set of specific techniques or steps, and (3) universal principles of design that are the underlying laws that cut across the various design approaches. These three aspects compose a conceptual metalevel mapping of design. There is interaction among the levels. As we use techniques, their use is guided by the general rules, and the general rules are formulated and supported by the underlying assumption of the universal principles. Furthermore, findings coming from the application of techniques might test the appropriateness of theories expressed by the rules and might lead us to revisit and change those rules. We might also find that the application of rules might suggest the need to revisit underlying universal principles. This circular and recursive process, then, might lead to rethinking and changing design techniques and methods. The approach portrayed here reflects a dynamic, open, and learning-focused approach to conducting systems design.

In my view (Banathy, 1991a), design is a creative, disciplined, and decision-oriented inquiry, carried out in iterative cycles. During the cycles we develop the design solution by repeatedly exploring organized knowledge as well as testing alternative solutions. We constantly integrate information, knowledge, insights gained, and the findings of testing into emerging design solutions. This process is not linear, sequential, or systematic. Design manifests dynamic interaction between feedback and feedforward, reflection and creation, and divergence and convergence. This dynamic process goes on until we develop confidence in the viability of one of the solution alternatives.

2.2.2. Design versus Science

Design is a continuous process of solution finding. It is concerned with what should be. Science, on the other hand, is concerned with what is. Design is a process of creating things and systems that do not yet exist. Science focuses on what already exists, and it aims at discovering and analyzing this existence. Science is predominantly analytical, whereas design is predominantly constructive. Simon (1969) elaborated the distinction between science and design. Science, says Simon, is organized in disciplines, such as physics, chemistry, biology, etc. These disciplines are interested in what things are and how they work. Thus, science develops knowledge about what is. Design uses knowledge to create what should be. Design is the core of all professional activities. This core is the intellectual activity of changing existing situations into desired ones. Professions such as "engineering, architecture, business, education, law, and medicine are centrally concerned with the process of design" (Simon, 1969, p. 56).

2.2.3. Design versus Art

Sless (1978), an art scholar, considering the differences between the arts and design, defines design as the "process of originating systems and predicting their fulfillment of given objectives." Art, on the other hand, is "unspecified experimental modeling" (p. 123). In his view, art and design share something in common in that they both create something that did not exist before. But the crucial difference between the two is in the realm of accountability. "The designer is accountable in terms of specified objectives. The artist's accountability is carried or smashed by the tide and waves of posterity." (Sless, 1978, p. 128). An assessment of the worth of design has to begin with the objective of the design. Art can be assessed only retrospectively.

Bevlin (1970) provides another perspective on the difference between art and design. She suggests that artists cannot find particular rules or formulas to help in their search for artistic expression, but rules and formulas are available to systems designers. Bevlin remarks that it would be rather difficult, if not impossible, to find a simple definition for design in art.

Reflecting on the two scholars' statements, it seems that their definitions talk to design as a rather closed system that does not reflect the characteristics of large-scale complex social systems. Furthermore, while design in art is an individual activity, the design of social systems is a collective venture.

2.2.4. Design and Other Modes of Organizational Inquiry

From an examination of the diverse literature on design, it becomes evident that design shares some characteristics with other disciplined inquiry activities, such as planning and problem solving. Shared characteristics of design, planning, and problem solving include (1) some form of disciplined thinking, rational behavior, or logical process; (2) sets of activities that one must go through in order to achieve a desired end; (3) the purposefulness of activities in all three modes of inquiry; (4) creativity, which is involved in all three modes; (5) the use of some form of disciplined and ordered methodology; (6) emphasis on collecting and evaluating information and knowledge; (7) choice and decision making; and (8) synthesis—an all-important activity in all three modes. These characteristics can lead to a blurring of differences between planning and design, design and problem solving, improvement and design, and restructuring and design.

2.2.4.1. Planning and Design

One aspect that sets planning and design apart is design's greater complexity. We would not say that we plan a social system; we design it. However, we usually say that we plan a meeting, which is of lesser complexity. The architect designs a building as he creates its blueprint. Based on the blueprint, the contractor develops a construction plan, which is a sequence of tasks organized in a time and resources frame. Planning is a set of steps that one takes toward a goal. The product of planning is a description of a sequence of activities to be accomplished in a time schedule, coupled with a scheme for the use of resources. Planners often operate in three- to five-year time frames. They revisit their plan at the end of the set period.

Designers, on the other hand, create a description of a system that has the capacity/capability to attain set purposes. The product of design is the model, the description of the system we designed. Once we have this descriptive representation of the new system, we can prepare and proceed with a plan that will bring the design to life. Designers revisit the design continually and create and implement new designs as time goes by.

In the social systems arena, once we recognize the need for change, in most cases we follow the traditional social planning approach (Banathy, 1991a). This approach reduces the problem to "manageable" pieces and seeks solutions to each. Planners believe that solving problems piece by piece ultimately will correct the larger issue of the problem situation. But "getting rid of what is not wanted will not give us what is desired," says the designers. In sharp contrast to the social planning approach, the systems design approach seeks to understand the problem situation as a system of interconnected, interdependent, and interacting problems and creates the design solution as a system of interconnected, interdependent, interacting, and internally consistent solution ideas. Furthermore, designers operate the specific levels of a multilevel system interactively and simultaneously. This requires coordination. Then they design for interdependency and internal consistency across all systems levels by integration. This simultaneous, all-over, whole-systems approach is the hallmark of systems design. It is totally different from the incremental, piecemeal, disjointed, and part-oriented approach of social planning.

2.2.4.2. Problem Solving and Design

Problem solving as a disciplined inquiry is concerned with (1) the finding/selection of the problem to address, (2) the analysis and structuring of the problem, (3) the selection of methods by which to address the problem, (4) the resolution of the problem, and (5) the presentation and evaluation of the resolution. In sharp contrast to problem solving, systems design as a disciplined inquiry is concerned with (1) the creation of an image of the desired state, (2) the selection of design approaches and methods by which to generate a set of solution alternatives, and (3) the selection and systemic description or modeling of the most promising alternative.

The difference between problem solving and design can be further explored and highlighted by presenting the perspectives of scholars who studied both problem solving and design. Newell and Simon (1972), in their epic work *Human Problem Solving,* suggest that we are confronted with a problem when we want something and do not know immediately what to do. In addressing a problem, we should engage in a series of rational actions. First, we translate the information about the problem into a problem formulation by rational analysis. Once the problem is formulated and analyzed, we should select a method appropriate to solving the particular problem. Then we apply the method, monitor it, and terminate it in case it is deemed to be inappropriate. When the method is terminated three courses of action are available: select another method, reformulate the problem, or abandon the attempt to solve the problem.

The problem-solving method described here is driven and dictated by the representation of the problem itself. This representation is the result of an in-depth analysis and structuring of the problem. The focus is on the problem. Solutions brought to the problem stay within the boundaries of the problem. It is assumed that the more time we spend on formulating the problem the more likely we will find the solution. A thorough formulation is expected to help us to select the method that best corresponds to the problem. The formulation of the problem is expected to lead us to the formulation of the solution. The problem-solving approach described here is appropriate in dealing with well-defined and well-structured problems.

But the problems designers face in designing social systems are anything but well defined. They are ill defined (Cross, 1984) or ill structured (Simon, 1984). Rittel and Webber (1984) call design problems "wicked" in comparison with "tame" problems that are the subject of rational problem solving. Ackoff (1981) holds that designers of social systems are always confronted with "messy" situations. In contrast to rational problem solving, designers cannot stay within the bounds of the problem situation. Their focus cannot be on the problem, but rather on the solution. Designers, therefore, have a solution-focused approach. They begin to generate solution concepts very early in the design process. Because design problems are "ill-defined, there is an inevitable emphasis on the early generation of solutions. An ill-defined problem is never going to be completely understood without relating it to a potential solution" (Cross, 1984, p. 172).

Focusing on the system in which the problem situation is embedded locks designers into the current system. But design solutions lie outside of the existing system. If solutions could be offered within the existing system, there would be no need to design. Thus, designers have to transcend the existing system. Their task is to create a different system or devise a new one. That is why designers say that they can truly define the problem only in light of the solution. The solution informs them as to what the real problem is.

2.2.4.3. Improvement and Design

Improvement focuses on how we can make what we have now better, more effective, and more efficient. In improvement efforts, the existing system becomes the unit of our analysis. Ackoff (1981) clarified a main difference between improvement and design. Improvement, he says, examines a current activity, process, product, or service and determines what is wrong with it. Improvement is concerned primarily with the removal of defects or deficiencies. Unfortunately, the removal of defects and deficiencies provides no assurance of reaching an end point that is more desirable than the point at which one started. In fact, the substitute for a defect or deficiency may be much worse than the defect or deficiency it is intended to correct. Design of a system should be directed at getting what we want, not getting rid of what we don't want. This observation points to one of the key differences between improvement and design.

The second major difference is that in improvement, the inquiry is usually focused on the specific parts of the problem situation. Thus, we take corrective measures problem by problem. This kind of change effort often results in piecemeal, fragmented, and discontinuous steps of correction; the "local consistency of its structure does not guarantee the consistency of the whole of it" (Sallstrom, 1992, p. 50).

We may very well fix parts and still find that the whole is not working. It is for this reason that systems designers often say that the excellence of a part does not prevent the bankruptcy of the whole. The frame of reference of designers is the whole system. Being conscious of the pitfall of focusing on specific parts, designers view a problem situation not as a heap of separate problems but as a system of interacting problems. They seek to create a system that manifests internal consistency and external viability as a whole. They know well that components and their parts get their meaning from their role in, and their contribution to, the whole.

Finally, in contrast with improvement, the focus of designers is not the existing system. They leap out from it and push the boundaries of the inquiry as far out as possible (Banathy, 1991a). They attempt to paint the largest possible picture within the largest possible context. An understanding of what "should be" emerges from an exploration of the larger (societal) environment. Designers have an expansionist orientation.

It should be noted, however, that some design scholars at times use the term "improvement" to mean design, in the sense it is used here (e.g., Drucker, 1989; Hood and Hutchins, 1993).

2.2.4.4.   Restructuring and Design

The term "restructuring" is often considered by its proponents to mean the same thing as design. But restructuring by any definition means taking a structure of a system and reorganizing it by rearranging its parts in a different configuration. What was discussed in comparing improvement and design applies to restructuring as well, with a minor difference. In restructuring we attempt to correct for the wrongs by changing relations among the parts, by redefining the role played by people in the system, or redefining the role of the components of the system. Having done so, we find that the boundaries of the system have not changed; neither have we changed purposes and functions. Restructuring might bring about change within the system, but it does not create a new system. Designers, on the other hand, focus on creating a new image of the system, define the purpose based on the image, and select the functions that attend to the purpose. Only then are they concerned with what components will be able to carry out the functions and how those components should then be arranged in a structure. An iron law in design is "form follows function."

Reflections

In this section we have continued the clarification of the meaning of design by differentiating it from other modes of intellectual inquiry. Systems design was juxtaposed with systematic design, science and the arts, planning and problem solving, and improvement and restructuring.

In discussing design with others—who are engaged in planning, problem solving, or restructuring—I am often told by them: "that is exactly what we are doing." It is of utmost importance that we make the kind of differentiation discussed here up-front in discussing design.

Activity #3

Review the juxtapositions and contrasting features. Note instances where you have experienced their use in systems in which you have worked. Report on your observations about working systematically rather than systemically, treating design as science, confusing planning with design or problem solving with design, and substituting improvement or restructuring for design.

2.3.   Design as Decision-Oriented Disciplined Inquiry

A main purpose of Chapter 2 is to develop an understanding of what systems design is. This section focuses on the relationship of design to other forms of disciplined inquiry (Banathy, 1989).

2.3.1.   The Domains of Disciplined Inquiry

Scholars and professionals have made the observation that the logic and methodology of the natural and behavioral sciences often have been adopted uncritically in the various professional fields. This has blurred the fundamental difference between the disciplines and professions, as Simon (1969) observed. The natural and behavioral sciences describe what things are and how things work. They form theories and make predictions based on those theories. Organized in compartmentalized disciplines, such as the physical, biological, behavioral, and social sciences, the salient intellectual process of the disciplines is analysis and their guiding orientation is reductionist.

In contrast, the professions and the various social service fields are concerned with what should be and how that is to be attained. Their focus is on design—they construct and reconstruct systems in specific contexts. In addition to analysis, their salient intellectual process is synthesis, their guiding orientation is expansionist, and their thrust is formulating and fulfilling purpose.

Both the sciences and the professions pursue disciplined inquiry, the quality of which distinguishes them from opinion and belief. Cronbach and Suppes (1969) suggest that disciplined inquiry is conducted in such a way that the argument can be examined. The report on the inquiry does not depend on its appeal or any surface plausibility, and the argument is not justified by anecdotes or fragments of evidence. They further observe that the investigator, using disciplined inquiry, institutes controls at each step in the exploring and reasoning process to minimize the possibility of error. To make their conclusions credible, their report displays their findings and the ideas entering into their argument—as well as the inquiry process by which the findings were developed. Disciplined inquiry is either conclusion oriented or decision oriented.

2.3.1.1. Conclusion-Oriented Disciplined Inquiry

Conclusion-oriented disciplined inquiry takes its direction from the investigator's commitment and hunches, by formulating questions about a particular issue. The aim is to understand the chosen phenomenon. Conclusion-oriented inquiry is the main domain of research. It produces knowledge. Its outcomes are technical reports. Results usually have little immediate consequence for practice, and the researcher has no obligation to transform newly found knowledge into actual applications.

2.3.1.2. Decision-Oriented Disciplined Inquiry

In the conventional mode of decision-oriented disciplined inquiry, designers are asked to provide information wanted by a decision maker such as a corporate executive, a government policymaker, an industrial manager, a board of education, etc. As a commissioned study, this type of inquiry is pursued against specific objectives and is expected to produce desired results and measurable changes when applied. The investigator is expected to create products, processes, or systems and produce information by which the users can evaluate or assess the effectiveness of what has been created. Often the task is to provide the decision makers or users with a range of alternatives or choices for their consideration.

Once people in self-organizing social systems develop competence in design, they can carry out their own inquiry and create their own system based on their image of the future they wish to create.

Figure 2.1 depicts the two modes of inquiry. As shown in the figure, conclusion-oriented inquiry contributes to, as well as draws on the findings of, decision-oriented inquiry. Conversely, decision-oriented inquiry uses knowledge developed by, and is a knowledge source for, conclusion-oriented inquiry.

2.3.2. *Placing Design in the Context of Disciplined Inquiry*

In placing design in the context of disciplined inquiry, one has to (1) locate it within the larger inquiry space of disciplined inquiry, (2) explore its relationship with other types of disciplined inquiries, and (3) identify the domains of design itself.

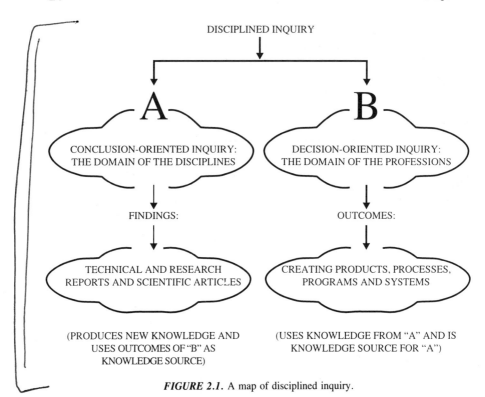

FIGURE 2.1. A map of disciplined inquiry.

2.3.2.1. Locating Design in the Larger Inquiry Space

Design is concerned with how things "out to be." It creates systems that are directed toward the attainment of aspirations and purposes. Designers set forth images of the desired future state, create alternative representations of that state, evaluate the alternatives, and select and describe the most promising alternative. Each and every activity described here implies decision making. Thus, in the larger space of disciplined inquiry, design fits into the domain of decision-oriented disciplined inquiry pursued by the various professions.

However, when design itself becomes a subject of scientific study, when we focus on what design is and how it works, when we conduct research on design to produce knowledge about it, then our inquiry becomes conclusion oriented. Thus, design as a purposeful human activity is decision oriented, but design as a subject of study is conclusion oriented. The relationship between the two modes of inquiry is explained next.

Research or conclusion-oriented study of design generates the knowledge base from which we can draw in conducting design. In addition to general knowledge about design, depending on the specific professional field in which design is applied, designers will draw on findings of the various disciplines and fields of inquiry that are relevant to the subject and the context of design. For example, in the design of social systems, relevant knowledge includes the theory and philosophy of systems in general and social systems in particular and design theory and methodology, creativity, organizational theory, ethics, communication, evaluation, and social evolution.

As we carry out design in a self-reflective and contemplating mode, we gain insight and knowledge about design itself. This knowledge informs both designers and social researchers.

2.3.2.2. Design and Other Modes of Decision-Oriented Inquiry

In addition to design, the space of decision-oriented disciplined inquiry is inhabited by other inquiry modes that operate in the various professions. These other types include:

- Representation, description, characterization, and modeling of systems and processes.
- Analysis, diagnosis, assessment, and evaluation.
- Planning, adjustment, improvement, and problem solving.
- Development, implementation, and institutionalization.
- Management of systems and management of change.

Earlier we made distinctions between design and some of the inquiry modes presented above. Now we explore how design is related to these various decision-oriented inquiry modes. As we shall see, all these inquiry modes are involved in design.

The front end of design involves the diagnosis of the design problem situation as well as the description of the existing system. The outcome of design is a description and modeling of the new system as well as its systemic environment. Throughout design we evaluate and assess the various alternatives of the future system. Once we decide that our design should "come to life," we formulate a plan for its development and implementation and then proceed with its development, implementation, and institutionalization. Finally, we are involved with continual adjustment and change.

2.3.3. Types of Designed Systems

The major types of systems we design include the following:

- Abstract, conceptual systems, such as theories.
- Physical systems, such as buildings, machines, and tools.
- Hybrid, machine–nature systems such as a hydroelectric plant.

- Man–machine systems, such as a car, a computer, a spacecraft.
- Human activity systems, such as organizations, and social systems. This last type is the subject of our interest. It may include all the above types.

Reflections

In this section we explored an understanding of design in the larger conceptual space of disciplined inquiry. We identified two major domains of disciplined inquiry: the conclusion oriented and the decision oriented. We understood that in the conclusion-oriented domain we seek to generate knowledge about design as we conduct research about design. We located design as a purposeful human activity in the domain of decision-oriented disciplined inquiry. We learned that as we carry out design we use design knowledge as well as gain knowledge about design. We explored the relationship of design with other modes of decision-oriented inquiry. In conclusion, in this section, we differentiated various design types. It is often said that we learn by making distinctions. In this section we advanced design learning by making some critical distinctions that have helped us to understand what design is and what it is not. The activity that follows will further help to enrich your understanding of design.

Activity #4

Task #1. Identify core ideas that help you to understand systems design as a decision-oriented disciplined inquiry.

Task #2. Select a social system of your interest and stipulate fields of knowledge that might contribute to the knowledge base of engaging in the (re)design of your selected system. Enter findings into your workbook.

2.4. The Nature of Design Problem Situations

We have now arrived at the midpoint of Chapter 2. In this section, the task is to define and understand the meaning of problems and the nature and characteristics of design problem situations.

2.4.1. Defining and Exploring the Meaning of Problems

Dictionaries define "problem" in a variety of ways: a substantive matter about which there is a concern; a matter raised for inquiry or solution; a source of perplexity or distress. A "concern" may be with uncertainty, difficulty, ambiguity, etc. The "matter" could be a situation, a phenomenon, an issue. Other definitions include a situation for which an organism does not have a ready response, a felt difficulty, or a difference between what we have and what we want. Design researchers provide some additional insights into the meaning of problems.

Ackoff and Emery (1972) define problem as a state with which a purposeful individual is dissatisfied and in which there is doubt about which of the available courses of action will change that state to a satisfactory one. A systems view holds (Banathy, 1992a) that problems are subjective images held by people. Different people perceive different problems in the same situation. One person's perception may not be more correct than someone else's, but some perceptions may be more useful than others. McWhinney (1992) differentiates problems from issues. He thinks of problems as well-bounded difficulties for which we have the requisite resources to solve (e.g., electronics engineers solve their design problems on a breadboard, isolating them from extraneous influences). An "issue," by contrast, is "an unbounded, ill-defined, and overwhelming complex of problems" (McWhinney, 1992, p. 63). The word "issue" (from the latin *exire*, "to go out") expresses, in McWhinney's words, a "turbulent intermingling of various streams" (p. 20). (In this sense, designers of social systems deal with issues.) And from Checkland (1981, p. 155): "a problem, relating to real world manifestations of human activity systems is a condition characterized by a sense of mismatch, which eludes precise definition, between what is perceived to be actual and what is perceived might become actual."

2.4.1.2. Problem Categories and Aspects

Earlier I differentiated structured problems from unstructured ones. Structured problems, on the one hand, can be explicitly stated and imply that a theory concerning their solution is available. These problems are in the domain of systems engineering or operations research. They can be approached with what we call "hard systems thinking." Unstructured problems, on the other hand, are manifested with a feeling of unease, and they cannot be explicitly stated without oversimplifying them (e.g., What should we do about educational reform or the reform of our health services?). Unstructured problems should be approached with "soft systems thinking" (Checkland, 1981) and are addressed by systems design.

Cartwright (1973) categorizes problems as simple, compound, complex, and metaproblems. Simple problems can be understood in terms of a specified number of calculable variables and are subject to analysis and optimization. Compound problems are defined in terms of an unspecified number of variables that are calculable. They are addressed by analyzing and optimizing a selected set of variables. Complex problems are definable in terms of a specified number of incalculable variables and are approached by exploring all variables and initiating systemic change in the overall situation. The least precise, metaproblems are made up of an unknown number of variables and are addressed by exploring those that are most relevant to contemplating solution to the problem situation. The design of social systems deals with both complex and meta-problems. Understanding the differences between the four types is of crucial significance. Forcing an inappropriate approach to a problem will ultimately change the nature of the problem. For example, we live next to a pristine canyon that is home to a rare California pine forest, with a rich fauna and wetlands. Considering the building of a freeway through this natural habitat as strictly an engineering problem transforms a highly complex problem of land-use, eco-logical, and environmental issues, and issues of societal values, into a simple problem.

Nadler (1981) considers three problem aspects: the substantive, the locus, and the value. Taking into account the substantive aspect of a particular problem ensures that an appropriate methodology will be used (e.g., a design issue will not be approached as a research problem). A problem's locus centers on the uniqueness of the situation, tailoring solutions to specific needs and resources. (Attempts to transfer America's agricultural solution to Third World countries underscore this point.) A problem's value aspect places it squarely in the context of human aspirations. Understanding the significance of a problem's aspects forestalls the unhappy tendency of defining problems by fitting them into the constraints of techniques that are available in the "toolbox" of a design profes-sional.

2.4.1.3.   Bounding Problems

There are problems that are well bounded, such as engineering problems. These have clearly defined approaches and requisite resources to solve them. In contrast, poverty is not a boundable problem. Any attempt to solve it must address many areas of social concern. Faced with the difficulty of "unbounded-ness," Majone and Quade (1980) say that often we are tempted to trim the questions we ask to the available data or frame of reference and bound the problem in accordance with our preconception. We often make our exploration conform to methods we know and prefer to use. Hall (1962) suggests that whereas unboundedness increases possible solutions, restrictions on bounding will decrease solution alternatives. Applying an unbounded approach, we search the total environment for new ideas, theories, and methods, then look for ways to use them in designing the system. Churchman (1971) calls this process "sweep-ing-in" as many relevant issues and ideas as we possibly can.

Using a bounded approach, we study existing operations and define prob-lems and plan for changes within the boundaries of the existing system. Recent educational improvement efforts are excellent examples of the "pitfalls" of a bounded approach.

2.4.2. *The Characteristics of Problem Situations in Design*

In designing social systems we are confronted with problem situations that compose a system of problems rather than a collection of problems. They are embedded in uncertainty and require subjective interpretations. Above all, design problems are ill structured and defy straightforward analysis. In design there is a continuous interaction between problems and solutions.

2.4.2.1. System of Problems

Social systems are unbounded. Factors assumed to be part of a problem are inseparably linked to many other factors. A technical problem of transportation becomes a land-use problem, linked with economic, environmental, conservation, and political considerations. Can we really draw a boundary? When we seek to improve a situation, particularly if it is a public one, we find ourselves facing not a problem but a cluster of problems, often called a "problematique." Peccei (1977, p. 61) says that

> Within the problematique, it is difficult to pinpoint individual problems and propose individual solutions. Each problem is related to every other problem; each apparent solution to a problem may aggravate or interfere with others; and none of these problems or their combinations can be tackled using the linear and sequential methods of the past.

Ackoff (1981) suggests that a set of interdependent problems constitutes a system, which he labels a "mess." Like any system, the mess has properties that none of its parts have. These properties are lost when the system is taken apart. In addition, each part of a system has properties that are lost when it is considered separately. The solution to a mess depends on how the solutions to the parts interact. A design problem situation should always be seen as a system of problems and not as independently obtained parts of a mess.

2.4.2.2. The Issues of Uncertainty and Subjectivity

The era of "quest for certainty" has passed. We live in an age of uncertainty in which systems are open, dynamic; in which problems live in a moving process. Ackoff (1974, p. 31) says: "Problems and solutions are in constant flux, hence problems do not stay solved. Solutions to problems become obsolete even if the problems to which they are addressed are not."

Lawson (1984) proposes that design problems cannot be comprehensively stated. We cannot ever be sure when all aspects of a design problem have emerged. Their features may never be fully uncovered and they are full of uncertainties. Design problems require subjective interpretations. Different people perceive problems differently and thus design problems are inevitably value laden. Furthermore, design problems tend to embrace several levels. A perceived design problem can be viewed as a symptom of higher-level problems. There is no logical way of determining the right level at which to state and tackle a design problem. "The decision remains largely a pragmatic one; it depends on the power, time, and resources available to the designer, but it does seem sensible to begin at as high a level as is reasonable and practicable" (Lawson, 1984, p. 87).

2.4.2.3. Transcending the Problem

Ulrich (1983) suggests that it is the task of designers to reflect critically upon "given" problems. He asks: How can we produce solutions if problems remain unquestioned? Design is a practical discourse. Such a discourse has to transcend problems as originally stated and critically explore the problem itself. In such a discourse those who represent the affected must not be required to submit to the rationality standards of the designers but "must be entitled to argue polemically" (Ulrich, 1983, p. 308).

Critical reflection leads us to propose another transcendence. We must be ready to transcend the standpoint from which we explore a design situation. Design problems should not be forced into the limits at which a specific standpoint or a worldview sets them. Only a perfect mind could foresee all the ramifications of a problem at the time of formulating the viewpoint from which to approach the problem situation. In systems design, as we explore a design problem situation, we must be ready to question and transcend both the problem and the perspective or worldview from which we approach the problem.

2.4.2.4. Design Problems Are Ill Defined and Wicked

From a review of the design literature, it seems that the first detailed exploration of the ill-defined nature of design problems was originated by Reitman (1964). He noted that well-defined problems are the kind in which the initial conditions, the goals, and the necessary operations can all be specified. Ill-defined problems are the kind in which the initial conditions, the goals, and the allowable operations cannot be specified or extrapolated from the problem. Rittel and Webber (1984) developed a detailed exploration of this difference. Their thesis is that science and engineering are dealing with well-structured or tame problems, and their paradigm underlies modern professionalism. But this paradigm is not applicable to open social systems. Still, many social science professionals have mimicked the cognitive style of science and the occupational style of engineering. They often approach problems in social systems the way scientists do their tame problems. But social problems are inherently wicked. Rittel and Webber (1984) characterize wicked problems as follows:

- It is not possible to provide a definitive formulation of a wicked problem. The information needed to understand the problem depends upon our idea for solving it.
- There is no stopping rule for wicked problems. The designer can always try to seek a "better" solution, but at a certain point he has to say, "This is good enough."
- Solutions to design problems are good or bad. They are not true or false. Judgments of "goodness of fit" of the solution may vary depending on people's interest, values, and perspectives.
- There is no ultimate test of the solution of a wicked problem. The waves of consequences over time are unbounded.
- There is no exhaustively definable set of solutions to wicked design problems. There is no way to consider all potential solution alternatives.
- Every design problem is essentially unique. Despite numerous possible similarities, there are always additional properties that distinguish seemingly similar problems.
- Every design problem can be considered to be a symptom of another problem. And the problem of which the identified design problem is a symptom might be a symptom of another problem.

In the design world of wicked problems, the aim is not to find the truth but to design systems that enhance human betterment and improve human quality.

2.4.2.5. Problem and Solution Are in Continuous Interaction

Rittel (1973) suggested that every formulation of a design problem is tentative and incomplete, and it changes as we move toward the solution. The design problem is formulated in view of the solution. And as the solution changes—as it is elaborated—so does the understanding of the problem. This "ill-behaved" nature of design frustrates all attempts to start out with an information phase and an analytical phase, at the end of which a clear definition of the problem is rendered and objectives are defined that become the basis for synthesis, during which in a monastic isolation a solution can be worked out. Design requires a continual recursive interaction—an interplay—between the initial state that triggers design and the final state, when the design is completed.

In a similar vein, Jones (1980) says that it is a mistake to begin design by focusing on the problem and leave thinking about the solution to later stages. He believes that "one's mind is best kept in a constant intermingling of both problem and solution so that the interdependency of each is evident throughout" (p. xxxiii). He also suggests that the pattern of the original problem may change so drastically over the course of design that the designers are thrown back to square one. "This instability of the problem is what makes designing so much more difficult and more fascinating than how it may appear to someone who has not tried it" (p. 10). Lawson (1984) suggests that many components of the design problem situation cannot be expected to be uncovered (if ever) until we generate design solutions.

Activity #5

The section above provided a picture of the nature and characteristics of design problem situations. You should now develop your own core ideas about the nature and characteristics of problem situations in design inquiry. You should reflect on the statements and explanations in this section and generate your own ideas and perspectives that can guide your thinking about design. Based on your experience of working with problem situations, describe one that is well-structured and one that is ill-structured. Enter your findings in your workbook.

Reflections

As I reflect on our discussion of the design problem, I am reminded of Vickers's (1970, 1983) notion of "appreciative world." Jantsch (1975) interpreted Vickers's notion in the context of social systems design. The appreciated world comes into being by self-reflection that encompasses the physical, cognitive, social, emotional, and spiritual aspects of our experiences. The appreciated world, says Jantsch,

> embraces our appreciation of what this world can do to and for us, and what we can do to and for it. It reflects our own place in the world as well as our responsibility toward it, the demands which we make on it, and the personal concept we have formed of it. Most importantly, it holds the difference between the world as we want it to be and the world as we actually perceive it. Thus, the appreciated world becomes the motor for change, induced by human action. (p. 106).

Reflecting on the words of Jantsch has opened up for me a new horizon for viewing the design problem situation. Without exception, the design scholars I quoted had a problem with the concept of a problem-focused approach to design. I know I have. In the context of designing social systems, I feel that the genesis of wanting to take action and changing our world by design is not driven by problems. It is guided by our visions and images of a better future. It is fueled by our aspirations, desires, beliefs, hopes, dreams, and expectations. I now feel that I have transcended the notion of problems being the genesis of design. I feel that I am now free from, I am liberated from, the "tyranny of problems." As I begin to imagine, on my own and with others, the world as I want it to be, it becomes possible to look at the world as it is and see—beyond the mess of wicked problems,—desirable and inspiring images of the future. The future is in my hands; it is in our hands. And I am—we are—responsible for its design. We are responsible for the design of the systems we inhabit.

2.5. The Three Cultures: Science, Humanities, and Design

In the course of societal evolution three intellectual cultures emerged: the cultures of science, humanities, and design. In the humanities we explore and portray the human experience. In science we study the world in which we live and describe it. In design we make things and create systems that do not yet exist and thus we change our world. The acquisition of these three cultures entails (Cross, 1990) (1) the transmission of knowledge about their specific domains, (2) education in the appropriate methods of their inquiry, and (3) internalization of their belief systems and values. But there are significant differences between these cultures in terms of what they study, why and how they study it, and what they value.

In this section, I briefly define culture and the culture of design. Then I juxtapose design culture with the cultures of science and the humanities. In conclusion, I explore the need for, and an approach to, building a design culture.

2.5.1. Culture and the Culture of Design

Culture includes social knowledge and understanding; ways of knowing, thinking, and doing; beliefs and dispositions; and customs and habits, shared by a community of people, and passed on though social transmission. Culture is learned and structured and it embraces every realm of human experience. We can speak of cultural maps that are generated in the minds of members of cultural groups. These maps regulate social actions, personal relationships, and attitudes toward the institutions of the culture. Cultural maps are drawn based on the shared values we hold, shared ideas about how the world works, and our perceptions of what our role is in the world. These maps are alive; they are created, confirmed, disconfirmed, elaborated, changed, and redrawn (Banathy, 1992c).

Design, in the most general sense, is purposeful creative action, the building of relations between us and our world. It is the conception and creation of novel phenomenon, the realization of what should be. It is the manifestation of knowledge, beliefs, and aspirations, translated into a great variety of what we want to bring about and make part of our way of life. Whatever we design expresses our culture and our designs are embedded in our culture.

Design culture is an integrated pattern of human behavior that is manifested in (1) design's own distinct ways of thinking; (2) the use of modeling, which is the language of design; (3) design concepts and principles that constitute the theory of design; and (4) the means and methods of design, by which creativity is applied in actions of inventing, making, assessing, and doing.

The culture of design is learned. It embraces every modality of human experience that aims at the creation of novel phenomena. Design culture is manifested in the human action of bringing to life what we believe should be. This action is based on the shared values we hold, our shared ideas about how the world should work, and the shared perception of what our role is in the world.

During the early eighties it was recognized that the capability to deal with increasing systemic complexities, rapid societal changes, and design decisions that affect our society cannot be left to the so-called design experts. This recognition led to the idea of a broad-based participation of the users of design in the design of their systems. It was proposed that to complement the expert culture of professional designers, we should build design cultures that include the general public.

2.5.2. Contrasting the Three Cultures

Figure 2.2 interprets Cross (1990), who contrasted the cultures of science, the humanities, and design. Science focuses on the study of the natural world. It seeks to describe what exists. Focusing on problem finding, it studies and describes problems in its various domains. The humanities focus on understanding

	SCIENCE	HUMANITIES	DESIGN
FOCUS	The Natural World	The Human Experience	The Man-made World
	Problem Finding	Understand the Human Experience and	Solution Finding
	Describe What "Is"	Portray It	What "Should Be"
PRIMARY METHODS	Experimentation	Analogy	Modeling
	Pattern Recognition	Metaphor	Pattern Formation
	Analysis	Criticism	Synthesis
	Classification	Valuation	Conjecture
	Deduction	Induction	Abduction
WHAT IS VALUED	Objectivity	Subjectivity	Practicality
	Rationality	Imagination	Creativity
	Neutrality	Commitment	Empathy
	Concern for "Truth"	Concern for "Justice"	Concern for "Goodness of Fit"

FIGURE 2.2. The three cultures.

and discussing the human experience. In design, we focus on finding solutions and creating things and systems of value that do not yet exist.

The methods of science include controlled experiments, classification, pattern recognition, analysis, and deduction. In the humanities we apply analogy, metaphor, criticism, and (e)valuation. In design we devise alternatives, form patterns, synthesize, use conjecture, and model solutions.

Science values objectivity, rationality, and neutrality, and it has concern for the truth. The humanities value subjectivity, imagination, and commitment, and they have a concern for justice. Design values practicality, ingenuity, creativity, and empathy. It has concerns for goodness of fit and for the impact of design on future generations.

Despite these differences, there is a strong relationship between design and the two other cultures. Design always uses the knowledge developed and the insights gained in both the sciences and the humanities in the pursuit of practical tasks. In turn, both the sciences and humanities use the creations of design. I suggest that the three cultures jointly constitute the wholeness of human intellectual affective and creative experience. A lack of any one of the cultures leads to a grave loss of substance and value, and a loss in the quality of human experience. Such a loss today is manifested in the paucity of design culture in the general human experience. It is clearly manifested in education by the fact that education focuses on literacy in the sciences and the humanities and neglects and is even unaware of the need for literacy in design.

2.5.3. Why Do We Need a Design Culture?

Our age has been described as the information/knowledge age, the postindustrial age, the age of complexity, the postmodern era, and the age of high technology. While these labels may all fit, our era can surely be called the age of design. Cross (1984) and Warfield (1987) suggest that in the age of design the building of a design culture is an inescapable necessity. Their ideas, discussed next, are helpful in answering why we need a design culture.

We design buildings, clothing, laws, processes, packaged food, power plants, all kinds of organizations, curricula, cars, and weapons—the list is endless. Take away designs and we strip the world of most of its enabling mechanism. We see around us daily evidence of designs of high and low quality. We recognize bad designs and their harmful consequences: faulty power plants, inadequate disposal methods, poorly designed legislation; designs of all kinds that are unsafe, uncomfortable, and even injurious. And there are many bad designs we don't even recognize as such. Nonetheless, we experience their harmful side effects. We are at the mercy of past and present designs, crafted for us by the experts. So we ask: Is it our fate to accept designs regardless of their quality? No, it is not. It is our choice not to accept, not to tolerate bad design. But how can we do that?

We have a choice. We can continue to be uninformed design illiterates. We can accept design decisions made by the experts, and continue to live in ugly and crowded block-houses, drive dangerous and polluting cars, use shoddy goods, work in unsafe factories, destroy nature so we can build more freeways, and suffer outdated educational designs. Or we can become design literate. A design literate population will be able to make judgments about the instrumental and social effectiveness of design decisions. It can make judgments about the socio-economic and political implication of design decisions. The contemplation of these two choices will help us now to define and explore literacy and competence in design and think about the ways and means of their development.

2.5.4.   Design Literacy

Design literacy enables us to understand what design is as a human activity, what design does and how it does it, what the role of design is in society and how it plays this role, what the impacts of design are on human quality and human betterment, what our role is in design and how we can play that role. But design literacy is only one side of the coin of design culture. Design literacy will create informed and judgmental users of products and systems in which the expertise in the design of technical products is a requirement. But when it comes to the design of social and societal systems of all kinds, including educational systems, it is the users—the people in the system—who are the experts. The primary right to design should be invested in the users of the system, provided they accept such responsibility and empower themselves by becoming competent in design.

2.5.5.   Competence in Design

Competence in design is the other side of the coin of design culture. We live in an age when change happens at all levels of society from local to global. Change is universal, it is ever ongoing, and nowadays it happens ever faster. We may fear change or we may want it; we may go against it, or go with it and direct it. The choice is ours. But having the intent to change is not enough. We should know how to design for change. We must have the will to initiate it and acquire the competence to do it. Building a design culture is the absolute prerequisite of taking charge of our future and shaping our individual and collective destiny.

2.5.6.   The Prime Directive: Building a Design Culture

During the relative stability of previous eras, piecemeal adjustments were able to bring our systems in line with the slow rate of change in the society. But in a time of ever accelerating and dynamic changes and transformations of the current era, piecemeal adjustments of systems that are still grounded in the design of the industrial machine age will create more problems than they solve. The constantly emerging new realities require continual design activity at all levels of society. They require the creation and re-creation of our systems so that they will coevolve with the new realities.

Faced with this requirement we have two choices. We can relegate authority and responsibility to others who represent us, as we do today. Or we can empower ourselves by acquiring design literacy and design competence so that we can assume responsibility for the design of systems in which we live and to which we are connected. The building of a design culture enables us to create participative democracy, about which we talk so much today, but which is not yet truly part of the human experience. The prime directive of building a design culture in the society requires resources, arrangements, and programs by which to build design literacy and design competence among our children and youth, and in the public sector.

Activity #6

Review the text of this section. Then (1) formulate your own core ideas about why we should build a design culture and (2) use your imagination to describe the state of affairs of a community that has developed a design culture. Enter your findings in your workbook.

Reflections

As I reflect on the relationship of democracy and design culture, I am reminded of Slater's (1991) notion of democracy as self-governance and the reinvention of ourselves and Follett's (1965) image of democracy as "self-creating coherence." Their ideas define participative democracy. But self-governance, reinvention, and true participation are not possible without competence in design. Design is "self-creating coherence." Participative democracy comes to life when we individually and collectively develop a design culture that empowers us to create, govern, and constantly reinvent our systems.

2.6. How People and Systems React to Change

> While an unchanging dominant majority is perpetually rehearsing its own defeat, fresh challenges are perpetually evoking fresh creative responses from newly recruited minorities, which proclaim their own creative power by rising, each time, to the occasion.
>
> —Arnold Toynbee

The fresh challenge that we face perpetually is the change that occurs around us constantly. It often dazzles us, leaves us confused, and generates uncertainty and ambiguity. Change happens to us, to the systems to which we belong, to other systems that surround us, and in the larger societal environment in which we and our systems are embedded. Change also flows from our systems to the environment. How we relate to change can make all the difference. We can be its spectators or its victims, or we can take charge by "evoking fresh creative responses" to it and, thus, become masters of change. The key characteristics of taking charge of change are "to proclaim our own creative power" and to engage in deliberate action by the purposeful design of our lives and our systems. But such action has to be responsible action. If we initiate change, we have to take responsibility for its results (Banathy, 1991a).

Before we engage in the design of our lives and our systems, we need to reflect on how we ourselves relate to change. The ideas and concepts about how people and systems relate to change, introduced here, offer a mirror for self-reflection and understanding of our attitudes toward change.

Ackoff (1981) describes four styles of working with change: reactive, inactive, preactive, and interactive. Ackoff suggests that the four "orientations are like primary colors; they seldom appear in pure form." Orientations, like colors, are mixtures, but "usually are dominated by one of the four pure types" (p. 53). Furthermore, orientations might shift over time. They might change depending on the situation and the preferences people have. Our orientation might change once we understand its implication to the challenges we face and to the responses we want to develop toward them.

In this section we explore these four ways people perceive and respond to change and consider their implications for organizational behavior. Each orientation is characterized in terms of (1) general attitude toward change, (2) the role of science and technology, (3) organizational mode and culture, (4) approach to planning and problem solving, and (5) attractiveness of orientation.

2.6.1. Reactive Orientation: "Back to the Future"

The general attitude. As reactivists, we are dissatisfied with the present, long for the past, and want to return to what was. We attempt to unmake changes and romanticize about the good old days when life was simple. We drive toward the future by looking in the rearview mirror and focusing on where we have been instead of where we want to go.

Perception of the role of science and technology. Consonant with the reactive orientation, we hold that experience is the best teacher. Faced with changes, we seek guidance from past experiences rather than use the knowledge offered by science and its applications. We consider technology the main cause of change and have little use for it. We rationalize this position by pointing to some of the bad consequences of technological applications.

Organizational mode and culture. Corresponding to the past experience-focused orientation, we have a tendency to rely on old, well-proven, and familiar organizational forms. In this mode, the current structure and operational procedures are protected from change, often at all cost. This mode is manifestly hierarchical, bureaucratic, and top-down driven. The vision of the organization is defined at the top. There is an official culture—the culture of management—and an employee culture, which is often very different from the official culture.

Approach to planning and working with problems. Planning is ritualistic and directed from the top. Given the official plan, other levels are to react to it and develop implementation plans level by level and pass it up to the next level for approval. Problems are addressed in a piecemeal fashion and so we fail to realize that problems are connected. We believe that if we get rid of what is not wanted we can achieve what is desired.

Attractiveness. Ackoff (1981) suggests that the reactive orientation has three main attractions. It maintains a sense of history from which we can derive guidance; it maintains continuity and seeks to avoid change; and it preserves tradition, protects familiar grounds, and maintains a feeling of stability and security.

2.6.2. The Inactive Style: "Don't Rock the Boat"

The general attitude. The inactivist is satisfied with things as they are. The label "inactive," however, is misleading. A great deal of energy and effort is spent on preventing change. The operating principle is preserving stability at all cost, and it takes a lot of work to keep things from changing. The inactivist says things may not be the best today, but they are good enough, or as good as can be expected. If nothing new is done, things will stay as they are, and that is what we want. "It is not really broken, so let's not fix it."

Perception of the role of science and technology. Having a dominant orientation toward the present, inactivists (1) want to preserve the status quo, (2) try to avoid changes, (3) rely on present practices rather than considering science as a guide, and (4) are reluctant to use new technology. Their attitude is best expressed in the saying: "We don't even use half the knowledge or the technology we now have."

Organizational mode and culture. Given the desire to keep things just as they are, inactivists' operational mode is bureaucratic. They rely on red tape as their instrument to slow things down and avoid change. They use committees and study groups in an endless process of gathering facts, pass on information from one group to the other, and revise positions and recommendations. This process goes on until they find that there is no longer a reason to change. In the organizational culture, conformity is valued more than creativity. The status quo is valued and "don't rock the boat" is the code word of the organization.

Approach to planning and working with problems. In view of a desire to maintain the current state, planning focuses on extrapolating from the present and operating within the boundaries of the existing system. If study groups recommend change that cannot be ignored, insufficient implementation resources and support are provided, virtually assuring failure. An approach similar to that of the reactivist is followed. Problems are treated piecemeal. Inactivists change as little as possible; they muddle through.

Attractiveness. Ackoff suggests that the inactive style is dominated by the perception that even if there might be situations that call for change, doing nothing is better than doing something. Inactivists believe that problems fade away if left alone and those who act cautiously seldom make mistakes.

2.6.3. The Preactivist Style: "Riding the Tide"

The general attitude. The preactivist anticipates change, prepares for it when it arrives, and exploits its opportunities. Since preactivists believe that change is brought about by external forces, they do all they can to guess where change might lead, and they ride its tide so that they can get on with it. They say: "When the future comes, we will be ready for it." They are not willing to settle for things in the present. They are eager to move into the future on the road that is projected by the experts.

Perception of the role of science and technology. Since the key to moving into the future is our ability to foresee it, preactivists rely on the science of prediction. They put a great deal of effort into finding out who are the best futurists or they use science-based methods of predicting. They agree with the reactivist that technology is the principal cause of change. But unlike the reactivist, they want to promote technology as a panacea.

Organizational mode and culture. Preactivists are more concerned about missing an opportunity than about making errors. For them, errors of commission are less costly than errors of omission. They value novelty rather than conformity. They build an organizational culture of anticipation of change and want to be the first to try out new things. They value growth; they want to become bigger, capture the largest share, and become "number one."

Approach to planning and working with problems. Planning relies on predictions and preparing for the future. Future environmental conditions are forecast, based on which broad strategy is formulated at the top and passed down to lower levels where plans for preparedness for the future are developed. Given several possible futures, plans are prepared for each. Preactivists seek solutions by searching for new techniques and want to be on the cutting edge of technology.

Attractiveness. Commenting on the attractiveness of preactivism, Ackoff (1981) proposes that its close association with modern science and technology accounts for much of its great appeal as well as its prestige. Accepting and advocating change give the preactivists a "progressive stance at the frontiers of the future. Their preoccupation with the future gives the impression that they have it well in hand" (p. 61).

2.6.4.   The Interactive Style: "Shooting the Rapids"

The general attitude. As interactivists, we believe that it is within our power to attain the future we envision and desire to bring about, provided we learn how to do it and have the willingness to do the steering. We place the past, present, and the future in an interactive relationship. We believe that the future depends more "on what we do between now and then than it does on what has happened until now" (Ackoff, 1981, p. 146).

Perception of the role of science and technology. Interactivists create the desired future state by engaging in the intellectual technology of systems design, which applies the concepts, principles, and models of systems and design inquiry. Interactivists believes that the value of technology is manifested in the way we make purposeful use of it as a tool.

Organizational mode and culture. Interactivists integrate their systems, operating at the various levels of the systems complex, through continuous and purposeful interaction. The organizational model, proposed by Ackoff (1981) for the interactive mode, is an interlocking and interactive system of design boards established across the various systems levels of the organization, ensuring information flow in both directions and promoting collective decision making. Interactivists seek idealized solutions. They want to create a future that is better than what we have now.

Approach to planning and working with problems. Interactivists engage in two major operations: designing the desired future and planning ways and means for implementing it. Interactivists believe that "we fail more often because of an inability to face the right problems than because of an inability to solve the problems we face." This orientation is the only one in which ideals play not only an important role "but they play the key role" (Ackoff, 1981, pp. 63–64).

Attractiveness. Interactivists feel empowered to create their own future. This orientation "provides the best chance we have for coping effectively with accelerating change, increasing organizational complexities and environmental turbulence. Moreover, it is the only one of the four orientations that explicitly addresses itself to increasing individual, organizational, and societal development and improving quality of life" (Ackoff, 1981, p. 65).

Activity #7

Review the four styles of working with change and assess and compare those styles to explore their usefulness, advantages, and disadvantages. Use the four styles as a mirror to ask: Which style is my dominant style? Then ask: (1) In what situations would a particular style be appropriate? (2) Why and how would a particular style work in my organization? (3) Which style do I aspire to develop as my preferred or dominant style? Enter your findings in your workbook.

Reflections

The above descriptions of the four orientations toward change might look to be biased toward the interactive style. I believe they are not. If one compares the four styles in a summative way, it becomes obvious that the interactive style is the only one that learns from the past, values what is good in the present, and takes responsibility for the future. It is the only orientation that focuses on harnessing individual and collective aspirations, creativity, and intelligence for the purpose of seeking the ideal, and, based on it, giving direction to change and shaping the future by design.

2.7. The Role of Design in a World of New Realities

Systems design in the context of human activity systems is a future-creating disciplined inquiry. People engage in design in order to devise and implement a new system, based on their vision of what that system should be. In this section we begin to contemplate the role of design in a changing world, how it affects our own lives, and how it affects our systems.

There is a growing awareness that most of our systems are out of sync with the new realities of the current era. Those who understand this and are willing to face these realities call for the rethinking and redesign of our systems. Once we understand the significance of these new realities and their implications for us individually and collectively, we will reaffirm that systems design is the only viable approach to working with and re-creating our systems in a changing world of new realities.

2.7.1. The New Realities

Around the middle of the twentieth century a new stage emerged in societal evolution. It has been defined by various labels: the "postindustrial society"

(Bell, 1976), the "postbusiness society" (Drucker, 1989), the "postmodern society" (Harman and Horman, 1990). The label "post" indicates the transformation of our society into something very different from what it had been. But we are not yet sure what name to give to the emerged age. Whatever the name, we know for certain that this new stage of societal evolution has unfolded new thinking, new perspectives, a new scientific orientation, and a new planetary worldview, and has brought about massive changes and discontinuities in all aspects of our lives. In his speech to the U. S. Congress, Czech President Václav Havel (1990) voiced his vision of a world in which history has accelerated, and that once again "it will be the human mind that will notice this acceleration, give it a name, and transform those words into deeds." The thrust of this book is to develop the understanding that it is systems design that has the potential to transform our words and intentions into deeds.

Drucker (1989) calls the emerged changes the "new realities." It is of primary importance that we individually and collectively understand what these new realities are, grasp their implications for the design of our lives and the design of our systems. But what are these new realities and what are their implications?

In the course of the last couple of decades, observers of the societal landscape have described a major societal transformation from the industrial machine age to the postindustrial information/knowledge age. Describing these transforming societal features, Bell (1976) distinguishes three discontinuous stages in societal evolution. Our current era, which focuses on processing with the use of knowledge-based intellectual technology, is vastly different from the earlier industrial age, which focused on fabricating with the use of machine technology. And the industrial era was vastly different from the preindustrial period of agriculture and mining, the nature of which was primarily extracting.

Bell (1976) describes societal change as a major shift from producing goods to generating information and knowledge. He suggests that this shift is characterized by (1) the centrality of knowledge, (2) the creation of new intellectual technologies, (3) the spread of a knowledge class, (4) a massive change in the character of work, (5) a focus on cooperative strategy, and (6) the central role of systems science.

These massive societal changes and transformations have crucial implications for our society at large, for our systems, and for all individuals. To deal with these emerged changes, Bell sees the need for a major shift from a trial-and-error, piecemeal tinkering with our systems to their radical transformation by design. Drucker (1989) says that the biggest change we experience in our postbusiness knowledge society is a shift from the industrial worker to the knowledge worker. This shift represents a sharp break with the past. Knowledge becomes the true capital of our age. The knowledge worker sets society's values and norms and defines what to learn and how to learn it. The dominant task of the day is by no means confined to high technology or technology in general. In fact, social innovation and its intellectual technology—social systems design—may be of greater importance and have much greater impact on life than any scientific or technical invention. Drucker's focus on social innovation through systems design confirms the need to build a design culture in our systems, in our communities, and in society. Drucker (1989) also says that to respond to the new realities, within the next decades education must "change more than it has changed since the modern school was created by the printed book over three hundred years ago" (p. 232). In the knowledge age only organized and purposeful learning can convert information into knowledge, which then becomes the individual's and society's most important possession.

Robert Reich (1991) points to a major shift in economic production from a "high-volume" focus to a "high-value" focus. His analysis indicates that most of our systems still mirror the high-volume production of the industrial economy. To maintain a viable society, we must shift to a high-value focus in all of our activities so that we can respond to the new realities and evolve continuously with our ever-changing society.

Maynard and Mehrtens (1993) suggest that fourth-wave organizations in the twenty-first century will serve as global stewards; leave valuable legacies for the future; promote economic and social justice; share leadership in local, national, and global affairs; focus on quality of life and align with the natural order; practice social and resource accounting and consensual decision making; practice freedom of expression for all, openness, flexibility, and lifelong learning; strive for seamless boundaries between work and personal lives; focus on integration of life and fulfillment of purpose; and integrate ethical concerns with all aspects of the organization.

In concert with the massive changes in general societal features, new realities have emerged in our various social systems and organizations. In their book *Creative Work,* Harman and Horman (1990) elaborate a new paradigm of organizations. They provide several features of this new paradigm by juxtaposing it to traditional organizational features. In Table 2.1, I present a modification of their paradigm as the "old" and "new" story of organization culture.

The table shows stark differences between the two organizational modes. Unfortunately, the organizational culture of many, if not most, of our current systems reflects the old story. A realization of this makes it essential that we accept the challenge of designing systems with an organizational culture that is reflective of the new realities and bring the new story to life.

2.7.2. Implications of the Emerged New Realities

The emerged new realities and societal and organizational characteristics of the current era call for the development of new thinking, new perspectives, new

TABLE 2.1
The Changing Organizational Culture

The old story	The new story
Fixed, bureaucratic structure	Flexible and dynamic structure
Status-laden and rigid	Functional and evolutionary
Power resides at top	Power shared by empowerment
Motivate, manipulate people	Inspire, care for each other
Compliance is valued	Value creative contribution
Focus on problems	Focus on creating opportunities
Blame people for failure	Support learning from failure
Short-term focus	Long-term perspective
Past regimen reinforced	Innovation and novelty nurtured
Work within constraint	Seek the ideal
Progress by increments	Progress by leaps
Technology and capital based	People and knowledge based
Linear/logical/reductionist	Dynamic/intuitive/expanding
Emphasis on high volume	Emphasis on high value
Insisting on "the right way"	Encouraging learning/exploring
Driven by survival needs	Desire to develop, fulfill self
Motivated by production	Personal/collective satisfaction
Need external acknowledgment	Acknowledgment comes from self
Adversarial and competitive	Cooperative and supportive
Goals are succeed, and to go ahead	Aim at having integrity and individual and collective identity

insight, and, based on these, the design of systems that will be in sync with our transformed society. In times of accelerating and dynamic changes, when a new stage is unfolding in societal evolution, inquiry should not focus on the improvement of our existing systems. Such a focus limits perception to adjusting or modifying the old design in which our systems are still rooted. A design rooted in an outdated image is useless. We must break the old frame of thinking and reframe it. We should transcend the boundaries of our existing systems, explore change and renewal from the larger vistas of our transforming society, envision a new image of our systems, create a new design based on the image, and transform our systems by implementing the new design.

2.7.3. *Systems Design: A New Intellectual Technology*

Systems design in the context of social systems is a future-creating disciplined inquiry. People engage in this inquiry in order to design a system that realizes their vision of the future society, their own expectations, and the expectations of their environment (Banathy, 1991a).

Social systems design is a relatively new intellectual technology. It emerged only recently as a manifestation of open systems thinking and corresponding soft systems approaches. This new intellectual technology emerged, just in time, as a disciplined inquiry that enables us to align our societal systems with the new realities of the information/knowledge age. Early pioneers of social systems design include: Simon (1969), Churchman (1971), Jantsch (1975, 1980), Warfield (1976), Sage, (1977) and Jones (1980). The watershed year of developing knowledge base for systems design was 1981. This year was marked by the works of Ackoff, Checkland, and Nadler. Then came the works of Argyris (1982), Ulrich (1983), Cross (1984), Warfield (1990; Warfield and Cardenas, 1994), Nadler and Hibino (1990), Checkland and Scholes (1990), Banathy (1991a), McWhinney (1992), Weisbord (1992), Flood and Jackson (1991), Jackson (1992), Mitroff and Linstone (1993), and Pinchot and Pinchot (1993).

As I noted earlier, prior to the emergence of open social systems design, the improvement approach to systems change manifested traditional social planning (Banathy, 1991a). This approach, still widely practiced today, reduces the problem to manageable pieces and seeks solutions to each. Practitioners of this approach believe that solving the problem piece by piece ultimately will correct the larger issue this method aims to remedy. But systems designers know very well that "getting rid of what is not wanted will not ensure the attainment of what is desired."

In sharp contrast with the traditional social planning approach, the systems design approach seeks to understand a problem situation as a system of interconnected, interdependent, and interacting issues and to create a design as a system of interconnected, interdependent, interacting, and internally consistent solution ideas. Systems designers envision the entity to be designed as a whole, as one that is designed from the synthesis of the interaction of its parts. A systems view suggests that the essential quality of a part of a system resides in its relationship with, and contribution to, the whole (Banathy, 1992b). Systems design requires both coordination and integration. All parts need to be designed interactively, therefore simultaneously. This requires coordination. The requirement of designing for interdependency across all systems levels invites integration.

Activity #8

(1) List the core ideas in this section. (2) Review the statements of Bell, Drucker, and Reich and seek examples of their views on societal changes, transformations, and new realities in the context of your own experiences. (3) Review the table that juxtaposed traditional and emerging organizational characteristics and describe how those are manifested in an organization you have observed. (4) Give your answers to the questions raised in the reflections below. Enter your findings in your workbook.

Reflections

Having reached the end of this chapter, we should look back on our journey toward understanding systems design and ask a few questions. The answers developed will help in synthesizing the findings of the journey. Answer in your own way the following questions: What is systems design? What differentiates it from other modes of disciplined inquiry? Why do we need it? Why do we need it now? What might be the role of design in a changing world? What might be the role of design in our own lives and in systems in which we live? Reflecting on these questions might lead to a realization that if we want to take charge of our own lives, if we want to give direction to change, if we want to participate in shaping the future of our systems and our communities, then it is not enough to understand what design is and why we need it. We must learn how design works and how to use design. The chapters that follow will generate a better understanding of design and guide us in the use of design.

3

The Products and Processes of Design

In Chapter 2, we explored two questions: What is design and why do we need it? In this chapter some additional questions about design are asked. Section 3.1 discusses two questions: When should we engage in design and what is the product of design? The main question of how we design social systems is addressed in Sections 3.2 to 3.5. Section 3.6 focuses on how we can present the product, the outcome of what we designed. One more question remains: How do we bring the design to life? This question is answered in Section 3.7.

3.1. When Should We Engage in Design? What Is the Product of Design?

Social systems are created for attaining purposes that are shared by people who are in the system. Activities in which people in the system are engaged are guided by those purposes. There are times when there is a discrepancy between what our system actually attains and what we designated as the desired outcome of the system. Once we sense such a discrepancy, we realize that something has gone wrong and we need to change either some of the activities or the way we carry out activities. The focus is on changes within the system, which are accomplished by adjustment or improvement.

There are times, however, when we have evidence that changes within the system would not suffice, when our purposes are not viable anymore and our system is out of sync with the environment in which it is embedded. We realize that we now need to change the whole system. We need either to redesign our system or to design a new system.

Changes to a system are guided by self-regulation, accomplished by feedback. Feedback is a process by which information introduced into the system either confirms the existing state or calls for change(s) within the system or the changing of the whole system.

Through feedback, the actual state of the outcome—the output—is continually compared with the expected and stated outcome. This comparison tells us if there is a discrepancy between what is produced and what is expected to be produced. If there is a discrepancy we take action to reduce the deviation. This kind of feedback is called "negative feedback." Negative feedback reaffirms the outcome as originally stated, but it tells us that we need to correct things in the system in order to attain the stated outcomes. This feedback is the dominant operating process in relatively stable environments, when adjustments and piecemeal improvements in the existing state of the system could bring the system in line with slow and gradual changes in the environment.

Massive, accelerating, and dynamic changes characterize the current societal scene, when adjustment and changes within the system will no longer suffice. A gap opens up between the evolution of the system and the evolution of its environment. Internal, piecemeal adjustments cannot keep the system in sync with its environment. The whole system has to change in order to make it compatible with its environment. Now we must increase the deviation from the existing norms and expectations. We need to define new purposes and introduce new functions, new components. The feedback that guides this action is called positive feedback. The system responds to positive feedback through "self-creation" and learns to coevolve with its environment by transforming itself into a new state. The process by which this coevolution and transformation comes about is systems design. The guidance and regulation of social systems, therefore, employ both negative and positive feedback, either to confirm the existing system or to create a new system. They are, on the one hand, means of guided self-governance and, on the other, guided self-creation and coevolution.

Activity #9

(1) Describe the core ideas you captured in this section. (2) Review situations that have occurred in your selected system and ask the following questions: (a) Have we in our system demonstrated an understanding that there are two kinds of feedback and consequently two kinds of actions that we should consider in working with changes in our system? (b) Have we introduced only improvements and adjustments when we should have initiated the redesign of our system? Enter your findings in your workbook.

3.1.1. The Product of Design

The above discussion identified a specific state of affairs in the life of social systems that calls for the redesign of a system or the design of a new system. We can say that the genesis of systems design is either a realization of the need for a comprehensive change of an existing system or the recognition that we need to create a new system. Now we make a quantum leap from the genesis as we ask:

What is it that we design? What does design produce? Answers to these questions start to define the major thrust of this chapter, which is the definition and characterization of the process of systems design.

The product of design is a comprehensive representation, a description of the system we design. This representation is called the model of the system. The process by which the model is produced is called model building. Consequently, knowledge about models and model building is a main concern to anyone who studies and practices design. This section defines and classifies models, discusses model building, and characterizes the model maker.

3.1.1.1. Models: Definitions, Classifications, and Characterizations

The term "model" has multiple meanings. Underlying all its meanings is the notion that it is a construct or description that represents or stands for something. It can be a small-scale model of a physical object. It can be an architectural blueprint, which is a model of a building. A model may be an exemplar, such as a model teacher or a fashion model. It can be a benchmark, an achievement standard, such as the current world record of the 100-meter dash. There are conceptual models, such as mathematical and theoretical models that furnish explanations. A deductive model illustrates known relationships and characteristics of existing systems. A normative model tells us how a system should behave. A prescriptive model is a representation of a system that does not yet exist.

In systems and design inquiry we work with both product and process models (Banathy, 1973). Product models describe the outcome of the inquiry. Process models set forth the processes, the activities, by which to conduct the inquiry. A product model is often called ontological because it answers the questions "what is" or "what should be." The process model is called epistomological. It addresses how to go about designing the product.

In systems design the ontological model represents either the desired future state of an existing system or the model of a future system. This model tells us what the system should be, what it should do, and how it should operate in its future environment. The process, or epistemological, model portrays, describes, or represents an inquiry system, a system of actions that will bring about a desired outcome. In systems design, a description of the process of designing is a model of designing as a disciplined human activity.

Model building is a most cost-effective and economical mode of disciplined inquiry. It allows one to speculate freely about potential design solutions by creating various alternatives that can be described and tested to arrive at the most promising solution without a large investment of resources. Thus, it is much cheaper to design and test several alternative solutions of a future system than to develop and implement one that we think might work and then eventually realize to our regret that it does not. Pressured by time, we often jump into the implementation of an idea, adopt an invented solution, put into place an armchaired proposal, and suffer the consequences of our haste. "Haste makes waste" is nowhere more relevant than in systems design.

Lippitt (1973) suggests that a model should have three major characteristics. It should (1) clearly describe the future system, (2) facilitate communication between the various groups involved in design, and (3) serve as a guide, enabling people to develop a course of action by which to implement the design. Jantsch (1975) describes modeling as a "brain process." Humans form models through which they can relate to the world. Thus, modeling is a conscious and creative interaction between the individual and his or her environment. The same process is manifested in systems design, which is a purposeful and creative interaction between the designers and their environment, in which they create a model of the future system.

3.1.1.2. Models and Model Building: Language, Utility, and Attributes

In the various types of model building we should use language that best expresses the specific content and context of what the model stands for and is most appropriate to the people who will use the model. Lippitt (1973) noted that one of the key elements of designing is perception, the process of becoming aware of relevant phenomena and their relationships within and outside of us. Models can both express and enhance perception. The model builder (Hanneman, 1975) abstracts the model from his perception of the world in order to express isomorphism between the model and what the model stands for. But a model of a social system cannot describe its full complexity. It will be a simplification of complexity, while still showing and displaying all the essential features of the system. Furthermore, inasmuch as the design of social systems is basically communication, the language used should be understandable to people in the system.

The utility of model building is manifested by several factors:

1. There is a time advantage of manipulating representations through language instead of working with the real system.
2. The cost of designing and testing alternative models is much cheaper than building and testing alternative systems.
3. The model enables contemplation.
4. It provides a framework for (1) structuring the effort, (2) acquiring the required knowledge base for design, and (3) carrying out evaluation.
5. Modeling offers potential for identification of additional questions not obvious initially to designers.
6. Models offer designers a structure for organizing, displaying, and examining the findings of their inquiry and testing them conceptually as well as in the real world.

The attributes of models are explored next by interpreting Mitroff's (1979) formulation:

1. The model should be complete. It should describe all of the important elements of the system and its relationships.
2. The model should be simple. It should depend on few assumptions, have few parameters, and have straightforward relationships.
3. It should be adaptable; its structure and parameters should be easily updated or modified.
4. The model should be fertile. It should provide new insights and its deductive properties should be rich.
5. The model should be believable and credible.
6. It should be economical, requiring a knowledge base that is available or economically attainable.
7. The model should be transparent. The assumptions and logic of the model should be easily communicable.

In building a model, Florian (1975) suggested that we should keep in mind several considerations:

1. Are the temporal effects of the model of prime importance, or is it robust enough to describe the state of the system through time?
2 What appropriate level of detail is to be employed in order to properly represent the system?
3. What precision is needed: do we need accurate solutions or are approximate solutions acceptable?
4. How will the user interact with the model; what preparation/education should the user have in order to benefit from the model and its potential?
5. What criteria may be used to evaluate the model, and how sensitive should the model be to the criteria?
6. What level of error is acceptable?
7. In what environment is the model to be implemented, or what environment will be required for the system?

These considerations are representative of the kinds of expectations that designers might have as they build models of social systems. To aid in the understanding of the term "model" in social systems design, it might be useful to read ahead and review briefly subsection 3.2.1 of this chapter.

3.1.1.3.   The Designer as a Model Maker

A discussion on model making should include studying the model maker. The model maker is the designer who constructs a representation, a model of the system to be created. Based on a contemplation and interpretation of the work of Lippitt (1973), a few of the salient attributes of the model maker are introduced here.

Confidence and a certain amount of courage are required to transcend an existing system or state and attempt to create a conceptual model as the design solution. Making a model of a desired future system requires a degree of confidence in one's assessment of the present and in one's commitment to a vision of the future, and it calls for willingness to take risk with conviction.

Situational sensitivity implies seeing things that others might not perceive, seeing things that are not obvious, stretching one's perceptual powers to capture and feel more about the situation than would ordinarily be the case, and becoming tuned in to the complexity of emerging design solutions.

Flexibility calls for adjusting quickly to emerging developments in the design situation, extending the boundaries of the design inquiry, experimenting with various design solutions, abandoning old assumptions and trying out new ones, and—most importantly—adding new dimensions to the solution rather than merely adjusting the old.

Tolerance for ambiguity and uncertainty means tolerating a certain amount of disorder in bringing meaning to contradictions and dynamic complexity, living with the uncertainty of emerging design solutions, and withstanding the pressure for immediate or quick solutions.

Moving back and forth between analysis and synthesis interactively is an essential requirement of the model builder. In synthesizing, we identify, create, combine, and enfold different elements into a holistic framework of the emerging process and structure of the system. At the same time, we constantly analyze what we create. Moving between synthesis and analysis is a key requirement of effective model building.

Managing design takes place in an environment of dynamic complexity that is unpredictable, ambiguous, and unique. Designers in such contexts cannot rely on standard procedures. They have to manage design with the use of methods tailored to the design situation while they seek solutions in the flow of all the processes that are manifested in social systems.

Activity #10

Review the section and ask yourself: (1) What core ideas and organizing perspectives can I formulate that will help me in building a model of a future system? (2) What are potential pitfalls of model building of which I should be aware as I attempt to build a model of the systems I design? (3) How do I rate myself as a model builder? (4) What specific attributes of a good model builder should I develop? (5) What additional learning should I undertake? Record your answers in your notebook.

Reflections

Two crucial aspects of designing social systems were discussed. First we answered the question of when we should engage in systems design as we characterized situations where the action, error detection, and correction process of initiating changes within the system are adequate. Then we discussed situations when correction within the system is no longer adequate and we need to change the whole system. Next we focused on the outcome of design. We discussed the nature of models and model building and characterized the designer as a model maker. An understanding of these three issues is deemed to be prerequisite to exploring the process of systems design.

3.2.   The Design Process

The main question of Chapter 3 is: How do we design social systems? The answer to this question is developed in the next four sections, where I progressively elaborate the process of designing social systems.

3.2.1.   *Research Findings on the Nature of Design Activity*

In chapter 2, we discussed the conclusion-oriented and decision-oriented modes of design inquiry. The conclusion-oriented inquiry is a research mode, aiming at producing knowledge about design. The decision-oriented mode uses knowledge about design in carrying out systems design. In Cross's compendium (1984), four design researchers report their findings on the general nature of design. I briefly review their findings.

Darke (1984) has found that contemporary designers have rejected the earlier systematic, objective, analysis–synthesis–evaluation approach to designing and replaced it with what Hiller *et al.* (1982) called "conjecture-analysis." The point of departure of this approach is not a detailed analysis of the situation, but the formulation of a conjecture, a contemplation of "what should be" that Darke has termed "primary generator." The primary generator is formed very early in the design process as a set of initiating concepts. (I later call this a system of core ideas: the first image of the system.) This primary generator helps designers make the creative leap between the problem formulation and a solution concept, as Cross (1984) noted. Broad design requirements, in combination with the primary generator, help designers arrive at an initial conjecture that can be tested against specific requirements as an interactive process. Conjectures and requirements mutually shape each other.

While earlier design approaches concentrated on design morphology as a sequence of boxes bearing preset labels, Darke (1984) finds that designers now fill the boxes with their own concepts and the sources of their concepts. An understanding of the subjectivity of designing reflects the diversity we find in human experience, which, in turn, should reflect the diversity in approaches to design.

Akin (1984) challenges earlier assumptions about design. As Darke did, he also takes issue with the analysis–synthesis–evaluation sequence in design. He says this approach was at the heart of almost all normative design methods of the past. He suggests that one of the unique aspects of designing is the constant generation of new task goals and the redefinition of task constraints. "Hence analysis is part of virtually all phases of design. Similarly, synthesis or solution development occurs as early as in the first stage" (Akin, 1984, p. 205). The rigid structuring of the design process into an analysis–synthesis–evaluation sequence and the tactics implied for these compartments are unrealistic. Solutions do not emerge from an analysis of all relevant aspects of the problem. Even a few cues in the design environment are sufficient to evoke a recombined solution in the mind of the designers. Actually, this evoking is more the norm than a rational process of assembly of parts through synthesis. Many rational models of design violate the widely used criterion of designers, namely, to find a satisfying, rather than a scientifically optimized, solution. No fixed model is complex enough to represent the real-life complexities of the design process. That is why designers select approaches that produce a solution that satisfies an acceptable number of design criteria. Social systems design, being an intuitive process, has to conform to the nature of the human experience.

Lawson (1984) conducted a controlled experiment between scientist and designers. He discovered that the scientist used processes that focused on discovering the problem structure, whereas designers used strategies that focused on finding solution structures. For the designers, the most successful and practical way to address design problem situations is not by analyzing them in depth, but by quickly proposing solutions to them. This way, they discover more about the problem, as well as what is an acceptable solution to it. By contrast, scientists analyze the problem in order to discover its patterns and its rules before proposing a solution to it. Designers seek solutions by synthesis, scientists by analysis. Accordingly, designers evolve and develop methodologies that do not depend on the completion of analysis before synthesis begins.

Thomas and Carroll (1984) carried out a wide range of studies on design that indicated similarities between the behavior of designers and their approaches to design, regardless of the particular subject of design. The authors said they changed their original assumption that design is a form of problem solving to the opinion that design is "a way of looking at a problem." They consider design as a dialectic interactive process between the participants of the design activity. In this process participants elaborate a goal statement into more explicit functional requirements, and from these they elaborate the design solution.

In reviewing the four research findings, Cross (1984, pp. 172–173) arrives at two major conclusions. The first is an inevitable emphasis on the early generation of solutions so that a better understanding of the problem can be developed. Second is that the earlier systematic approaches tend to focus on an extensive problem analysis, which seems unrealistic to ill-defined problems.

Reflections

The above discussion highlights the differences between systematic and systemic approaches to design as a recurring issue. The term "systematic" was in vogue in the 1950s and 1960s. During that period, closed, engineering systems thinking dominated the scene. The term implied regularity in a methodical procedure. In design, it means following the same steps, in a linear, one-directional causation mode, and adhering to the prescribed design method, regardless of the subject and the specific content and context of the design situation. Designers of the 1970s and 1980s have learned the confining and unproductive nature of the systematic approach. Once we understood the dynamic complexity and nonlinear nature of the open system, and the mutually affecting nature of social systems, we developed a "systemic" approach that liberated us from the restrictive and prescriptive rigor of a systematic approach. Systemic relates to the dynamic interaction of parts from which the integrity of wholeness of the system emerges. Systemic also recognizes the unique nature of each and every system. It calls for the use of methods that respect and are responsive to the uniqueness of the particular design situation, including the unique nature of the design environment. A systematic approach in not sensitive to such uniqueness.

Activity #11

The four research reports are rich ground from which to draw and describe core ideas about design. Describe examples that demonstrate the difference between a systematic versus a systemic approach in disciplined inquiry. Record your findings in your workbook.

3.2.2. Models for Building Social Systems

Until the 1970s, design as a disciplined inquiry was primarily the domain of architecture and engineering. In social and sociotechnical systems, the nature of the inquiry was either systems analysis, operations research, or social engineering. These approaches reflected the systematic, closed, hard systems thinking discussed in the previous section. It was not until the 1970s that we realized that the use of these approaches was not applicable, and, in fact, they were counterproductive in working with social systems. We became aware that social systems are open systems, that they have dynamic complexity, and that they operate in turbulent and ever-changing environments. Based on this understanding, a new orientation emerged, based on "soft systems" and "critical systems" thinking. The insights gained from this orientation became the basis for the emergence of a new generation of designers and the development of new design models applicable to social systems. In the last section of Chapter 2, I listed systems and design researchers who made significant contributions to the development of open social systems design. Of this group, I selected three scholars, namely, Ackoff, Checkland, and Nadler, who developed comprehensive process models of systems design. Their work has set the trend for continuing work in social systems design.

3.2.2.1. Ackoff: The Design of Idealized Systems

The underlying conceptual base of Ackoff's (1981) design model is a systems view of the world. He explores how our concept of the world has changed in recent time from the machine age to the systems age. He defines and interprets the implication of the systems age and the systems view to systems design. He sets forth design strategies, followed by implementation planning. At the very center of his approach is what he calls "idealized design."

Design commences with an understanding and assessment of what is now. Ackoff calls this process "formulating the mess." The mess is a set of interdependent problems that emerges, the problems being identifiable only in interaction. Thus, the design that responds to this mess "should be more than an aggregation of independently obtained solutions to the parts of the mess. It should deal with messes as wholes, systemically" (Ackoff, 1981, p. 52). This process includes systems analysis, a detailed study of potential obstructions to development, and the creation of projections and scenarios that explore the question: What would happen if things would not change?

Having gained a systemic insight into the current state of affairs, Ackoff proceeds to the idealized design. The selection of ideals lies at the very core of the process; "it takes place through idealized design of a system that does not yet exist, or the idealized design of one that does" (p. 105). The three properties of an idealized design are that it should be technologically feasible, operationally viable, and capable of rapid learning and development. This model is not a utopian system but "the most effective ideal-seeking system of which designers can conceive" (p. 107). The process of creating the ideal includes selecting a mission, specifying desired properties of the design, and designing the system. Ackoff emphasizes that the vision of the ideal must be a shared image. It should be created by all who are in the system and those affected by the design. Such participative design is attained by the organization of interlinked design boards that integrate representation across the various levels of the organization.

Having created the model of the idealized system, designers engage in the design of a management system that will (1) guide the system, (2) identify threats and opportunities, (3) identify what to do and how to have it done, (4) maintain and improve performance, and (5) guide organizational learning. The next major task is the design of the organization that will have the capability to carry out the functions, an organization that is "ready, willing, and able to modify itself when necessary in order to make progress towards its ideals" (p. 149). The final stage is implementation planning. It is carried out by selecting or creating the means by which the specified ends can be pursued, determining what resources will be required, planning for the acquisition of resources, and defining who is doing what, when, how, and where.

3.2.2.2.   Checkland: Soft Systems Methodology

Checkland (1981) and Checkland and Scholes (1990) create a solid base for a model for systems change by reviewing science as human activity, the emergence of systems science, and the evolution of systems thinking. They differentiate between hard systems thinking, which is appropriate to working with rather closed, engineered types of systems, and soft systems thinking, which is appropriate to working with social systems. They try to make systems thinking a conscious, generally accessible way of looking at entities. Based on soft systems thinking, they formulate a model for working with and changing social systems.

Their seven-stage model generates a total system of change functions, leading to the creation of a future system. Their conceptual model of the future system is similar in nature to Ackoff's idealized system. Using their approach, during the first stage we look at the problem situation of the system, which we find in its real-life setting as being "unstructured." At this stage our focus is not on specific problems but the situation in which we perceive the problem. Given the perceived "unstructured situation," during stage two, we develop a richest possible structured picture of the problem situation. These first two stages operate in the context of the real world.

The next two stages are developed in the conceptual realm of systems thinking. Stage three involves speculating about some systems that may offer relevant solutions to the problem situation and preparing concise "root definitions" of what these systems are (not what they do). During stage four, the task is to develop abstract representations—models of the relevant systems—for which root definitions were formulated at stage three. These representations are conceptual models of the relevant systems composed of verbs denoting functions (what these systems do). This stage consists of two substages: the conceptual model is first described and then checked against a formal model of systems, such as Churchman's (1971) model of assessing the systemic nature of an entity.

During the last three stages, designers move back to the realm of the real world. In stage five, the conceptual model is compared with the structured problem situation that was formulated during stage two. This comparison enables identification, in stage six, of the feasible and desirable changes in the real world. Stage seven is devoted to taking action and introducing changes into the system.

3.2.2.3. Nadler: The Planning and Design Approach

An early proponent of designing for the ideal, Nadler (1967) is the third systems scholar who developed a comprehensive model (Nadler, 1981) for the design of sociotechnical systems. During phase one, his strategy calls for the development of a hierarchy of purpose statements, which are formulated so that each higher level describes the purpose of the next lower level. From this purpose hierarchy, the designers select the specific purpose level with which to create the system. The formulation of purpose is coupled with the identification of measures of effectiveness that indicate the successful achievement of the defined purpose. During this phase, designers explore alternative reasons and expectations that the design might accomplish.

During phase two, "creativity is engaged as ideal solutions are generated for the selected purposes within the context of the purpose hierarchy" (Nadler, 1981, p. 9). Nadler introduces a large array of methods that remove conceptual blocks, nurture creativity, and widen the creation of alternative solution ideas.

During phase three, designers develop solution ideas into systems of alternative solutions. Here designers play the believing game as they focus on how to make ideal solutions work, rather than on the reasons why they won't work. They try ideas out to see how they fit.

The solution is detailed in phase four. Designers build into the solution specific arrangements that might cope with potential exceptions and irregularities while protecting the desired qualities of solutions. As Nadler says, "Why discard the excellent solution that copes with 95% of the conditions because another 5% cannot directly fit into it?" (p. 11). As a result, design solutions are often flexible, multichanneled and pluralistic.

Phase five involves the implementation of the selected design solution. In the context of the purpose hierarchy, we set forth the ideal solution and plan for taking action necessary to install the solution. But we have to realize that the "most successful implemented solution is incomplete if it does not incorporate the seeds of its own improvement. An implemented solution should be treated as provisional" (p. 11). Therefore, each system should have its own arrangements for continuing design and change.

Activity #12

(1) Review the three design models and capture the core ideas of the three authors about conducting design. (2) Contemplate the use of the core ideas in the context of the system of your interest. Enter your findings in your workbook.

3.2.2.4.   Banathy: Social Systems Design

The three design models introduced above have been applied primarily in the corporate and business community. Still, we can learn much from them as we seek to formulate an approach to the design of social services and other societal systems. I outline a process model of social system design, which has been informed by the work of Ackoff, Checkland, and Nadler, and is a generalized version of my earlier work (Banathy, 1991a). The sections that follow discuss this model in some detail.

The process of design that leads us from an existing state to a desired future state is initiated by an expression of why we want to engage in design. I call this expression of want the genesis of design. Once we formulate this genesis, we initiate five major processes:

- Transcending the state or the existing system and leaving it behind.
- Envisioning an image of the system that we wish to create.
- Designing the system, which, when implemented, transforms the existing state to the desired future state.
- Presenting/displaying the model(s) of the system we designed.
- Planning for the implementation of the design.

Transcending, envisioning, and designing for transformation; displaying the model of the system we designed; and planning for bringing the design to life by developing the system are the major strategies of designing social systems. A brief outline of these processes follows.

3.2.2.4a.   Transcending.   If we want to change the existing system or create a new one, we are confronted with the task of transcending the existing system or the existing state of affairs and creating a framework that can serve as a guide to mapping out alternative boundaries for the design inquiry and major solution options. The exploration of options leads us to make a series of decisions that give direction to the design program.

3.2.2.4b.   Envisioning the First Image.   Systems design creates a description (a representation, a model) of the future system. This creation is grounded in the designers' vision, ideas, and aspirations of what that future system should be. As the designers draw the boundaries of the design inquiry on the framework, described above, and make choices from the options they created, they collectively form core ideas about the desired future. They articulate their shared vision and synthesize their core ideas into the first image of the system. This image becomes a magnet that pulls designers into designing and describing the system that will bring the image to life.

3.2.2.4c.   Designing for Transformation.   The image expresses an intent. One of the key issues in working with social systems is: How does one bring intention and design together and create a system that transforms the image to reality? The image becomes the basis that initiates the strategy of transformation by design. The design solution emerges as designers (1) formulate the mission and purposes of the future system, (2) define its specifications, (3) select the functions that have to be carried out to attain the mission and purposes, (4) organize these functions into a system, (5) design the system that will guide the functions and the organization that will carry out the functions, and (6) define the environment that will have the resources to support the system.

3.2.2.4d. Presenting the Model. This process involves the conceptual representation of the future systems and the description of the systemic environment of the new system. (The systemic environment is that part of the general environment with which the system regularly interacts.)

3.2.2.4e. Planning for Implementation. This process includes the preparation of a plan for the development of the system that brings the design to life, the development of the system and its implementation in the context of the system's environment.

Reflections

In this section a major step has been taken toward the understanding of systems design by exploring some research findings about design, examining a set of comprehensive design models, and proposing a process model for the design of social systems. The sections that follow build on the material introduced here by extending the horizon of understanding of both the process and the products of the design of social systems.

3.3. The Process of Transcending

A realization that we should change the existing system as a whole or that we want to create a new system is the genesis of design. Once we make the decision to redesign our existing system, we have to leave the system behind, transcend it, and leap out from its boundaries. This leaping out and transcending are probably the most troublesome parts of systems design. This difficulty could be reduced by following Ackoff's (1981) idea of speculating about what would happen if we were not to change the system.

Even if we decided to design a new system, we still have to transcend the existing state of affairs and leap into a sphere of conceptualization where we can envision and create the new system. The ease or difficulty of leaping out and transcending depends a great deal on our attitude toward change. (Various modalities of responding to change were reviewed in Chapter 2.) In this section, I introduce a framework that enhances the process of transcending. Using the framework, designers can create an option field, consider major design options, and explore the implications of the use of the option field.

3.3.1.   The Creation of a Framework that Offers an Option Field

I introduce here an example of a framework that enables designers to transcend the existing system, establish boundaries of the design inquiry, and create some major design options of a desired future system. The framework is constructed by first identifying dimensions that create the option field. As an example, Fig. 3.1 depicts an option field for the design of learning and human development systems (Banathy 1991a).

The option field in Fig. 3.1 comprises four dimensions for the framework: the focus of the inquiry, its scope, relationship with other systems, and the selection of systems type. On each of the dimensions, several possible options are plotted. The options gradually extend the boundaries of the inquiry. Using the framework, designers can consider and draw alternative and expanding boundaries of the design inquiry (in the figure, options are marked A, B, C, and D). By contemplat-

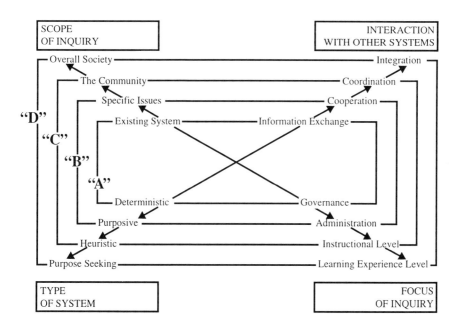

FIGURE 3.1. A framework for creating an option field.

ing various possible boundaries, designers map out option fields within which they can explore a range of major solutions. Their exploration leads to tentative choices. The synthesis of those choices will help designers envision what is possible and eventually help them create the first image of the new system.

The advantage of the framework is that it creates a place where designers can land as they leap out from their existing system. The framework creates a "comfort zone" in which the anxiety of leaping out melts away and the excitement of entering into the space of envisioning, imaging, and creating replaces troubling uncertainty and the fear of the unknown.

3.3.2. Considering Options

The option choices of the four dimensions outlined in Fig. 3.1 are described next.

3.3.2.1.   The Focus Options

The focus options mark the four levels at which the primacy of a system level can be conceptualized. In Fig. 3.1 these levels are governance, administration, and the instructional and learning experience levels. The question is: Which system level is in focus? In societies where the state or the church operates education, governance dictates. In the United States, the public education system is built around administration. In some recent restructuring programs authority has been shifted to the instructional level. The fourth option calls for designating the learning experience as the primary level and building the whole systems complex around it. When learning comes into focus, the learner becomes the key entity of the system, and the primary task is to provide resources, arrangements, and opportunities for learning and human development. Choosing learning as the focus will require a major boundary change and, consequently, it calls for the design of a new system.

3.3.2.2.   The Scope Options

The scope options include (1) staying within the boundaries of the existing system, (2) considering certain issues in the environment that might be addressed by education, (3) making the community the context of design, and, finally, (4) defining the whole society as the context of the inquiry. Choosing scope options (1) and (2) means that we do not change boundaries; therefore, we cannot talk about design. Option (3) extends the boundaries into the community and contemplates the inclusion of all systems in the community that can provide resources, opportunities, and arrangements for learning and human development. Option (4) means a further extension of boundaries into the overall society as the context of design inquiry. Considering this option, we are exploring the educational implica-

tions of (1) the emerged characteristics of the postindustrial, informa-
tion/knowledge era and (2) the massive changes and transformations that have
emerged in the sociocultural, socioeconomic, sociotechnical, scientific, and organizational spheres of the society. Considering the educational implications of
emerged characteristics will guide us in (1) the redefinition of the societal functions of education, (2) the development of new learning agendas, and (3) the
creation of new organizational arrangements for learning and human development. Choosing this fourth option calls for a major boundary change and therefore
for the design of a new system.

3.3.2.3. Relationship with Other Systems

The dimension of relationship with other systems offers four options:

1. Information exchange with other systems in the environment, which is the
   typical practice today.
2. Occasional cooperation with other systems in the community.
3. Coordination with other systems that might offer educational and human
   development resources and opportunities.
4. Integration with all other social service, health, human development, and
   community service systems.

Option (4) will create a "new educational species," a community-based
system of learning and human development. Options (1) and (2) do not transcend
the existing system but might involve some minor boundary adjustments. Option
(3) and (4) imply new boundaries and call for the design of a new system.

3.3.2.4. Selecting the System Type

Social systems, such as education, are purposeful creations. People in these
systems select, organize, and carry out activities to attain a purpose. We can
classify these systems by the degree to which they are open or closed, mechanistic
or systemic, unitary or pluralistic, and complex or simple. Based on these dimensions, we can differentiate four system types (Banathy, 1988c).

- Deterministic systems are rather closed. They have clearly assigned goals;
  thus, they are unitary. People in these systems have a limited degree of
  freedom in selecting methods. The systems operations are rather mechanistic. Their complexity ranges from simple to detailed. Examples are industrial bureaucracies, centralized national educational systems, and government bureaucracies.
- Purposive systems are still unitary. Their goals are set at the top. They are
  somewhat open and react to changes in their environment in order to
  maintain their viability. People in the system have freedom in selecting

operational objectives and methods. Examples are corporations, social service agencies, and our public education system.

- Heuristic systems, such as R & D agencies and innovative business ventures, formulate their own goals under broad policy guidelines; thus, they are somewhat pluralistic. They are open to changes and often initiate them. Their complexity is dynamic and their internal arrangements and operations are systemic.
- Purpose-seeking systems are ideal seeking, guided by their vision of a future. They are open and coevolve with their environment. They exhibit dynamic complexity and systemic behavior. They are pluralistic in that they constantly seek new purposes and new niches in their environments. An example is communities seeking to establish an integration of their educational, social services, an human development systems.

Deterministic and purposive systems mean no change in boundaries. But changing to heuristic and purpose-seeking system types calls for changing inquiry boundaries and purposes, and therefore designing new systems.

3.3.3. Implications of Using the Framework

The implications of choosing the options marked in Fig. 3.1 are explored next. If we stay within the boundaries marked A and B, we might attend to problems in the existing system and somewhat improve or restructure it. Pursuing options A and B, we are not engaged in design. Options C and D, on the other hand, create two gradually expanding design inquiry spaces. The further we go out in selecting options on any of the dimensions of the framework, the more novelty we create, the more complex the system will be, and the more involved and time–resource consuming is the design. As we use the options framework, tackle decision points, and make choices among the options offered by the four dimensions, a conversation unfolds around several questions:

- Why did we select a particular option? As we discuss reasons for selecting options, we begin to articulate, share, and harmonize our values, assumptions, preferences, and ideas that underlie and support our choices.
- Is there internal consistency between the choices we made? Exploring this question, we begin to probe the alternative solution ideas we are considering within the larger picture of the whole context of our inquiry.
- Are our choices compatible with our stated values and ideas? Are they supportive of each other? Answering these questions, we explore how particular choices affect each other and test their compatibility with stated core values and core ideas.

Going back to our educational example:

- Selecting the learning experience level as focus we recognize that we need a much larger base for learning resources and opportunities than what is available in the classroom. We need to create learning territories beyond the walls of our buildings. This will require at least coordination and possibly integration with other systems in the community.

- If the system we aspire to create is expected to coevolve with the societal environment in which it is embedded, and if we wish to work continuously with ongoing changes, then we should design a purpose-seeking system and select the overall society as the functional context for design.
- If our aspiration is to integrate learning with human and community development, establish lifelong learning and bring to life a learning society, then we shall seek the integration of all those systems in the community that offer opportunities, resources, and arrangements for learning and human development.

3.3.4. Generalizing the Example

The educational system context is an example of using a framework for transcending an existing system, setting inquiry boundaries, and considering some solution configurations. Generalizing the example into the context of other social systems, it is suggested that in most cases several dimensions of the example would be applicable. In designing social systems we must define what the focus of the inquiry is. We should establish the scope of the inquiry, consider relationships with other systems, and make a choice of a system type. But there might also be some other dimensions that are unique to the particular social system. The option markers on the dimension of system type might be generalizable across social systems. But the option markers on the dimensions of scope, focus of inquiry, and relationships with other systems will be unique and specific to the particular system in question.

Activity #13

First describe core ideas of transcendence. Then, choose a system for (re)design. (1) Create a framework from dimensions that define an option field. (2) Mark options on each of the dimensions. (3) Draw alternative inquiry boundaries and contemplate option configurations. (4) Answer the question: What values might support boundary choices and choices of design options? Note your findings in your workbook.

Reflections

Transcending the existing state is the first task of designers who wish to bring their system in sync with the new realities of our age or who wish to design a system that responds to their desires, aspirations, and expectations. They must transcend old ways of thinking and reframe their thinking, shift from a problem focus to a solution focus, unload the baggage of the past, unlearn past habits and practices, and learn new ones. These changes "within" are the most difficult part of design. As I reflect on transcendence, I am reminded of Einstein's admonition: we cannot address a problem from the same consciousness that created it. We must think anew.

3.4. Envisioning the Image of the New System

> A young nation is confronted with a challenge for which it finds a successful response. It then grows and prospers. But as time passes the nature of the challenge changes. And if a nation continues to make the same once successful response to the new challenge, it inevitably suffers a decline and eventual failure.
> —Arnold Toynbee

Standing at the threshold of the twenty-first century we are faced with the challenge of a new era. This challenge requires new societal responses. It requires the envisioning of new images of our social and societal systems, and, based on the images, it requires the transformation of our systems by design.

The framework discussed in the preceding section enables designers to transcend the boundaries of the existing system and explore alternative inquiry options. In the course of this exploration a conversation develops in the designing community that leads to a collective clarification of shared core values and core ideas and to an integration of these into an image of the new system. It is important to note that the use of the framework, the consideration of options, the articulation of core values and ideas, and the creation of the image constitute a dynamic interacting, interdependent, and integrating process. To better understand this process and its impact on design we must now ask: What is the meaning and significance of the notion of societal images in general and the creation of an image in the design of social systems?

3.4.1. The Meaning and Significance of Image

"Images of humankind which are dominant in a culture are of fundamental importance, because they underlie the ways in which the society shapes its institutions, educates its young, and goes about whatever it perceives its business to be" (Markley and Harman, 1982, p. 201). Our society is now undergoing a transformation that is more profound, more intensive, and more dynamic than any in human history. This transformation shapes emerging societal images that are vastly different from the image of the by-gone industrial era. But we now realize that despite these massive changes, many of our social systems are still grounded in the images of the industrial machine age.

When I study the notion of image, I go to Boulding's (1956) classic booklet *The Image*. His first proposition is that behavior depends on the image. The behavior of designers of social systems depends on their image of their society and the image they have about the societal function of the system they wish to create. Boulding's second proposition is that our experiences provide us with messages that might produce changes in our image. The majority of the messages do not bring about change. There are some, however, that call for some adjustments in the image. However, when the message hits the nucleus of the image, "the whole thing changes in quite a radical way" (p. 8). Images are resistant to change and we often reject a message that challenges our prevailing image. But if the message persists, we begin to have doubts. "Then one day we receive a message which overthrows our previous image and we revise it completely" (p. 9). It is precisely such a message event to which systems design responds.

"The basic bond of any society, culture, subculture or organization is a public image, that is, an image the essential characteristics of which are shared by the individuals participating in the group" (Boulding, 1956, p. 64). Our concern here is such a public image, an image of the system we design that is shared by the members of the designing community. The making and remaking of such an image are linked with our value system. In the development of the image "the factual and valuational images are inextricably entwined" (p. 174). We experience this entwining as we integrate our core values, which guide our design choice decisions, with the core ideas that emerge as we explore design options.

In their study of societal evolution, Markley and Harman (1982) suggest that if an inherited image is used to guide the society at a time when a new evolutionary stage is emerging, the old image retards development. Its continued use creates far more problems than it solves. It opens up a developmental gap. By contrast, when a new image leads sociocultural development, it provides direction to the design and redesign of social systems.

3.4.2. Sources in Creating Image

There are a variety of sources that designers can use in their quest to create an image of the society they wish to have and, based on it, an image of the system they design. There are some invariant image elements that are universal in nature. Other elements are specific to the unique context of the design. Huxley (1945) calls invariant sociocultural images the "perennial philosophy." These images can be found in the traditional lore of cultures as well as in advanced philosophy and in the ancient and contemporary forms of religion. These invariants include the notions of harmony, balance, and wholeness; the universe of individual, collective, and divine consciousness; an awareness of a higher self and quest toward its realization; the unlimited human potential; motivation toward creativity; and a quest toward higher levels of consciousness that enables us to participate in conscious evolution. Churchman's (1982) invariant value is the "human within," which he considers sacred. For him this implies the moral law that "we should undertake to design our societies and their environments so that the people in the future will be able to design their lives in ways that express their own humanity" (p. 21). Wilson's (1993) invariant aspects of "moral sense" include sympathy, fairness, self-control, and duty.

The kinds of invariant images mentioned above, and others that designers select, will be integrated with sets that are unique to the system, sets that designers articulate as they collectively define the core ideas and values that guide their design inquiry. The outcome of this process is the creation of a value-based and value-driven image of the future system. The more intensive the exploration of the option field, and the more core values and core ideas formulated and integrated, the more compelling and powerful the image.

In Table 3.1, I introduce the example of an image (Banathy, 1991a) that reflects "invariant" values as well as core values that might be sources for designing systems of learning and human development.

TABLE 3.1
An Image Example

The image of a desired future system	The image of the existing system
Education should reflect and interpret society through coevolutionary interaction, as a future-creating, innovative, and open system.	Education is an instrument of cultural and knowledge transmission, focusing on maintaining the existing state, operating in a rather closed systems mode.
Education should be coordinated with other social and human service systems, integrating learning and human development systems.	Education is an autonomous social agency, separated from other social services and human service systems.
Education should provide resources, arrangement for learning, and life-long experiences for the full development of learners.	Education now provides instruction to individuals during their schooling years.
Educational experiences should embrace the sociocultural, ethical/moral, economic/occupational, physical/mental/spiritual, political, scientific/technological, and the aesthetic.	Education now focuses on the basics and limited preparation for citizenship and employment.
Education should be organized around learning experiences; arrangements should be made in the environment of the learner by which to master the learning task.	Education is now organized around instruction; arrangements are made that enable teachers to present subject matters to students.
Use a variety of learning types; self-directed, other-directed, cooperative team learning, social and organizational learning.	Today teacher–class and teacher–student interactions are the primary means of the educational experience.
Use the large reservoir of learning resources and arrangements available in the society to support learning and human development.	The use of learning resources and arrangements are confined within the school.

We may consider the image to be the "DNA" of an education system. The DNA is imprinted in all parts of the system so as to help create an internally consistent and viable living organization (the metaphor of a holographic image would also apply). But the image is not yet the system. It represents only an intent, and "intentions are fairly easy to perceive, but often do not come about. Design is hard to perceive. But it is design and not intention that creates the future" (Boulding, 1985, p. 212). One of the key issues in working with social systems is how to bring intention and design together and create the system that we intend to bring about.

Activity #14

Keeping in mind the choices you made in Activity #13 concerning your framework, inquiry boundaries, and option configurations, (1) identify the core values and core ideas that led you to make those choices and (2) create a tentative image of your system. Describe your findings in your workbook.

Reflections

Most likely, an image of a new system will be revolutionary, but the move toward it can be evolutionary. The image is speculative and serves to trigger further exploration and refinement. Its components should be compatible with each other, composing a system of internally consistent markers of the future system. The image becomes the basis of engaging in the design of the desired future system.

3.5. Transformation by Design

Systems design in the context of social systems is a future-building, decision-oriented, disciplined inquiry. It aims at the creation of a system that will bring to life our aspirations. Once we have transcended the existing system, explored major design options within the boundaries of the design inquiry, and envisioned an image of the future system, we are ready to proceed with the transformation by engaging in the design or redesign of our system. Design inquiry involves the conceptualizing and testing of potential solution alternatives, and the selection of the most desirable alternative.

3.5.1.   The Design Architecture

Systems design is a conceptual process that takes place in five spaces (Banathy, 1991a). The relational arrangement of these spaces is the design architecture, as shown in Fig. 3.2.

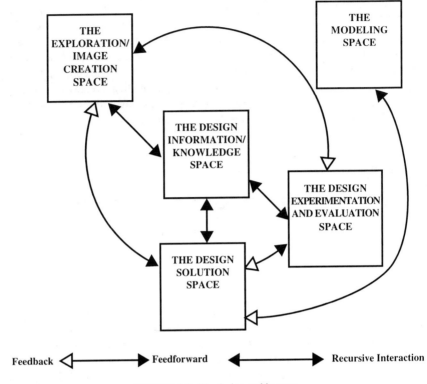

FIGURE 3.2. The design architecture.

3.5.1.1. Exploration and Image Creation Space

In this space we report on the findings of the inquiry process described in Section 3.4, namely, we (1) establish inquiry boundaries, (2) explore design options, (3) generate core ideas and core values, and, based on them, (4) create an image of the future system.

3.5.1.2. Design Information and Knowledge Space

In this space we display knowledge that we generated in the exploration/image creation space—knowledge about (1) the content and context of the design, (2) the characteristics of the system's environment, and (3) the various design models, methods, and tools from which to select.

This process of information/knowledge development is not a one-shot deal. We constantly need new knowledge to support the creation and testing of alternative design solutions.

3.5.1.3. Design Solution Space

In this space, we (1) formulate the core definition of the system that responds to the image we created, (2) select systems specifications based on the core definition, (3) design the system of functions that have to be carried out to attain the desired outcomes, (4) design the enabling systems that have the capacity and human capability to carry out the functions, and (5) design the systemic environment that will support the system.

3.5.1.4. Evaluation and Experimentation Space

In this space we (1) experiment with and evaluate solution alternatives and (2) evaluate them by using the criteria that are based on the image, and the core definition and specifications formulated in the design solution space.

3.5.1.5. Modeling Space

In this space we display (1) the model of the system and (2) the model of the systemic environment.

3.5.2. Design Dynamics

The dynamics of design are shown in Fig. 3.3, which portrays the process flow of systems design through the various spaces of the design inquiry.

The lines that connect the spaces stand for the various spirals of the inquiry process. Each spiral is a complex of several component spirals that represent alternatives that we construct and evaluate. The arrows indicate the recursiveness of the process. The blackened arrows show the direction of spiraling and the white arrows show the direction of the feedback process.

In the course of spiraling through the design spaces, we create and evaluate various potential design solution alternatives. The arrows on the spiraling line show two-way interactions between the design spaces. The spirals represent design inquiry accomplished in the substantive domains of systems design: the formulation of the core definition, the definition of specifications, the selection of functions, and the design of the enabling systems and the systemic environment. As we spiral through the spaces, the interaction is never one-directional, but recursive and mutually affecting. Feedback and feed-forward are always ongoing, two-pronged operations.

The second type of dynamics is the interplay of divergence and convergence, as shown in Fig. 3.4. This dynamic operates already in the exploration and image creation space of design. In the context of the framework, we first diverge as we consider a number of inquiry boundaries, a number of major design options, and sets of core values and core ideas. Then we converge, as we make choices and create an image of the future system. The same type of

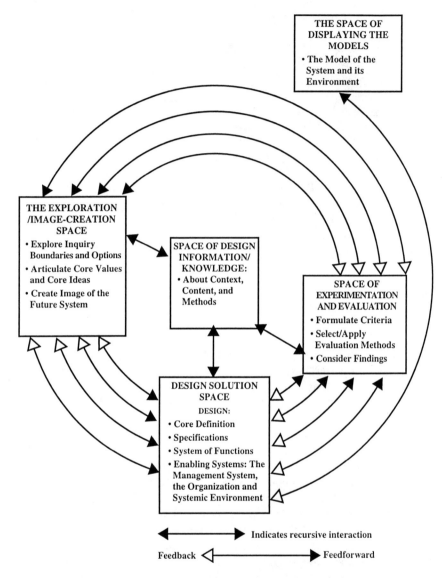

FIGURE 3.3. The dynamics of design.

divergence-convergence operates in the design solution space. For each of the substantive design domains (core definition, specifications, functions, enabling systems, systemic environment), we first diverge as we create a number of alternatives for each, and then converge as we evaluate the alternatives and select the most promising and most desirable alternative.

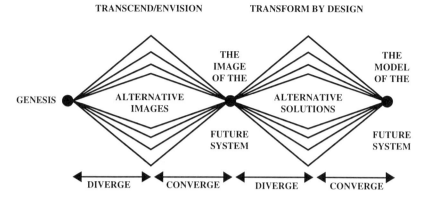

FIGURE 3.4. The dynamics of divergence and convergence.

3.5.3. The Spirals and Their Operations through the Five Spaces

Returning to Fig. 3.3, we see that it portrays the flow of the process of systems design as we move through the various design spaces of the inquiry. The figure also shows how the five spaces are related. Our journey through the spaces is accomplished in five spirals.

3.5.3.1. Spiral One: Formulating the Core Definition

During spiral one, we formulate the core definition of the new system. The core definition interprets the image we created in the exploration space in terms of the mission of the system. In formulating the mission of a social system, we ask the overall question: What is the new system about? And more specifically: What is our goal in serving humanity, the larger society, our community, the clients and the stakeholders of the system, people who serve the system, and all those who are affected by the system, most importantly future generations? What is the shared commitment that can give everyone in the system clear direction and guidance? A synthesis of answers provides us with a comprehensive definition of the future system. We then elaborate the mission in terms of specific purposes. This description reflects Ackoff's (1981) idea on formulating mission statements, as well as Nadler's (1981) notion of formulating purposes at several systems levels, and Checkland's (1981) formulation of root definitions.

The core definition spiral is connected with the knowledge base that informs the decisions we make. It spirals through testing, where we test the alternative core definitions based on the criteria we displayed in this space. This evaluation will guide us in converging on a specific core definition. Then, we spiral back to the exploration space, where we might reformulate our design ideas and the image in view of the core definition.

3.5.3.2. Spiral Two: Developing Specifications

Our design work during the second spiral produces the specifications of the new system (Ackoff, 1981). The bases of formulating specifications include the image, the mission, and the statement of purposes. Specifications interpret these as we ask questions such as: Who are the clients of the system? What services should we offer to them? What characteristics should those services have? When, where, and how should we provide those services? Who should own the system? How should ownership be distributed? What rights and responsibilities should owners have? What rights and responsibilities should clients and stakeholders have? How should the system relate to other systems in the environment?

Specifications are checked against the core definition. In the forward-moving spiral the selected specifications are evaluated in the space of experimentation/evaluation. The spiral now proceeds to the exploration space, where we review the compatibility of the specifications with the core values and core ideas and the image of the system. The specifications spiral connects with the knowledge base upon which we draw in formulating specifications.

3.5.3.3. Spiral Three: Selecting Functions

During spiral three, we ask such questions as: What are the key functions that have to be carried out that enable the system to attain its mission and meet its specifications? How do these functions interact to constitute a systems of functions? What are the subfunctions and how do they integrate into subsystems of the key functions? What are the component functions of subfunctions? How can we organize the subfunctions in subsystems of functions? By answering these questions we unfold the systems complex of functions that constitutes the first systems model of the new system. This system model of functions is a system of verbs, as noted by Checkland (1981), who calls this the conceptual model of the system. In most organizational work, once the mission and purposes are stated, people move on to establish the structure of the organization. In systems design there is an iron law: Form follows function. We cannot design a social system without first specifying the functions that have to be carried out.

The functions spiral connects to the knowledge base and is tested against core definition and specifications. Next it moves into the experimentation/evaluation space and then back to the exploration/image creation space.

3.5.3.4. Spiral Four: Designing the Enabling Systems

This spiral comprises three subspirals. First, we design the management system (Ackoff, 1981) by asking: What design will enable it to guide the functions, energize and inspire people in the system, ensure the availability of needed information and resources, and provide for continuous organizational learning?

Next we design the organization that carries out the functions by asking: What systems components, and what kind of people and resources in those components, have the required capabilities? How should we integrate the selected components in relational arrangements? What resources should be allocated to whom? And finally, the third subspiral produces the design of the systemic environment in which the new system is embedded and which is capable of providing the resources/information required by the system.

Each of the enabling systems draws on the knowledge base and is tested against the functions, specifications, and the core definition. The designs of the enabling systems are then tested in the evaluation/experimentation space. Then we move into the exploration and image creation space where we compare and validate the designs. Furthermore, each spiral goes through several iterations as we explore, test, and select solution alternatives. Moving with a spiral, we are initially in a divergent mode as we create a number of alternatives. But gradually we converge, as we evaluate the alternatives and select the most desirable solution.

3.5.4. Modeling the Design Solution

The term "model," as used here, is a mental or conceptual representation or description of the future system. We model the design solution by first constructing the functions/structure model, then the process/behavioral model, and the systems-environment model of the new system (Banathy, 1992a). These models jointly provide a comprehensive characterization of the new system. The use of these models is described in the next section.

Activity #15

First, describe core ideas that you identified in this section.

To get a "feeling" for designing a system, select a system you wish to (re)design and go through the spirals just once. This will be a kind of design simulation. Enter your findings in your workbook.

Reflections

The process of systems design, introduced in this section, might be viewed as a journey that took us through the various spaces that make up the territory of design. I mapped the design territory in the figures of this section. A decision to embark on the design journey was our point of departure. Then, we contemplated the questions of where we want to go and how we want to proceed. Having made those decisions, we set out on the course to complete the journey. In reflection, we can consider the design of social systems as a journey to create systems through which we can shape our individual and collective futures.

3.6.   Three Systems Models that Portray Design Outcomes

The product of design is an organized description, a representation or model of the system we created. Models of social systems are built in view of systems concepts and principles that represent the context, the content, and the process of a particular system. The relational arrangement of these concepts and principles can be organized into systems models. A couple of examples highlight the above statements. Input, transformation, and output are familiar systems concepts. Their relational arrangement gives us the following systems principle: input is transformed into output. Feedback and adjustment are also systems concepts. Their relational arrangement gives us two principles: feedback informs us about the adequacy of output, and based on feedback, adjustments might be introduced in the system. The three systems principles formed above can be arranged into a simple image of a very low resolution general model of systems, depicted in Fig. 3.5.

3.6.1.   Building Models of Social Systems

By observing various types of social systems and studying their behavior, we recognize characteristics that are common to them. Once we have identified and described a set of systems concepts that are common to social systems, and observed and discovered between them certain relationships, we can construct systems principles. A systems principle emerges from an interaction and integration of related systems concepts. Next, we can organize related principles into certain conceptual schemes, called systems models.

Following the line of reasoning described above, I constructed three systems models (Banathy, 1973, 1992a). The first, the systems-environment model, defines social systems in relationship to their environment. The second, the functions/structure model, focuses on what the system is. And the third, the

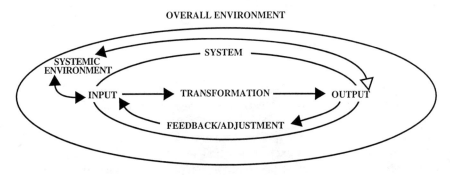

FIGURE 3.5. A general model of systems.

process/behavioral model, portrays how the system operates through time. These three models help us to understand and work with social systems.

We can use these models as lenses. Using the systems–environment lens, we can see the relational arrangements and observe the dynamics between the system and its environment. The functions/structure lens helps us to see the system at a given moment, to see what it is and what it does. It is a snapshot or a still picture of the system. The process/behavioral lens projects a motion picture of the system. It shows the behavior of the system through time. None of the lenses gives us a whole picture of the system. Only as we superimpose the three images can we capture a comprehensive view and description of the system. A metaphor that reflects this "superimposing" well is Galileo's invention. By using multiple lenses in his telescope, he gave his instrument much higher power.

Models are frames of reference we can use to examine and talk about the system that the models represent. We work with models all the time. When we exchange ideas about something, we usually do so by using conceptual models. In a discourse, it is helpful to have a common frame of reference of what we talk about, so that we have some assurance that everybody is on the same wave-length. The three models that I introduced are most beneficial in systems design for the reasons just presented. If we describe the outcome of our design in terms of models, we can examine our product, and everybody else will be able to understand what we have designed.

3.6.2. A Description of the Three Models

The following description of the three models gives us only a broad-brush picture of modeling social systems. For a detailed presentation of the models consult my earlier work (Banathy, 1992a).

3.6.2.1. The Systems-Environment Model

This model is an organized and relational arrangement of systems concepts and principles that enables us to describe and represent relations and interactions between a system and its environment. This model describes and presents the system we designed in the context of its environment. As a "lens," it projects a bird's-eye-view of the landscape in which the system operates. The systems concepts and principles that are pertinent to this model help us to define systems-environment relationships, interactions, and mutual interdependence. The model provides the following:

- A detailed description of the environment—the geopolitical, socio-cultural, and economic systems that influence the system and establish its requirements. It defines those systems in the immediate (systemic) environment that affect the system and are affected by the system we designed, upon which the system depends for its support and into which it sends its output.

- A description of the boundaries of the system, such as the spatial/physical, geopolitical, economic, social, sociocultural, sociobiological, psychological, temporal, ethical, technological. We also define the nature of boundary judgments to be made: who makes them and how they are made.
- A definition of all input entities that enter the system as well as the output entities that the system sends to its environment. (The source of some of the input–output entities, such as expectations, include not only the environment but also the system as a whole and people in the system.)
- A definition of the relationship and interaction patterns between the system and its environment, including characterization of how these patterns operate and the open or closed nature of the system.
- A description of the processes by which differentiation is made between adjustments within the system and changing the whole system. A description of the arrangements whereby organizational learning is assured in interaction with the environment.

The primary source of description of the systems-environment model is the exploration and definition of the systems environment, which is accomplished in the course of the last spiral of the design process.

3.6.2.2. The Functions/Structure Model

This model provides us the lens through which we can focus our attention on what the system is and what it does. The source of describing this model comes primarily from the resolution of design solutions we generated in the course of working through the spirals of the core definition, specifications, systems functions, and the enabling systems. This model is a snapshot of the system, projecting a still picture image of the system at a given moment in time.

In presenting this model, we describe (1) the mission and the purpose of the system, (2) the specifications that characterize the system, (3) the functions that the system carries out to attain the mission and the purpose, (4) the components and their parts that are engaged in attending to functions, and (5) the relational integration of the components into the structure of the system.

3.6.2.3. The Process/Behavioral Model

This model provides the designers with a lens that focuses on what the social system does through time. It projects a motion picture image of the system we designed. It helps to represent how the system behaves as a changing, living social system in the context of its environment. The process/behavioral model describes how the system does the following:

- Processes input, and more specifically, how the system identifies, receives, screens, and assesses incoming information and resources and how it sends the input on for transformation.
- Transforms input to output, and more specifically, how the system carries out transformation production, transformation facilitation, and transformation guidance.
- Processes output; how the system develops and applies the output model, how it facilitates and guides the output process and dispatches the output into the environment.
- Carries out the assessment of output by testing for, and collecting evidence of, the relevance and adequacy of the output; how the system analyzes and interprets evidence, and, if indicated, how it constructs models of adjustment and how it introduces adjustments. This process also describes operations that should be activated in case there is a need for changing the whole system.

The above description of the three models gives us only an orientation in the use of the three systems models.

Activity #16

Identify core ideas that might guide you to model the design of a system of your interest. Then, use those ideas in sketching out an imaginary model. Enter your findings in your workbook.

Reflections

People who begin to work with the three models often ask: Which model is a true representation of a social system? Which is the most important? The answer is: No single model or even a combination of two is sufficient in portraying social systems. Each has its own function. The systems-environment model maps the space in which the system lives. It portrays systems-environment relations and tells us why the system exists. The functions/structure model depicts what the system is at a given moment in time and how it is structured. The process/behavioral model tells us how the system operates and lives through time. Only if considered jointly, as if superimposed on each other, do these models tell us the real story of a social system. Only if we integrate them do they reveal a system's true nature.

3.7. Developing a Plan for Bringing the Design to Life

The design program reaches its climax in the display of the models of the system. These models are descriptions, conceptual representations of the desired future system we designed. The task is now to bring the design to life by developing, implementing, and institutionalizing it. This section outlines implementation approaches developed by Ackoff (1981), Checkland (1981), Checkland and Scholes, (1990), Nadler (1981), and Nadler and Hibino (1990) and presents additional ideas about implementation and institutionalization.

3.7.1. Ackoff's Means and Resources Planning

Once a comprehensive version of the idealized design has been completed and accepted by consensus, it should be compared with the reference scenario. (The reference scenario is developed prior to the initiation of design. It is extrapolated from the present state and stipulates what will happen if we do not change our system.) The task of planning is to close the gap between the extrapolated image and the ideal image. At the far side of the gap is the ideal image, and the task of planning is to realize that image or at least to move toward it. The process of doing so, according to Ackoff, involves two kinds of planning: means planning and resources planning.

3.7.1.1. Means Planning

Ackoff's means planning focuses on the development and evaluation of alternative means. "A means is behavior that either produces a desired outcome or brings one closer to its attainment" (Ackoff, 1981, p. 170). Means are of different types. They include courses of action; procedures and processes; projects that integrate actions and processes; programs that are systems of projects, directed at the attainment of the ideal image; and polices that guide the application of means.

Planners develop and evaluate alternative sets of means. The choice of means is based on comparative evaluation. A judgment of how much effort should go into this comparison is guided by an assessment of the potential cost-regret of selecting less than the best set of means, the relative effectiveness of the alternative, and the comparative cost of a less or more intensive evaluation program. The use of means should be continually evaluated as the conditions of their use might significantly change over time.

3.7.1.2. Resource Planning

Resource planning proceeds once means are selected and it is determined what resources are required to implement the means. Resources include materials, supplies, energy, services, facilities, equipment, personnel, and money. Planners ask such questions as: How much, when, and where each of these types of resources will be required? What resources will be available when required? What is the gap between what is required and what will be available? How should the gaps be resolved? How much will it cost to fill the gaps? Ackoff is particularly concerned about ensuring the availability of appropriate personnel. This will require thoughtful inquiry and even experimentation. However time-consuming and costly it may be, detailed assessment and experimentation is usually more than justified to ensure the efficient and effective availability and use of human resources.

3.7.2. Checkland's Implementation Approach

The fifth stage of the model of Checkland and Scholes (1990) involves the comparison of the conceptual model (stage four) with the problem situation expressed (stage two). This comparison shows the difference between what is and what should be. Applied in different settings, this approach yielded four variations. In one setting, the conceptual model was used as a basis for an ordered questioning of the problem situation. It initiated debate about changes that might be implemented. In another case, designers reconstructed the past events that lead to the emergence of the problem situation and asked what would happen if the conceptual model was actually implemented. In a third case, they asked: What features of the conceptual model are especially different from the present reality and why? The fourth application involved the conceptual model-ing of the existing situation and its comparison with the conceptual model of the designed system. (This approach is similar to the one proposed by Ackoff, when the ideal system is compared with the reference scenario.) The authors also mention a "greenfield situation" when a new system is designed. In this case, we compare the conceptual model with defined expectations. This is similar to the design approach I described in the last part of Section 3.5.

The purpose of stage five, the comparison stage, is to generate debate about changes that should be implemented. Stage six focuses on a determination of changes that should be implemented. Stage seven is the implementation of feasible and desirable changes. Checkland says that three kinds of change are possible: changes in procedures, in structure, and in attitudes. Structural changes may include organizational groupings and structures of functional responsibility. Procedural changes are changes in the dynamics of functions, interactions, and activities. Changes in structures and procedures are relatively easy to specify and implement. This is not the case with changes in attitudes. These involve many important, but intangible, characteristics that reside in the individual and in the collective consciousness of the design group.

3.7.3. Nadler's Approach

We have discussed Nadler's five-phase design approach. Nadler's last phase is the installation of the workable design solution. Nadler warns us that any solution will need modification regardless of how well detailed and documented it is. The installation phase proceeds as we (1) test, simulate, or try out the solution and establish installation schedules; (2) create procedures for presenting the solution; (3) prepare operational resources that include equipment, location preparation, job descriptions, organizational specifications, and personnel training and allocation; (4) install the solution and provide close and continuous monitoring; (5) establish operational performance measures; (6) evaluate the installed solution; and (7) schedule the planned improvement of the solution. Nadler suggests that we should design a separate system that will implement the selected solution. He suggests that the design of an implementation system must start at the very beginning of the design effort.

Achieving the four societal and organizational values, namely, greater effectiveness, better quality of life, human dignity, and individual betterment, "is always so dynamically uneven that redressing the balance requires a continuing search for change and improvement" (Nadler, 1981, p. 269). That is why we say that design never ends.

3.7.4. Systems for Institutionalizing Change

In agreement with Nadler, I (Banathy, 1986) proposed that the implementation and institutionalization of design requires the establishment of a system that will bring about the change implied by the design. Designers should set forth a set of functions that the implementation systems should carry out. These functions include the definition and display of the system we designed, the contemplation of the implication of the implementation of design, and the definition of the implementation functions.

- In defining and displaying what we have designed we shall explain the rationale for the design; characterize its expected impact; specify its functional, structural, and operational characteristics; specify resources that are required to implement the design; and make arrangements by which information on the items above can be introduced to those who are affected by the design.
- In contemplating implementation we shall assess the short- and long-range institutional and environmental impact of the design, consider potential barriers and plan for their removal or mitigation, design and test alternative arrangements for implementing the design, select the most promising arrangement, and develop an implementation plan based on the four items above.
- Implementation functions include the following: developing staff capability in carrying out implementation, establishing management capacity and capability to guide the accomplishment of those functions, developing or acquiring resources required for implementation, commencing im-

plementation and continuously monitoring it, and initiating change in both the implementation program and in the design of the system, if indicated.

Activity #17

(1) Capture the core ideas of the four implementation approaches discussed. (2) Compare the four approaches introduced above and identify their differences and similarities. (3) Make an assessment in the context of your selected system of the feasibility and relevance of the use of the various approaches. (4) Speculate about an implementation system that would synthesize the best features of the four approaches. Enter your findings in your workbook.

Reflections

We have now reached an important milestone in our journey toward understanding the application of systems design in the context of social systems. This milestone marks a point in the journey where, having worked with the first three chapters and completed activities embedded in the text, you should have developed what I would call an orientation-level competence in systems design.

The journey toward understanding systems design continues as we review the evolution of design as a human activity and learn more about the approaches, methods, and tools of system design.

The Design Landscape

In the previous two chapters an orientation was provided about design as a human activity and about approaches to the design of social systems. Three questions were explored: What is design, how does design work, and why do we need it? In this chapter we build on the experience gained in the two previous chapters and develop a more advanced understanding of the nature of design, and the approaches, methods, and tools of design.

Imagine that we are high above a landscape. Our first view is a bird's-eye-view. But as we move closer to the landscape, we get higher-resolution images; we see more and more details. This chapter works very much like this. We view the "universe" of design, examine major design approaches, and then move closer to the design landscape and focus on specific processes, methods, and tools of social systems design.

In Section 4.1, we explore the history of design as an activity of "man the maker" and arrive at the here and now of man, "the maker of systems." In Section 4.2, we have a glimpse of the universe of design as a disciplined inquiry. Following a definition of key concepts, a map is developed that depicts this universe. In Section 4.3, we explore various generations of design approaches and recently emerged approaches to social change and social systems design.

In the next three sections, a synthesis of various design methods is developed within the framework of the three major phases of social systems design. In Section 4.4, the question of when to engage in design is explored, and available approaches that help to transcend the existing systems are introduced. Section 4.5, reviews useful methods for envisioning a desired future state and creating the first image of the new system. Section 4.6, reviews methods that help to transform the image into a comprehensive design of the system. In the last section, Section 4.7, design tools are reviewed that facilitate design inquiry.

4.1. The Evolution of Design: The Historical Landscape

In this section, we travel back through time. Tracing the evolution of man and society, we observe three major transformations. The first was marked by the emergence of "homo faber," man the maker of things. Then the maker of the machine age came on the stage. The third transformation was lead by "homo *gubernator*," man the designer and steersman of large, complex, interactive systems.

4.1.1.   The First Transformation: The Emergence of Homo Faber

> Man came silently into the world. As a matter of fact he trod so softly that, when we first catch sight of him as revealed by those indestructible stone instruments, we find him sprawling all over the world from Cape of Good Hope to Peking. Without doubt he already speaks and lives in groups: he already makes fire.
>
> —Pierre Teilhard de Chardin

Teilhard de Chardin (1959) says that it was the Neolithic Age, the last period of stone age, when civilization was born. This stage of human social evolution marks one of most solemn events of all the epochs of the past. After a period long enough for the selection and domestication of all the plants and animals on which we are still living today, we find a sedentary and socially organized human in place of the nomadic hunter. In a matter of twenty thousand years, man divided up the earth and stuck his roots in it.

The birth of civilization was marked by a transformation from hunting and gathering tribes to communities of agricultural societies, from speech communication to writing, from survival technology to fabricating technology (Banathy, 1991a). This transformation and all the activities that became associated with it required conscious and thoughtful design. From this time period on, design as a human activity stands for the whole of sociohistorical activity, in the course of which people changed their natural and social environment to adapt it to their needs and aspirations. Design in its application as a conceptual activity means the thinking out and formulating of mental images, mental templates of all the means and techniques of designing.

4.1.1.1.   Homo Faber's Design Ventures

Design is a manifestation of homo faber's—"man the maker's"—cognitive venture. Since the Neolithic Age, design tasks and design activities have included:

- Creating habitat for the family and the community: building shelters, houses, storage places, etc. (architecture), and roads and waterways (civil engineering).
- Crafting instruments for everyday use and for warfare, and developing the technology of their use (industrial design).
- Designing environments for growing food, and providing water systems (environmental design).

- Extracting raw materials from the earth (mining engineering), and developing the techniques to process them for craft and fabrication.
- Using materials to craft aesthetic pieces, creating beauty and a great variety of art (arts and crafts design).

The design realms described here have accompanied humankind through the various stages of evolution. Through the ages they became expanded in scope and depth and have become ever more complex in their applications. Competence in design was passed on from the master builder-grower-engineer-artist to his apprentices. In some realms of designing the knowledge of the craft was considered a secret, possessed only by those who belonged to a guild. Until recently, a formal cognitive interest in design—the formal description of the design process—was seldom in evidence.

4.1.1.2. The First Design Scholar: Vitruvius

Still, we have one significant and monumental statement about design, written over two thousand years ago. The series of ten books by Marcus Vitruvius Pollio was titled *De Architectura Libri Decem* (Vitruvius, 1955). This fascinating work was more than a rendering on architectural design. At the time it was written, the notion of architecture embraced almost all of what we now call technology and engineering. Commenting on the nature of the work, Gasparsky (1984, p. 20) says: "Vitruvius showed how to design and implement particular technical objects known to himself. The work contains reflections of a general character as expressed in the preface of each of the ten books."

Vitruvius's specifications of the architect's competence—his profile of the architect—is remarkably similar to what we can say about such a profile today. Gasparsky juxtaposed quotations from Vitruvius's work with R. D. Hall's specifications of the ideal systems engineer. Just a brief quote helps us to appreciate the quality of Vitruvius's thinking. "The science of the architect depends upon many disciplines and various apprenticeships which are carried out in other arts," and "technology sets forth and explains things wrought in accordance with technical skills and methods" (Gasparsky, 1984, p. 22).

4.1.2. The Second Transformation: The Designer of Machines

> In every epoch man has thought himself at a 'turning point of history.' And to a certain extent, as he is advancing on a rising spiral, he has not been wrong. But there are moments when this impression of transformation becomes accentuated and thus particularly justified. When did this turn begin? (Teilhard de Chardin, 1955, p. 213)

Teilhard de Chardin places the first vibrations of this change as far back as the Renaissance. And it is clear that by the end of the eighteenth century the course had been changed. "Since then, in spite of our occasional obstinacy in pretending that we are the same, we have in fact entered a different world" (p. 213).

The sources of this epic change are multiple. First is economics. Even two centuries ago our civilization was based on the soil and its partitions. Gradually, the "dynamisation of money and property has evaporated into something fluid and impersonal, so mobile that already the wealth of nations has almost nothing in common with their frontiers" (p. 214). Second is industry/technology. Despite the many improvements through the ages, up to two hundred years ago the only source of chemical energy was fire and the source of mechanical energy was the human and animal muscle, enhanced by tools. Then, "man the designer of machines" came on the scene and our physical power became multiplied by the machine. Social changes awakened the masses. Teilhard de Chardin, observing this transformation, quotes Henri Breuil: "We have only just cast off the last moorings which held us to the Neolithic age" (p. 214).

4.1.3. The Third Transformation: The Age of Homo Gubernator

In quick succession, we passed through the industrial machine age, the age of oil and electricity, and arrived at the atomic age, the space age of cybernetics and high technology: the systems age, the age of complexity. The burden of this transformation is well described by Stafford Beer (1975, p. 18):

> Man is a prisoner of his own thinking and of his own stereotypes of himself. His machine for thinking, the brain has been programmed to deal with a vanished world. The old world was characterized by the need to manage things—stone, wood, iron. The new world is characterized by the need to manage complexity. Complexity is the very stuff of today's world.

We are entering the postindustrial, postmodern, information/knowledge age that, as Stafford Beer (1975) says, marks the end of homo faber.

> Today we live in a revolutionary ethos. Do not let us have our revolution the hard way, whereby all that mankind has successfully built may be destroyed. We do not need to embark on the revolutionary process with bombs and fire. But we must start with a genuinely revolutionary intention: to devise wholly new methods for handling our problems. The methods become clear once the stereotypes are overthrown, and the need to design viable systems is accepted—whole books are available about how to proceed. That is not the difficulty. The difficulty is how to replace Homo Faber with a new kind of man. He will not be "man the maker" any longer. He will be "man the steersman"—of large, complex, interactive systems. I call him Homo Gubernator."
> (p. 36)

4.1.3.1.   The Task of the Steersman: Guiding Evolution by Design

> We humans are integral agents of evolution: we spearhead it on our planet and perhaps in our entire solar system. We are evolution and we are—to the extent of our power—responsible for it.
>
> —Erich Jantsch

A quick review of the markers of the stages of societal evolution will help us to appreciate the burden and the task of steering societal evolution. Adapted from the work of Curtis (1982), Fig. 4.1 depicts the historical landscape of evolution as it displays three transformations, marked by three vertical lines.

It was around the middle of this century when we crossed the third vertical line, shown in Figure 4.1, as we entered the fourth stage of societal evolution. This new era has brought about massive changes and transformations in our scientific, technological, economic, cultural, and organizational spheres, affecting all aspects of our lives. Bell (1976) characterized the three transformations of our evolutionary journey. He distinguished our current era—which focuses on processing information and creating knowledge with the use of intellectual technology—from the industrial era, which focused on the design and use of machine technology, and the preindustrial period, which was primarily fabricating and extracting through agriculture and mining. The transformation into the industrial era extended our physical powers through machine technology and the transformation into the postindustrial era extended our cognitive power by cybernetic/systems technology.

As we look at Fig. 4.1 we readily recognize the great disproportion of the time span of the various evolutionary stages: five hundred thousand years, ten thousand years, five hundred years, and fifty years. As we consider the synergic effect of the speed and the intensity of developments during stage four, we understand why this effect has resulted in a perilous evolutionary imbalance (Banathy, 1987b). In earlier times, when societal evolution was rather slow and gradual, there was time for the various systems of the society to respond to the overall changes in their environment. The mechanisms for such change were adjustment and adaptation. But adjustment and adaptation do not work for us anymore.

STAGE ONE	STAGE TWO	STAGE THREE	STAGE FOUR
hunting gathering	agricultural society	industrial society	post-industrial society
half million years	ten thousand years	five hundred years	fifty years
speech	writing	print	electronic communication
wandering tribes	communities city-states	nation states	regional/global societies
magico-myth paradigm	logico-philosophical paradigm	deterministic scientific paradigm	cybernetics/systems paradigm
survival technology	fabricating technology	machine technology	intellectual technology

FIGURE 4.1. A historical view of societal evolution.

During the last fifty years mankind has experienced the unleashing of unprecedented scientific and material advancement and a technological revolution. While this revolution has given us power—earlier unimagined—it has accelerated to the point where we have lost control over it. We have failed to match the advancement of technological intelligence with an advancement in sociocultural intelligence and wisdom. Only such intelligence and wisdom can guide technological advancement for the benefit of all mankind (Peccei, 1977).

In evolution, the most advanced state of existence is human consciousness (Banathy, 1991a). It is expressed in its highest form in those who are most developed mentally/spiritually/aesthetically as well as in terms of their relationship to others, and in terms of their ability to interact harmoniously with all else in their sphere of life. They have the greatest capacity to shape change. Evolutionary consciousness empowers us to collaborate actively with the evolutionary process and use the creative power of our minds to guide our systems and our society toward the fulfillment of their potential. Salk (1983) remarked that evolutionary consciousness can motivate action toward giving direction to our future by consciously guiding evolution, provided we have a clear vision of what we wish to bring about. Conscious evolution, says Jantsch (1981), provides a sense of direction for cultural and social development by illuminating it with guiding images. And the faster we go—as we do at our current evolutionary stage—the further we have to look for images to guide us in our evolutionary journey. I am reminded of Csanyi (1982, p. 427), who said: "Evolution on earth, and within it the history of mankind, is a unique story . . . man can create his own evolution, choose his own history, and this is his freedom."

The human race has profoundly changed the parameters of the evolutionary process. Our unlimited capacity for learning and the explosive rate at which we produce knowledge and design complex systems have an extraordinary, and often unintended, impact on societal evolution. The question that confronts us now is: For what purpose are we going to use this limitless capacity for learning and our creating power? We can use this capacity and power to guide wisely societal development and create a better future for all. This, however, depends on meeting four conditions (Banathy, 1989): (1) the development of evolutionary consciousness, and based on it (2) the creation of guiding images for the future; (3) the aquisition of the competence needed to design our systems based on those images; and (4) the application of this competence in designing our systems. It is through this process that evolutionary images can be transformed into reality in societal development.

Reflections

Faced with a massive evolutionary transformation, with a change in the nature of change itself, we must recognize that incremental adaptations or restructuring of our existing systems are not working for us. We must realize that we have to transform our systems—as the society has been transformed—and we have to learn to change and coevolve with our constantly changing environment. The mechanism for this kind of change is social systems design applied on a broad and comprehensive scale. To be able to accomplish this, however, we individually and collectively have to develop competence in design so that we can begin to give direction to our evolution. We have to become homo *gubernator* so that we can steer our fate and shape our future. This present work aims to be in the service of enhancing the emergence of homo *gubernator*, the steersman of evolution.

Activity #18

Describe the core ideas of this section. Then, based on your thinking and ideas, construct a statement (of 2 to 3 pages) that you could use in discussing, with any individual or group of your choice, the necessity of developing design competence as a means to guide individual and collective evolution.

4.2.   The Universe of Design Inquiry

In this section, the idea of the universe of design inquiry is developed and the notions of "generic" and "specific" design inquiry are explored. At an abstract level, I map "generic design inquiry" and, at a less abstract level, I map design inquiry types that are specific to various professional fields or to various classes of systems (e.g., social systems). Then I map design inquiry as it operates in a specific, real-world functional contexts.

4.2.1.   Design Inquiry

In Chapter 2 we discussed design as a disciplined inquiry having two operating modes: (1) the conclusion-oriented mode of producing knowledge about design, and (2) the decision-oriented mode of conducting design. Here I describe the three branches of design inquiry: design philosophy, design theory, design methodology, and the functional context in which these are applied.

4.2.1.1.   A Definition of Design Inquiry Domains

Design inquiry incorporates three interrelated domains: design philosophy, design theory, and design methodology. *Design philosophy* seeks a general understanding of values, core ideas, and beliefs that guide design inquiry. It explores assumptions that express fundamental beliefs about design. *Design theory* articulates interrelated concepts and principles that apply to design. It seeks to offer plausible and reasoned general principles that explain design as a disciplined phenomenon. *Design methodology* has two domains. It studies methods in design situations in order to generate knowledge about design methods in general. It also identifies and describes strategies, approaches, methods, and tools that are applicable in conducting design.

4.2.1.2. Interactions between the Domains

Design philosophy, design theory, and design methodology come to life as they are used and applied in the functional context of design situations. It is in the context of use that design philosophy, theory, and methodology are confirmed, changed, modified, and reconfirmed. Thus, the *functional context* in which design is used becomes the fourth domain of design inquiry.

Design philosophy presents us with underlying assumptions, beliefs, and perspectives that guide us in defining and organizing in relational arrangements the concepts and principles that constitute design theory. Design philosophy and theory then guide us in developing, selecting, and organizing approaches, strategies, methods, and tools into the scheme of design methodologies. Methodologies are used in the functional context of design situations. But this process is not linear or one-directional. It is recursive and multidirectional. One domain continuously confirms and/or modifies the other, as described next.

As design theory is developed and used, it gets confirmation from two directions: the underlying assumptions of philosophy and its application through methods used in functional contexts. Methodology is confirmed or changed by testing (1) its "faithfulness" and relevance to its philosophical/theoretical foundations and (2) its appropriate use in design situations. Philosophy is enlightened, confirmed, and changed as it is applied in developing theory and methodology and as it guides design in functional context situations. The functional context—the society in general and systems of all kinds in particular—is the primary source that places demands on design inquiry. For example, it was the emergence of complex social systems and their increasing interdependence that brought about the realization of the need for a special mode of design: social systems design. Figure 4.2 shows the interactions described here.

Design inquiry comes to life and develops as a disciplined inquiry as the four domains interact recursively (and not in a linear fashion). In the course of this interaction, the two modes of disciplined inquiry, the decision oriented and the conclusion oriented, blend into each other. We learn from the ongoing design activity and thus create new design knowledge. We then use this knowledge in conducting design. Thus, as we move up on the spiral in accomplishing various design programs, design philosophy, theory, and methodology become ever more refined for us and gain increasingly more power of application in a variety of functional contexts.

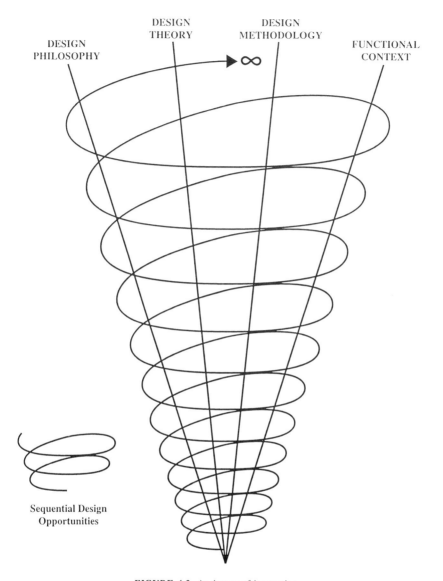

FIGURE 4.2. An image of interaction.

Exploring this influencing issue further, Rowland (1995) suggests that the question becomes: Are the philosophical and theoretical links (and links between theory and methodology) made explicit or are they left implicit and unexplored and, thus, open to contradiction by those who blindly apply them?

We can find a representation of the interaction of these four domains in

Warfield's (1990) generic theory of design. The key lesson from the ideas developed here is that only design methods that are grounded in philosophy and theory are acceptable to be used in systems design.

4.2.2. Mapping the "Universe" of Design Inquiry

The universe of design inquiry can be conceptualized on a continuum having various degrees of generality versus specificity. At the most general or abstract level, we can find "generic design inquiry." At the next levels of abstraction we have field-specific, subfield specific, or systems class specific design inquiry. Within a subfield that is identified at the lowest level of abstraction, we have real-life, concrete, functional context applications of design inquiry as we design unique systems.

4.2.2.1. Levels of Generic Design Inquiry (GDI)

Generic design inquiry operates in the conclusion-oriented disciplined inquiry mode, as it produces knowledge about design. Generic design (Christakis, 1987, p. 19) is "derived from the observation that no matter what is being designed, certain kinds of creative and organizational efforts are necessary" that cut across design situations. Defining GDI as generic design science, Warfield (1990) says that it "relates to those characteristics, attributes, phenomena and conditions that (a) are common to all design situations or (b) would be common to them if recognized by designers" (p. 22). Thus, generic design inquiry deals with matters that are common across the entire spectrum of design inquiry. The function of GDI is not to provide a recipe appropriate to all design situations. Each design situation is unique. Whatever is deemed to be generic is defined based on our observation of those elements of design that are invariant across design situations. The generic does not aim to—and cannot—account for those elements that are variant and unique to a specific design situation.

4.2.2.2. Field-Specific Design Inquiry (FSDI)

Field-specific design inquiry refers to those aspects of design inquiry that designers, operating in a specific field, use as specialized knowledge and activities applied in design situations. "Specific design knowledge and activities are not generally of interest to—or readily understood by—people outside the specific specialized area" (Christakis, 1987, p. 19). Discussing field-specific design, Warfield (1990) says that specific design inquiry (he calls it science) is typically restricted to a single field and has a well-defined set of applications linked to it.

Field-specific design inquiry is pursued in a conclusion-oriented disciplined inquiry mode. We seek knowledge about design attributes, characteristics, perspectives, beliefs, values, approaches, methods, and tools that are common—or generic—to a specific field of interest. Field-specific areas of design inquiry can be organized in various configurations. Exploring the question: Who should be primarily involved in design: the expert or those affected by the design? provides us with an opportunity to define two distinguishable field configurations of design inquiry. We label these fields "A" and "B."

Field "A" configuration would include architecture, habitat (e.g., urban) and environmental design, the various fields of engineering, law, medicine, economics, etc. These fields are the domains of professions where expert knowledge in design is of primary importance. But it should be noted that the involvement of those affected by the design becomes increasingly emphasized. Earlier, we discussed the need for all of us to acquire design literacy so that we can become informed "consumers" of systems designed by the experts.

Field "B" configuration would comprise the great variety of social systems, including education, human and social service agencies, government, community and volunteer agencies, religious organizations, and nonprofit corporations. In these fields the emerging trend is to give the primary role in conducting design to those who serve these systems, who are served by them, and who are affected by the design. As we noted earlier, all these groups should become "user designers." It is therefore imperative that we provide means, opportunities, and arrangements for everyone to develop design culture.

We can characterize the difference between field "A" and field "B" via their generic approach toward design inquiry by exploring the observations made by Nadler and Hibino (1990). They differentiated between the "doubting game" of the design experts and the "believing game" in social systems design.

On field "A," technical experts play the "doubting game." They focus on an in-depth problem diagnosis and definition that leads them to a detailed problem analysis. This is followed by the formulation and evaluation of alternatives and the display of the preferred solution. Their approach requires detachment, objectivity, and rationality. They ask tough and piercing questions. They "poke holes in ideas, tear apart assertions, probe continually." They put "people on the defensive, eliciting reactions that protect previous and current positions" (Nadler and Hibino, 1990, p. 66). "The doubting game, a specific result of rationality, supposedly makes a person feel rigorous, disciplined and tough-minded." They consider those who do not follow these methods as "unintellectual, irrational, and sloppy" (p. 67). The authors suggest that this approach is followed by Western corporate executives, academic intellectuals, and political leaders. This is "the prevailing mind set of the contemporary power elite, who, not coincidentally, have generally failed to find effective solutions to the common problems we face today" (p. 67).

On field "B", the social systems designers play the "believing game." They are committed to openness in exploring new ideas and in searching for the ideal. Their approach is subjective and flexible. They seek deep experiences and ever larger and more encompassing purposes. They work readily with other people, listening to and incorporating their ideas. They accept all assertions and their first and foremost rule is to refrain from doubting. They hold that no proposed solution—however seemingly impossible—is to be abandoned at the outset. This approach

> causes people to seek ways of achieving desired ends, thus putting forth positive reactions to questions. The believing posture produces responses such as, How could it work? What are the larger ends that a particular solution will also achieve? "All men's gains are the fruit of venturing," the way Herodotus put it in 450 B.C., is an early version of the believing game. (Nadler and Hibino, 1990, p. 67)

4.2.2.3. Functional Context Design Inquiry

Within a particular special field or subfield, we have real-life and concrete manifestations in functional context applications of design inquiry as we design specific and unique systems. In these applications designers can draw upon the design knowledge in the "generic" design domain as well as the relevant "field-specific" domain. Figure 4.3 presents an example of mapping "fields within fields" in the overall domain of design.

4.2.2.4. The Continuum of Generality-Specificity

As we formulate "generic" knowledge about design at various levels of abstraction, we face a difference in degree of the amount, the substantiveness, and the detail of what we can say. At the highest level of abstraction of generic design inquiry (Level I in the figure) we can say that design produces "novelty," and it is concerned not with what "is" but with what "should be." It involves such activities as transcending the current state; developing a knowledge base for design; synthesizing, analyzing, and selecting alternatives; and communicating, displaying, and implementing the design solution.

If our general interest is designing various types of systems, (e.g., Level II), then substantive statements, coming from systems philosophy, systems theory, and systems methodology might apply in systems design. In designing social systems, in addition to the general systems inquiry, we shall draw on theories such as organizational theory, the theory of social evolution and communication theory, and disciplines such as sociology, psychology, and anthropology.

The closer we get to specific unique design situations in a specified design field of interest (Levels III, IV, and V), we will be able to generate more substantive and more detailed design knowledge to be offered to designers. For example, my *Systems View of Education* (Banathy, 1992a) provides generic statements about education (Level IV), whereas my *Systems Design of Education* (1991a) provides a set of propositions of the design of systems of K–12 education at Level V.

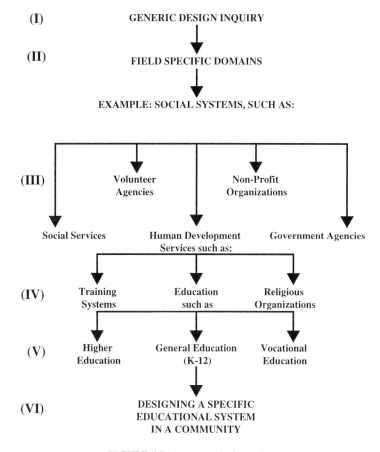

(I)   GENERIC DESIGN INQUIRY

(II)   FIELD SPECIFIC DOMAINS

EXAMPLE: SOCIAL SYSTEMS, SUCH AS:

(III)   Volunteer   Non-Profit
                   Agencies   Organizations

Social Services   Human Development   Government Agencies
                     Services such as:

(IV)   Training   Education   Religious
                  Systems   such as   Organizations

(V)   Higher   General Education   Vocational
                 Education   (K-12)   Education

(VI)   DESIGNING A SPECIFIC
                  EDUCATIONAL SYSTEM
                  IN A COMMUNITY

FIGURE 4.3. An example of mapping.

Activity #19

(1) Given the example in Fig. 4.3 and using your field of interest, develop a specific configuration of design inquiry that will range from the "generic" to a real-life application. (2) Based on (1), formulate some statements you could make about design at various levels of generality. (3) Does "A" or "B" or a combination of the two best reflect the approach to your field? Please explain. Enter your findings in your workbook.

Reflections

In this section, we started out with a set of definitions and mapped out various levels of abstractions of design inquiry. We compared two major types of field-specific design and reinforced our conviction that social systems are different from all other types. In the sections that follow, we continue this exploration as we examine approaches, models, methods, and tools applicable to the design of social systems.

4.3. Approaches to Systems Design and Social Change

In this section, several generations of approaches to design that have been documented in the design literature are reviewed. Then, some recently emerging approaches to social change and systems design are discussed. Reflecting on the theme of this section, I ask the questions: How do designers think about and approach their design tasks? What is their overall scheme or framework that expresses their design philosophy, theory, and strategy?

4.3.1.   Generations of Design Approaches

Reviewing the state of the art in design, several design scholars discuss—in Cross's (1984) compendium—generations of design approaches that have marked the design field during the sixties and seventies. Broadbent (1984) says that during the early sixties design scholars discussed the nature of design as if it would be traditional science, and developed methods that reflected a "Cartesian view" of designing, namely, "breaking the problem down to fragments and solving each of these separately before attempting some grand synthesis" (p. 337). A fundamental tenet of this approach was that (1) designers should abandon any attempt at preconceived design solutions and (2) design is the domain of the design expert (the design scientist). By the late sixties this approach, which was labeled by Rittel (1984) as the "first generation," lost much of its credibility. Alexander (1971) suggested that the first generation approach had become irrelevant and its continuing practice became a pointless excuse for people with a fear of engaging in real design activity. According to Rittel, the first generation approach was based on the systems engineering techniques of mechanical thinking that were wholly inadequate to the "wicked" problems of design. Rittel proposed new design principles that marked what he called the "second generation." Accordingly, the design process should be based on an "argumentative" structure and expertise and relevant knowledge should "be distributed amongst a wide range of participants." The role of the "designer is that of a midwife or teacher rather than the role of one who designs for others" (p. 305).

Christopher Jones, another leader in the field of design and a noted developer of the first generation approach, radically changed his perspective on design during the early seventies. He was prepared to go even further than Alexander and Rittel in the search for a new approach to design (Jones, 1977). He had a "continuing concern with trying to resolve the apparent conflicts between rationality and intuition, logic and imagination, order and chance" (Cross, 1984, pp. 305–306). Jones reacted strongly against machine language, behavioralism, and the use of preset logical frameworks. Like Alexander and Rittel, Jones was an antiexpert—against planners and designers who decided how everyone should live—and his view dominated through the development or second generation design methods during the 1970s (Cross, 1984, p. 306).

Broadbent (1984) suggested a synthesis of the first two generations. Taking the best from the first two, this third generation approach was based on Popper's science-driven conjecture and refutation model, proposing that the designer—the expert—make conjectures that should be open for refutation by the users of the design. Archer (1984, p. 34–8) took issue with this approach, saying that the science-based,

> logical models,—however correctly they may describe the flexibility, interactiveness, and value-laden structure of the design process—are themselves the product of an alien mode of reasoning. There exists a designerly way of thinking and communicating that is both different from the scientific and scholarly ways of thinking and communicating, and is as powerful as the scientific and scholarly methods of enquiry, when applied to its own kinds of problems.

Confirming that design problems are ill-defined, Archer further remarked that "an ill-defined problem is one in which the requirements, as given, do not contain sufficient information to enable the designer to arrive at a means of meeting those requirements simply by transforming, reducing, optimizing, or superimposing the given information alone" (p. 348). Archer, then, remarks that most of the issues we face in our everyday life are ill-defined, but in the course of human evolution we have found effective ways of dealing with them, and this coping behavior is rooted in human nature. Thus, any design approach must be based on innate human capabilities. Cross (1984), commending on Archer's paper, says that "there is a new confident view, expressed by Archer, that design must not try to ape the methods of the sciences or the humanities but must be based on the ways of thinking and acting that are natural to design" (p. 307).

Reflections

The last observation in the paragraph above was reflected in Section 2.5 of Chapter 2, where I developed the notion of the need to develop a design culture complementing the cultures of the sciences and the humanities. The deep appreciation of "humanness" in design thinking, manifested by Archer, Jones, and others during the 1970s and 1980s reveals a remarkable insight. Note that the design scholars quoted above are from the field of architecture, but their message has been heard by many of us—and should be heard by all of us—who work in the field of social systems. They were influenced, as we have been, by the emergence during the 1970s and 1980s of soft systems thinking, which changed the metaphors and the substance of systems inquiry away from the hard systems, social engineering approach. This new way of thinking, this shift in worldview, has moved us toward an open systems, coevolutionary, dynamic, value-laden and ethics-based approach to the design of social systems.

In this section, I continue a review of the social systems design field by introducing recently emerged approaches and models of working with social systems. I include in this review only those approaches that are grounded in explicitly stated philosophical and theoretical positions. I will not discuss some currently popular approaches of "management gurus" that lack firm conceptual grounding.

As a background to the review of recently emerged approaches it is appropriate to note that the impetus of this "emergence" has been the recent major historical transformation from the industrial to the postindustrial knowledge society. This transformation has called for the capturing of a new image of humanity and the implementation of this new image by the design and transformation of our social systems (Markley and Harman 1982).

The emergence of this new image is grounded in several factors:

1. An awareness of the demonstrated inability of the industrial/machine age paradigm to address the emerged new realities of our era.
2. The need for a guiding vision of a workable society, which is to be built around the new image of humanity and a corresponding image of social systems.
3. The guiding vision has to include provisions for the full and valued participation of people in the change and design of their systems.
4. A heightening of a sense of stewardship and public responsibility, which will support a moral perspective of economic and social justice and will nurture self-realization and ecological and social ethics.
5. The embracing of various cultural and racial diversities, while seeking unification and integration of the human community.

4.3.2. Total Systems Intervention (TSI)

This approach, developed by Flood and Jackson (1991), is grounded in critical systems thinking and liberating systems theory. It aims to liberate systems theory from the tendency of self-imposed insularity, from the delusions of objectivity and subjectivity, and to emancipate people and groups from domination and subjugation evidenced in social situations. TSI accepts the value of the diversity of available systems methods and the richness of methods in dealing with the changing complexities of the social world.

TSI draws on a grid of method types, organized the way social reality can be conceived metaphorically (e.g., like a machine, an organism, a culture). If a metaphor can be found that is appropriate to the system type of the design situation, then we can draw on the methods that the metaphor labels and it can lead to a reflection between the situation and the methods. This, then, can guide participants toward the selection of an appropriate method for intervention. TSI organizes tried and tested systems design methodologies and approaches and it reveals why each works in some situations but not in others. It shows the strengths and weaknesses of the various approaches and enables users to make an informed choice.

TSI is based on five positions, derived from critical systems thinking and liberating systems theory: "critical awareness, social awareness, complementarism at the theoretical and methodological levels, and commitment to human well-being and emancipation" (Jackson, 1992, p. 272). It encourages creative thinking about the nature of the design situation before a decision is made about the character of the issues to be addressed.

> Once the decision has been taken, the approach will steer toward the type of methodology most appropriate toward resolving problems of the kind identified as being most significant. As the intervention proceeds, however, the nature of the problem situation will be continually reviewed, as will the choice of appropriate methodology. (p. 227)

TSI might lead to nominate one methodology as the dominant one in a particular context and others that might play a supporting role. It harnesses the richness and diversity of methodologies. Its task is to make sure that practitioners have access to and can make informed choices about the use of appropriate methodologies.

4.3.2.1. The Three Phases of Total Systems Intervention

Principles that are embedded in TSI (Jackson, 1992) include the following. Organizations are too complex to approach with one model or to tackle with quick fixes. They should be investigated with a range of metaphors. Metaphors that highlight the system can be linked with appropriate systems methodologies and different methodologies can be used in a complementary way in addressing different organizational characteristics. TSI should be applied in cycles, with back-and-forth interactions among its three phases. Stakeholders should be engaged in all phases of TSI. The three phases are creativity, choice, and implementation.

4.3.2.1a. Creativity. The creativity phase brings forth systems metaphors that assist the inquirers to think creatively about their system. Examples of metaphors (Morgan, 1986) include the organization as a machine, as an organism, as brain, as culture. (In Section 6.6 of Chapter 6 these metaphors are discussed.) Questions that we might ask in the creativity phase are: Which metaphor would we appropriate to guide thinking about the existing system? What metaphor might throw light into the design situations? What metaphor can guide thinking about the system we wish to create? The outcome of this phase is a "dominant" metaphor as a basis for choosing an appropriate intervention methodology.

4.3.2.1b. Choice. The task in this phase is to choose a systems-based intervention methodology—or a set of methodologies—that match the characteristics of the system we wish to create. TSI provides here the tools that are needed to make reasoned choices. The choice making is guided by the propositions of critical and social awareness, theoretical and methodological complementarity, and commitment to human well-being and emancipation. The outcome of this phase is the emergence of a dominant methodology.

4.3.2.1c. Implementation. This phase implements the particular intervention methodology, or a system of methodologies selected during the choice phase. This intervention translates "the dominant vision of the organization, its structure, and the general orientation adopted to concerns and problems into specific proposals for change. The outcome of the implementation stage is coordinated change, brought about in those aspects of the organization currently most vital for effective and efficient functioning" (Jackson, 1992, p. 275).

4.3.3. Guiding Deliberate Social Change

The core question that Etzioni (1991) asks in *A Responsive Society* is: How can a society influence its future, rather than being subject to the whims of historical and environmental forces beyond its control? Etzioni refers to a 1987 conference on institutional design that asked the questions: How free is the society to choose the path of its change? Is it free to design or is it dominated by historical and evolutionary forces? Etzioni suggests that the answer lies between the malleable social system (or a very powerful overlay) that is free to design at will, and the determinist who assumes little or no power to guide. He suggests that "the productive question is: Under what conditions is the ability to guide enhanced?" (p. 23). Guidance reflects a combination of "downward" control and "upward" flow of consensus, which sustains and limits control. There is a trade-off here. "The greater the consensus the less need for control. When both control and consensus are relatively high, more change can be guided than when both are lower, without an increase in alienation" (Etzioni, 1991, p. 39).

The "mixed scanning approach" was developed by Etzioni in contrast to rationalist models of change and piecemeal incrementalism. His metaphor for the approach is satellites with two lenses, the wide-angle and the zoom. The wide-angle lens gives us the big picture of the ideal and fundamental change and the zoom lens focuses on incremental steps toward the ideal. "This approach is less demanding than the full search of all options that rationalism requires, and more 'strategic' and innovative than incrementalism" (p. 49). In the incremental mode, there is no overall guidance. The changes are random and scattered. In mixed scanning the fundamental image of change provides the guidance for directed and gradual change.

4.3.3.1. Create Alternatives

Etzioni suggest that a list all relevant possible alternatives should be developed, including those that do not appear to be very feasible. Examine alternatives briefly, and set aside those against which there appear to be utilitarian objections (means not available), normative objections (violate our basic values), or political objections (of others whose support is important). Further elaborate the remaining alternatives and screen them as proposed above. Develop in more detail remaining alternatives and acquire enough information and criteria to differentiate among them. Continue until we arrive at the preferred alternative.

4.3.3.2. Prepare for Implementation

The next step is to develop implementation plans. Arrange implementation in serial steps, commit assets for steps, but keep reserves. More costly and less reversible steps should come later. Schedule continuing information collection and more encompassing scanning that help in implementation decisions (be prepared for unanticipated delays).

4.3.3.3. Reviewing While Implementing

As implementation proceeds, introduce "semi-encompassing" scanning as sets of steps are implemented. If things work, continue scanning at longer intervals, and conduct overall review even less frequently. Scan more encompassingly if difficulties arise. Make sure to scan in full overall review at set intervals even if everything seems all right.

The approach described here is most compatible with a progressive viewpoint and innovative spirit. The approach seeks to avoid both the overly rationalistic mode and the excessively pragmatic approach to social systems design. A deep sense of right and wrong should influence efforts to spur social change. A deeply felt moral sense of what is right will guide us to make ethical choices that serve the greater good of the human community.

4.3.4. Breakthrough Thinking

Nadler and Hibino (1990) offer a design approach that "consists of focused principles that you can learn and apply to master your own environment, which, like that of any human being, is one of constant change" (p. xii). Rather than addressing purposes, most people ask: What's wrong? What is the problem? The answer given to such questions will be "culturally skewed, historically biased, conceptually constrained, and ultimately limited—precisely because it focuses on problems, not solutions" (p. 73). Such conventionally reasoned approaches worked well when we had unlimited resources and all the time needed. "Today, the distilled analysis of what exists is the breeding ground of defeatism." Today demands "breakthrough solutions providing maximal improvements" (p. 73). Today we are faced with issues of fundamental importance we never faced before, which are not only new but come at us faster than at any time before. The methods and tools we used so successfully in the past no longer work. Faced with this world, we must change our perspectives and thinking, change our heart and habit, and develop new ways of approaching our ever changing world. Nadler and Hibino (1990) propose seven principles that guide design thinking:

- The *uniqueness principle*. No two situations are alike; each problem situation is unique and is embedded in a unique array of related problems, requiring a unique approach. "Solution to a problem in one organization will differ from the solution to a similar problem in another organization" (p. 101).
- The *purposes principle*. Focusing on purposes helps "strip away" nonessential aspects of the problem situation. It opens the door to the creative emergence of larger purposes and expanded thinking. It leads to an increase in considering possible solutions. An array of larger purposes guides long-term development and evolution.
- The *ideal systems principle*. Having an ideal target solution puts a time frame on the ideal system to be developed, guides near-term solutions, and infuses them with larger purposes. "Even if the ideal long-term solution cannot be implemented immediately, certain elements are usable today" (p. 140).
- The *systems principle*. Systems thinking helps us to understand that every problem is part of a larger system of problems. It helps us to see "not only relationships of elements and their interdependencies, but, most importantly, provides the best assurance of including all necessary elements, that is, not overlooking some essentials" (p. 168).
- The *limited information collection principle*. This principle guards us against excessive data gathering that may make us an expert in the problem area when we should become experts in designing solutions. Too much focus on the problem prevents us from seeing new ways to create excellent alternatives.
- The *people design principle*. This principle says that those who carry out the solution should be intimately involved in its development. "We all

want to be involved in making decisions that influence our lives. And we accept and feel good about implementing a solution we help to devise" (p. 225). In some instances, the benefits of participation in creating solutions can be more important than the solution itself.

• The *betterment timeline principle*. In order to preserve the vitality of the solution, we should build into it the potential of continual change. "A real breakthrough is not only the 'big bang' or major change solution, but also the assurance of continual change and improvement in the area of concern" (p. 258). "Don't get stuck in the mud of assuming the continued viability of a system installed even a day ago" (p. 276).

In applying the principles of the breakthrough thinking approach we must "remember that problem solving is synergistic: The whole is greater than the sum of its parts" (p. 283). A holistic view leads us to apply the breakthrough principles not one by one, but comprehensively, in an integrated fashion. The only way to preserve the vitality of design solutions is to build in a program of continual change. Breakthrough thinking seeks solutions that have built within them the seeds of future changes.

4.3.5. Rethinking Soft-Systems Methodology (SSM)

Checkland's (1981) design model was introduced in Chapter 3. Over the last decade SSM was used extensively. Special focus was placed upon creating new insights about the relevance, efficiency, and effectiveness of SSM. This "learning about" SSM has led to some important discoveries.

Checkland and his coauthor, Scholes (1990) refocused on the distinction between the hard systems and soft systems approaches. They suggest that the single concern of the systems engineer who uses the hard systems approach is how to meet a need that was expressed "in the form of a named system with given objectives," such as "a system to build a supersonic aircraft meeting a defined specification within a stated time and to a stated budget." Thus, "systems engineering looks at 'how to do it' when 'what to do' is already defined" (p. 17). This approach is the "Achilles heel" of systems engineering when applied to ill-defined problem situations in social systems. In the context of social systems the genesis of initiating design is "no more than a feeling of unease, a feeling that something should be looked at, both from the point of view of whether it is the thing to do and in terms of how to do it" (p. 17).

Under the label "gathering and learning the lessons," the authors give an account of a set of what they call "new constitutive rules" of the soft systems approach. I review their formulations from the perspective of the designer of social systems as follows:

• The approach represents a structured way of thinking that focuses on real-world issues, which we perceive to be problematical. An issue emerges in the unfolding and interacting flow of events in the context of a social system. In this context there is a sense of unease, a sense that things could be better than they are.

• The structured thinking of the approach is based on ideas coming from a conscious reflection on the real world by using systems ideas. The organized expression of these ideas yields the epistemology of the approach. The front-end part of the epistemology includes the following:

1. An exploration of the social (cultural) characteristics of the design situation that includes the exploration of values. This activity reflects the "expansionist" systems orientation and the recognition of the importance of values.
2. An exploration of the power-related aspects of the design situation. (Activities (1) and (2) were not included in Checkland's earlier (1981) model.)
3. A diagrammatic representation of the entities, the structures, processes, relationships, and issues of the design situation.

• In the main body of the epistemology aspects of the root definitions have not changed, but the criteria for judging the transformation process are defined, and they include efficacy (does it work?), effectiveness (are goals achieved?), ethics (is it the moral thing to do?), and elegance (is it aesthetically pleasing?). The rest of the epistemology—(1) developing the conceptual model, (2) comparing it to the perceived real world, (3) defining desirable changes and (4) implementing changes—basically follows the earlier model (Checkland, 1981).

• A constitutive rule is a claim that the use of the approach is dependent on the following guidelines:

1. If a part of the real world is selected as a system to be designed, then that is done by conscious choice.
2. A distinction should be made between unreflective involvement in the everyday world and conscious systems thinking about it. The user of the approach moves back and forth between the two.
3. In the conceptual phase of constructing purposeful social systems (which embody emergent properties, communication, and control), the constructions used initiate dialogue, debate, and discourse; they are aimed at defining changes in the real world.

• The approach will be used differently by different people in different situations. Conscious thought should be given to adapting the approach to a particular design situation.

• The approach should be taken as a methodology and not a technique. "Every use of it will potentially yield methodological lessons in addition to those about the situation of concern" (p. 287). The lesson may be about the approach's framework of ideas, its processes, the way the approach was used, or all of the above. The lesson awaits discovery by conscious reflection on the use.

The constitutive rules, described above, were not articulated explicitly in the earlier description of the soft systems approach. It is important for us to note the last paragraph, which is an explicit call for research on the approach while it is applied.

4.3.5.1. Research on the Approach

In Chapter 2, I presented two modes of design inquiry: (1) the decision-oriented mode, which is the formal application of a design model or methodology in the design of social systems, and (2) the conclusion-oriented mode, which develops knowledge about design by reflecting on the use of an approach, and by learning from specific contextual design applications. In Section 4.2 of this chapter, I developed a statement about the specific relationship between these two modes of inquiry.

The authors of soft systems methodology have arrived at the same two modes of use. They propose that the experiences of the use—and reflection on the use—led them to the recognition of a spectrum from "a formal stage-by-stage application of the methodology" to "an internal mental use of it as a thinking mode" (p. 281). The authors refer to Vickers's (1983) much used metaphor: "the two-stranded rope of ideas and events." Namely, as the events of the intervention unfold, ideas about their use are generated through the research mode.

4.3.6. The Future-Search, Whole Systems Approach

In *Discovering Common Ground,* Weisbord (1992) presents a future-search, whole systems approach that brings people together to achieve break-through innovation, empowerment, shared vision, and collaboration. His main theme is that the world is moving from experts designing for people toward everybody, experts included, designing whole systems. He suggests that getting everybody involved is "the best strategy if you want long-term dignity, meaning and community" (p. ix). He says that a new paradigm is sweeping the postindustrial world. The paradigm is characterized by the ideas of learning, empowering, democratizing, partnering, and wanting to bridge gaps of culture, class, race, gender, and hierarchy. The search is now on to find approaches that are equal to these values. The purpose of such a method should include (1) the discovery of common grounds and the imagining of ideal futures, (2) the involvement of all sectors of the society, (3) the provision of such task-focused techniques as discovery and analysis, (4) the broadening of global perspectives and the promotion of self-management that leads to committed action, and (5) the building of democratic values and a higher quality of life.

An approach that aims at implementing the guiding concepts described above is called the Future Search Conference (FSC), designed by Weisbord (1989, 1992). The FSC integrates the past, present, and future, as these are experienced and desired by the participants of the conference. (This is similar to Ackoff's interactive design approach.) Each segment requires the building of a knowledge base, the joint examination of the knowledge base and its interpretation, and the drawing of conclusions. This process uncovers shared values and leads to congruent action. The FSC approach is based on seven assumptions, interpreted as follows (Weisbord, 1992, p. 13):

- The real world is knowable to ordinary people and their knowledge can be collectively organized by them. "In fact, ordinary people are extraordinary sources of information about the real world."
- It is believed that people can create (1) their own individual future and (2) collectively, the future of their systems.
- People want to have the opportunity to engage their heads and hearts in determining their future. They want to—and are able to—join the creative process of doing so rather than leaving such creation as the sole domain of the "organization's elite."
- The nature of participation is egalitarian. Everyone is equal and has an equal voice and right to make a contribution.
- Given the chance, people are much more likely to cooperate than fight. The task of those who coordinate (facilitate) the process is to structure opportunities to cooperate.
- The process should empower people to feel more knowledgeable about and in control of their future.
- Diversity should be appreciated and valued.

Weisbord (1992, p. 66) sets forth a "minimum critical specifications" for the success of FSC. These are as follows:

- "Get the whole system in the room." This means to involve in the process the broadest possible community that represents maximum variety and diversity of interdependent people.
- Have this community look at itself in a "global context." Explore all the system-relevant events, trends, relations within the wider world, and the institution/issues in focus. This means the exploration of the broadest possible knowledge base and common ideals before zeroing down to the issue at hand.
- Ask people to be task focused and "self-manage" their work and by so doing, reduce dependency, conflict and task avoidance.

All the above conditions must be present, says Weisbord. Mapping the whole system cannot be done by the views of management or of any specific stakeholder group. It has to be done by all the people in the system. Neither can the task be accomplished by focusing on a restricted field of inquiry.

4.3.7. Interactive Management (IM)

Interactive management (Warfield and Cardenas, 1994) is a decision-oriented disciplined inquiry that enables organizations "to cope with issues and situations whose scope is beyond that of the normal type of problem that organizations can readily solve" (p. 1). IM is grounded in the science of general design (Warfield, 1990). The concept of IM was developed by Warfield and Christakis at the University of Virginia in 1980 and its practice spread to many places. It is the first design inquiry approach that has made significant use of software. In each case of application, the outcomes of IM are defined to suit the specific requirements of the design situation. The three main phases of IM are definition, the design of alternatives, and choice of design. IM is a highly complex process, overviewed here only very briefly.

4.3.7.1. Definition

The context of the design situation is defined and components involved are identified. Patterns are constructed that show how the components are related and the patterns are interpreted to gain an understanding of design issues and insights into requirements of the design of alternatives. Based on the interpretation and insights gained, the definition phase might produce statements of context, a set of objectives, sets of triggering questions, and general questions that might guide the inquiry. In the course of this process "participants will have introduced their concepts about the situation being explored. Through dialogue, the collective best ideas of participants will have emerged as part of the definition. Participants will have become aware of critical relationships among factors, and take these into account as they design alternatives" (Warfield and Cardenas, 1994, p. 19).

4.3.7.2. Creating Alternatives

The second phase creates alternative designs for resolving complex issues, or conceiving the creation of a system. The generation and clarification of design solution options are prompted by triggering questions, devised by the participants, using nominal group technique (see Section 4.7). Once the options are clarified, they are processed for significance, and categories of options are organized in an option field. Now the question is raised whether each category is essential as an alternative. If yes, it becomes a design dimension. Dependent dimensions are clustered and structured. Attention is now turned to select design options.

4.3.7.3. Choice of a Design

Given the structural representation, developed during the previous phase, options will be chosen from each dimension. The result is an "option profile" that displays one design alternative. Other design options are developed in the same way. Now the design alternatives that have been created are described and evaluated with the use of the tradeoff analysis methodology (TAM) (described in Chapter 7). Before TAM is applied, selection criteria will have to be developed by the use of idea-generating methods.

Other relevant outcomes of IM are (1) participants are involved in significant learning about the issue at hand and ways of addressing it, (2) their involvement creates commitments to the choices made, and (3) careful and detailed documentation enables wide distribution of findings as well as having available knowledge for continuing design.

Reflections

The approaches to social systems design introduced in this section provide us with a very rich picture of a variety of design orientations that have emerged in the course of the last few years. It is particularly significant and rewarding to recognize significant growth in understanding the richness of systems design and the new insights that have emerged. We can easily become aware of this growth and the insights gained when we compare the earlier writing of Checkland, Flood, Jackson, Nadler, and Warfield with their recently published works, discussed here. It seems to me that the disciplined inquiry of social system design is a dynamic, vibrant, and powerfully evolving field that provides us with ever more sophisticated and meaningful approaches and an ever more enlightened understanding and practicing of systems design.

Activity #20

(1) Review the description of design approaches in this section and identify core ideas of design that appear to be significant for you. (2) Arrange the selected core ideas as follows: (a) ideas that provide insight about what design is and how it works, (b) ideas that might give guidance on how to think about design, and (c) new approaches to design and new methods of conducting design. (3) If you wish, out of the core ideas you might formulate a design approach that would be

most useful to the design of your selected system. Enter your findings in your workbook.

4.4. Initiating the Design Inquiry and Transcending the Existing State

In the next three sections, a broad-scoped knowledge base is presented that encompasses strategies and methods of the three phases of systems design: transcending, envisioning, and transforming systems by design.

More specifically, strategies and methods and tools are considered that are applicable to (1) the transcending of the existing state and initiating the design inquiry (Section 4.4), (2) the envisioning of the future by creating an image of the new system (Section 4.5), and (3) the designing of the model of the new system and modeling the systemic environment (Section 4.6).

This section is focused on the following questions: Why do people initiate change and what kind of change can be defined as design? What strategies and methods are applicable to transcending the existing system and initiating the design of a new system?

4.4.1. Why Change? What Change Can Be Defined as Design?

We can speculate about a host of reasons why people in systems initiate change. I present here a set of "reasoned configurations" presented as a three-dimensional model of understanding why people engage in change. The model is both comprehensive and inclusive. It is comprehensive in that it aims to embrace all possible configurations of change initiation. It is inclusive in that it aims to account for (1) changing part(s) of the system, (2) changing the whole system, and (3) creating a new system—a system without a history. Therefore, the model accounts for change that is the adjustment or the improvement type, as well as change that is carried out by systems design.

4.4.1.1.   The Three Dimensions

The three dimensions of the model are (A) the context of the change, (B) the "trigger" of change, and (C) the focus of the change inquiry. Each dimension has multiple options (Fig. 4.4).

- Dimension "A": the context of change can be (a) a specific part or a specific component(s) of the system, (b) the system as a whole, (c) the system in conjunction with other systems, and (d) a novel context.
- Dimension "B": the change trigger might be (a) negative feedback that calls for change within the system, (b) positive feedback that calls for

changing the whole system, (c) an aspiration or intent to extend systems boundaries, or (d) the intent to create a new system ("Greenfield situation").

- Dimension "C": the focus of the change inquiry can be (a) a problem focus, (b) a solution focus, or (c) a search for "novelty."

Given the three dimensions, and options within those dimensions, one can construct a set of "reasoned configurations" that can be considered to be the

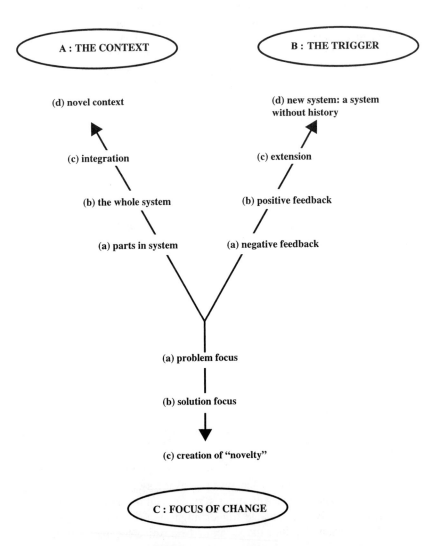

FIGURE 4.4. Change dimensions.

"genesis" of change. These configurations are constructed by considering possible and permissible combinations of options that are offered by the three dimensions. It is important to take into account the permissible, in that there are some configurations that are mutually exclusive. An exploration of the various configurations will enable us to differentiate between change that calls for design and change that does not.

4.4.1.2.   Potential and Permissible Configurations of Initiating Change

What follows is a set of reasoned configurations that may or may not call for design. Combinations of markers such as **A** (a), refer to dimensions and options displayed in Fig. 4.4.

Change by Considering the Context:

(I)   Configuration [**A**(a), **B**(a), **C**(a)]: The context is a part in the system. The trigger is negative feedback, calling for change/adjustment/ improvement within the system. The focus is on a specific problem(s) in the system. This change configuration does indicate adjustment/improvement but it does not call for design. Most organizational change efforts are dominated by this configuration even if the overall situation would call for systems design.

(II)   Configuration [**A**(b), **B**(b), **C**(b)]: The context is the whole system. The trigger is positive feedback, calling for changing the whole system. Thus, the focus is on (re)designing the whole system. This is an emerging configuration in organizational change, when there is a realization that the system is out of sync with its environment, or there is an aspiration within the system to search for a (more) ideal state.

(III)   Configuration [**A**(c), **B**(c), **C**(c)]: The context of inquiry extends into the environment. The trigger is positive feedback and it indicates integration with other systems. This type calls for the creation of a new system. This configuration indicates the design of a new system and prevails when systems intend to merge. Example: designing a community-based integration of educational, health, and social service systems.

(IV)   Configuration [**A**(d), **B**(d), **C**(d)]: The context is new. We design a system without a history. The focus is on the creation of novelty. This configuration calls for the design of a new system.

"Trigger" Initiates Change:

(V)   Configuration [**B**(b), **A**(b), **C**(b)]: The trigger of change is positive feedback, calling for the change of the whole system. The focus is on finding a design solution. This configuration indicates systems design and is similar to configuration (II) above.

(VI)   Configuration [**B**(c), **A**(b), **C**(c)]: The trigger is a desire/intent to extend the systems's boundaries, and create a new system by integration with other systems. This configuration calls for systems design and is similar to configuration (III).

(VII) Configuration [**B**(a), **A**(a), **C**(a)]: Negative feedback drives change, meaning that something is wrong within the system. This leads us to identifying this wrong and attempt to correct it. Here we have a problem focus, leading to intensive problem analysis. This configuration indicates improvement or adjustment in the system and does not call for design. This configuration is similar to configuration (I) above.

The "Focus" Calls for Change:

(VIII) Configuration [**C**(a), **B**(a), **C**(a)]: We focus on a problem in a part or in an operation of the system. This configuration, like (I) and VII), indicates adjustment in the system and does not call for systems design.

(IX)   Configuration [**C**(b), **A**(b)]: We seek solution to a design situation, which affects the whole system and is triggered by positive feedback calling for changing the whole system. This configuration calls for systems design and is similar to (II) and (V) above.

(X)   Configuration [**C**(c), **A**(b)]: We wish to create a new system in a new environment ("Greenfield situation"). This configuration calls for the design of a new system.

Reviewing the 10 configurations, we can define four major types of change as follows:

- Type one configurations are (I), (VII), and (VIII). These configurations, albeit for various reasons, indicate change within the system. They call for adjustments in parts or in operations of the system, or call for improvement. Type one cannot be considered systems design.
- Type two configurations are (II), (V), and (IX). These configurations, for a variety of reasons, indicate whole systems change, which calls for the (re)design of a system.
- Type three configurations are (III) and (VI). These configurations indicate the intent to integrate with other systems that are relevant to the subject system. They call for systems design.
- Type four configurations are (VI) and (X). These call for the creation of a new system, to be designed in a new environment.

We can now answer the question: What kind of change constitutes design? In view of the discussion developed above, we can disregard type-one configurations, in as much as these focus on improving or "fixing" the existing system. On the other hand, we can consider types two, three, and four as change types that can be defined as systems design. These types call for changing the whole system or creating a new one. Whole system change can be accomplished only through systems design.

Activity #21

First, capture the core ideas of design, implied by the text in this section. Next, review the four change types and propose or construct situations that would exemplify the various types. Then look at your own selected system and contemplate the change type that would apply to it. Enter your findings in your workbook.

4.4.2. Strategies and Methods Applicable to Transcend the System

In the text above, conditions that lead to a decision to (re)design a system were reviewed. If those conditions are present, people in the system realize that they should change the whole system. Such realization could lead to a decision to commence with design, in which case the decision can be considered the "genesis" of systems design. The insights we gained from most of the design literature and research indicated that once the realization to change the whole system was reached, we should leave the existing system behind, transcend it, and leap out from its boundaries. This leaping out and transcending often create anxiety, a good deal of uncertainty and hesitation, and possibly even denial. I have found over the years that when people are faced with a choice of changing their system and engaging in design or proving that there is no need to do so, they almost always find reasons for leaving things as they are.

I suggest that the ease or difficulty of transcending the existing state depends a great deal on our attitudes toward change. We reviewed four modalities of attitudes toward change in Section 2.6 of Chapter 2. Those who have a "reactive" (reactivating the past) or "inactive" attitude toward change will have the highest degree and intensity of anxiety and even denial. Having a "preactive" attitude, on the other hand, would lead people to rush into change whenever certain "trends" indicate, or "forecasters" predict, the likelihood of change. People with a "proactive/interactive" attitude will be ready to transcend their existing state and seek to engage in purposeful design inquiry. We can differentiate two major strategies—types "A" and "B"—for initiating design.

Design scholars who advocate type "A" are very much aware of the possibility that, for some people, leaping out might be troublesome. They approach the reduction or mitigation of the anxiety to transcend with plausible strategies and methods. Strategy type "A" suggests (1) an exploration of the problem situation (an analysis of what is wrong), (2) the creation of a "rich picture" of the existing problem situation, or (3) the addressing of such questions as: What are obstructions to our development and what would happen if we did not change?

Strategy type "B," on the other hand, suggests that it is a waste of time to analyze the problems of what is wrong now; therefore, this strategy sets forth an approach for quickly transcending the existing system and focusing on finding solutions.

4.4.2.1. Strategy "A": Situational Analysis Prior to Transcending

Strategy type "A" accommodates those who are not sure that they should proceed with design. Thus, they want to get some assurance by engaging in an analysis of the current situation. There is a set of strategies offered by design scholars that might help in reducing anxiety about changing the whole system.

Three strategy "A" approaches are introduced below, as defined by leading design scholars.

4.4.2.1a. Checkland's Approach. In his soft systems approach, Checkland (1981; Checkland and Scholes, 1990) presented an elaborate strategy for developing a rich picture of the problem situation. The essence of his strategy is a two-staged approach. During the first stage, we explore the problem situation in the real-life setting of the system of interest and find it "unstructured." At this stage our focus is not on specific problems but on the situational context in which we perceive the problem. We quickly realize that this unstructured problem situation does not help us. Thus, we move on to the second stage, where we develop the richest possible picture of the problem situation, represented as an organized and structured system of problems. This picture will help us to realize that we should seek out and create some new systems configurations that might be relevant to address and respond to our situation. Thus we are now ready to leap out from and transcend the existing state and seek a solution model.

4.4.2.1b. Weisbord's Approach. Future search conferences offer several examples of type "A" strategies. Described by Weisbord (1992), in his compendium, future search conferences (FSC) cannot be considered to be comprehensive systems design, but their strategies may be appropriate for the transcending and envisioning phases of design. As a rule, an FSC starts out with an exploration of the present and past situations and an analysis of the problem setting. In Weisbord's compendium (1992) I found several statements relevant to transcendence, including those reported by Haugen (1992), Frank, and Rehm (1992), Franklin and Morley (1992), Smith (1992), and Borbuloglu and Garr (1992). In these reports, I have found variations among the descriptions of future search conferences, but in most cases the front-end strategy follows three steps: (1) Explore changes in the world around us and explore anticipated future trends,

(2) review the history of the systems and (3) analyze the present system. Some of the events started out with a focus on the past, followed by focus on the present. This sequence was also described as "changes in the past" and "predicted changes." In one event, the front-end strategy calls for only two activities: (1) creating an image of current realities and (2) exploring their effects on the participants and their system. In another case, following the exploration of the environment and trends that affect us, the event quickly moves on to the design of the future system. In each of the FSC events described here, the initial exploration of the present and past and the analysis of problems is followed by the search for an ideal future and the setting forth of activities that bring that future about. But the approach does not include provisions for the design and modeling of the future system.

4.4.2.1c. Ackoff's Strategy.   Ackoff's (1981) front-end strategy is elaborate. He calls this strategy formulating the "mess" we are in. His three-pronged strategy—systems analysis, an exploration of obstructions to development, and making reference projection—is summarized as follows:

- The systems analysis stage provides a comprehensive description of the state of the existing system and its environment. It describes how the system operates, whom and what the system affects, and by what the system is affected.
- The obstructions to development phase explores constraints on the organization. These are constraints on growth and development. According to Ackoff, growth refers to increase in quantity and development refers to gain in quality. Constraints on growth are found primarily in the environment, while constraints on development are found within the organization. The within-the-system constraints are usually self-imposed and are often unconscious. They are manifested in discrepancies and conflicts within the system.
- A reference projection extrapolates the performance characteristics of the system from its past into the future. It assumes no significant change in the environment or in the behavior of the system. It stipulates the continuation of the systems's recent history and asks: What would happen if we would not change? A reference projection is often shocking and reveals the important characteristics of the "mess" that a system faces. It leads to exploring other futures, transcending the existing system, and engaging in systems design. In Ackoff's strategy, the three-pronged front-end exploration is followed by the ideal design of the desired future system.

The three approaches, described above, provide a representative sampling of what I labeled type "A" strategies. These are basically analytical in nature and could lead to a decision to leap out from and transcend the current state of the system and engage in systems design. Before I turn to describing some type "B" strategies, an activity is in order.

Activity #22

Review the three major front-end strategy types and identify their core ideas. Next, synthesize the strategies by creating various "strategy scenarios" for the front-end exploration of the system of interest. Then, make a judgment of which of the scenarios would be useful in the redesign of systems in which you are involved.

4.4.2.2.   Type "B": Transcending the System and Focusing on Solutions

Designers who define themselves as favoring the type "B" approach start out by transcending the existing state and by focusing on defining/finding solution alternatives rather than engaging in an analysis of problems or exploring the past and present. Four type "B" strategies are described next.

4.4.2.2a.   Nadler and Hibino's "Breakthrough" Approach.   Nadler and Hibino (1990) suggest that rather than engaging in an analysis of the current state we should focus on envisioning purposes of the future system. The purpose principle manifests a full range of motivations and makes results possible by changing what exists. Purposes challenge existing assumptions, thinking, and restrictions. "Breakthrough thinking" opens the door to many possible solutions. The authors say that "visions are started with big purposes" (p. 108), and "the purpose principle gives you a mechanism for seizing the opportunity to transform the problem into productive change" (p. 111).

If people accept a problem as presented, it leads them to "obvious solutions" and eliminates the opportunity for breakthrough solutions. "A purposes orientation helps you to avoid being sold a solution to the wrong problem" (p. 111). Accepting a problem as stated means the automatic acceptance of constraints that are associated with the problem. "Thinking purposes is an important defense against the analysis-first—and technology traps—of conventional approaches" (p. 113). An array of interrelated, embedded purposes provides a guide to the long-term development of the system.

Nadler (1981) proposed the development of an expanding array of purposes in his earlier work. He asked for the formulation of several hierarchical levels of ever larger purposes and a determination/selection of the primary purpose level around which to design the ideal system, the system that connects to all other purpose levels of the hierarchy,. Purpose expansion helps to eliminate "functional fixedness." It calls for leaping out and transcending the "fixedness" of the existing system. People and organizations need an understanding of purposes and not problems so as to enable themselves to move out of trenches and go ahead.

4.4.2.2b. Hammer and Champy. Hammer and Champy (1993) suggest that the structure, management, and performance of American business throughout the nineteenth and twentieth centuries have been shaped by principles laid down two centuries ago. It is time, they say, that we retire those principles and start over by developing a new set. This "starting over" and the development of new principles and practices does not mean analyzing current problems and trying to fix them. It does not mean tinkering with our existing systems and making incremental changes or jury-rigging them so that they can work better. It means asking this question: "If we were re-creating this company today, given what we know and given current technology, what would it look like?" For Hammer and Champy, redesign means "tossing over old systems and starting over" (p. 31).

American corporations do not perform badly because our workers are lazy or management is inept. Their performance over the last century is proof enough that this is not the case. They "now perform so badly precisely because they used to perform so well" (p. 10). But today, "the world in which they operate has changed beyond the limits of their capacity to adjust or evolve. The principles on which they are organized were superbly suited to the conditions of an earlier era but can stretch only so far" (p. 11). The basic organizing principles of our corporations are sadly obsolete.

Hammer and Champy are today the most vocal advocates in the corporate world of the type "B" strategy of transcending and leaping out from the existing system. For them this strategy means not only leaving behind the past and present but even unlearning the principles and techniques that brought success to our systems in the past. These principles and techniques do not work anymore in the information/knowledge age. As presented earlier in this section, the authors' reengineering model calls for the fundamental, radical, and dramatic redesign of our systems.

4.4.2.2c. Banathy's Leap Out. Banathy's (1991a, 1993a) proposals for a type "B" strategy are based on a recognition that when a new stage emerges in the evolution of a society, such as when the postindustrial knowledge age emerged around the midpoint of this century, the continued use of old "cognitive maps" (that guide our thinking and actions) creates increasingly more problems. Many, if not most, of our social systems still operate based on the outdated cognitive maps of the industrial society of the by-gone machine age. They are losing their viability. As I said in one of my columns on systems design (Banathy, 1991b): We cannot improve or restructure a horse and buggy into a spacecraft regardless of how much money and effort we put into it. Today our horse-and-buggy-style systems, born in the last century, operate in a continuous crisis mode. They surely face continuing decline and eventual termination unless they (1) understand the new realities of our transformed era and grasp the implications of what those realities mean to them, (2) leave their old ways of thinking and doing behind and learn new ways, (3) transcend their existing system, (4) envision a new image of their system, and (5) based on the image, design the future system and bring the new image to life.

Once we accomplish (1) and (2) and understand the new realities of our society and their implications for our system, and if we are prepared to learn new ways of thinking, we face the task of transcending what now is. Transcending is probably the most troublesome aspect of systems design. In addition to the anxiety and ambiguity that transcending the system generates, the crucial issue that we face is: Do we have the will and the capacity to leave the past behind and enter into the territory of the not yet known? As used here, "capacity refers to the open-armed attitude we need as we delve into something that's different" (Hawley, 1993). Transcending the familiar, that "what is now and what is known," requires that designers have, or develop, such individual and collective capacity. In the last section of Chapter 3, I reviewed the strategy for offering designers a "safe" space to land when they leap out from and transcend the existing system. This space was defined as the "option field," where designers can begin to think about solutions.

Metaphors/stories often help in encouraging us to leap out from and transcend the here and now. In the Native American story of the "jumping mouse," the mouse, hearing an enticing noise (of a stream), left the well-known home ground and, encouraged by the frog, dared to jump high (transcending) and capture the image of the sacred mountains (envisioning). He then embarked on an arduous journey (the design journey), during which he had to give up his sight (leaving the past behind) in order to be guided to the top of the sacred mountain, where he became transformed into an eagle (the transformation phase of design).

4.4.2.2d. Bridges's Transition. Bridges (1991) suggests that "change is the name of the game today, and organizations that can't deal with it effectively aren't likely to be around long" (p. ix). The effective way of dealing with change requires "transition," which I call transcendence and transformation. "Unless transition occurs, change will not work. And the starting point of transition is the ending that you will have to make by leaving the old situation behind" (p. 4). We have to let go the old reality, the old setting, the old thinking. The first step is "letting go," leaving behind and transcending. The second step is envisioning what we should become, which, as Bridges says, happens in the "neutral zone" of creativity, renewal, and design. Then comes the third step, the transition into the new reality (transformation by design).

For the first step, the "letting go," Bridges proposes a set of strategies, described as follows:

- "Mark the endings." Do not just talk about "the leaving behind" but create activities that dramatize the ending of what has been. Bridges tells the story of Cortés. When they came ashore at Veracruz, Cortés and his men were aware of the great uncertainty and adversity ahead of them. Many of the men wished that they never came. Cortés burned the ships . . . this dramatic mark of an ending became a new beginning, the beginning of a new journey.

- "Treat the past with respect." Never denigrate the past. Designers are tempted to denounce the past to distinguish it from a promising future. What we should say is: "What we accomplished is what enables us to stand on the brink of the new beginning. Honor the past for what it has accomplished" (p. 30).

- "Let people take a piece of the old way with them." "Endings occur more easily if people can take a piece of the past with them. You are trying to disengage people from it, not stamp it out like an infection. And, in particular, you don't want to make people feel blamed for having been part of it" (p. 31). Bridges tells the story of Western Airlines. When it was sold to Delta, the employee store sold out in a few hours all the items with the big red "W" company logo.

- "Show how endings ensure continuity of what really matters." The endings we seek to honor represent the only way to protect the continuity of what is truly important for us, the continuity toward something bigger. The old ways have to be relinquished before new systems can be created. "Conservatism is the worship of dead revolutions," said Clinton Rossiter. "Yesterday's ending launched today's success, and today will have to end if tomorrow's changes are to take place (p. 32).

Commenting on his strategy for "leaving behind," Bridges says that he is not urging to let it go slowly piece-by-piece, but rather, just the opposite: "Whatever must end must end. Don't drag it out."

Reflections

"One does not discover new lands without consenting to lose sight of the old shore" (French novelist André Gide). In this section, I focused on two questions: (1) Why people initiate change (and what kind of change can be defined as design) and (2) What strategies/methods are applicable to transcending and leaping out from the existing system and initiating the design of a new system?

We reviewed conditions that are to be present in order for us to recognize the need for change. We differentiated change that aims at the improvement of the existing system from change that qualifies to be called design. We reviewed two types of major strategies that designers use in transcending the existing state: The first might lead to a decision to engage in design through the analysis of the problem situation, through the analysis of the existing state. The other strategy is to leave the existing system behind and focus on the solution rather than the problem situation.

Which of the two strategies is the "right one?" I can't say. It depends on the particular design situation and our attitude toward change. Many design scholars speak for a type "B" strategy by saying: don't waste time to analyze what's wrong; rather, focus on seeking solutions. I am reminded of Bernard Shaw's remark: "You are looking at what *is* and ask, *why?* I dream of things that never were and ask, *why not?*

But we have found that in many design situations there are people who have a high level of anxiety when they are challenged to leave the existing system behind. They are rather fearful in the face of uncertainty and the unknown that leaping out brings with it. Thus, our best approach in such a situation is to be sensitive to the anxieties that emerge as people in an organization are faced with "leaving the old shores of the homeland behind." The type "A" strategies reviewed here aim at reducing those anxieties and enhancing the likelihood of developing an informed judgment that might lead to a commitment to engage in design.

I close my reflections with a quote from the essays of Chris Jones (1984).

> Designing, as I see it now, is, or could be, the process of unlearning what we know of what exists, of what we call the "status quo," to the point where we are able to lose our preconceptions sufficiently to understand the life, and the lives, for which we design, and where we are aware of the ways in which new things, added to the world, can change the ways we see it. (p. 172)

Activity #23

(1) Describe core ideas that may guide your design thinking and actions.

(2) Review the first part of this section and contemplate the difference between change for improvement and change that is implemented by engaging in design. Looking at systems that you are involved in, or are knowledgeable about, develop a rationale for engaging in improvement in some situations and initiating design in others.

(3) Consider the design or the redesign of certain systems of your choice in which people have a rather anxious (even fearful) attitude toward change. Review the strategies introduced as type "A" strategies and select and synthesize those which appear to be to you the most useful in reducing anxiety and leading to a decision to engage in design. Explain the reason for your selections.

(4) Review the strategies introduced as type "B" strategies and select and synthesize those that you would prefer to use in a design situations when people are ready for change. Explain your reason for your selections. Enter your findings in your workbook.

4.5. Approaches, Strategies, and Methods for Envisioning and Creating an Image of the Future System

In this section, I introduce various strategies and methods that might be useful in exploring and envisioning the future system and creating its first image.

The outcome or the product of systems design is a detailed description, a model that represents the future system. The design journey that leads from the leaping out from the existing system to the final comprehensive description of the future system is a long, adventurous journey. It begins with a fleeting vision of the "what should be" and it proceeds from there through a series of "pictures" of the desired future that is ever more developed, ever more elaborated, and ever more specified and differentiated as it becomes increasingly clearer and detailed through the process of design inquiry.

The front part of this journey is the envisioning of an image of the future system. It calls on our creative intelligence, our unrestricted imagination, our unbounded thinking. This freedom of envisioning becomes possible only if we free ourselves from the here and now, if we leap out from and transcend the present state that would only lock us into what now exists and would constrain and restrict our creative envisioning and imaging. In the course of this journey, we consider first the overall strategy of the exploration of a desired future. In this section the major activities of this exploration are described and their relationships and their flow are shown. Then I discuss methods we might apply in the process of envisioning and creating the first image of the system.

4.5.1. Initial Exploration: Envisioning the Future

> "Where there is no vision, the people perish"
>
> —Proverbs 29:18

For the term "vision" we find a two-pronged definition in Webster's New Collegiate Dictionary (1979): "the act or power of imagination" and "the act and power of seeing." In systems design, using the power of imagination and of seeing the future stands for the process of our "vision-quest" in search of the ideal. Our vision-quest commences as we leap out from the existing system and capture a vision of the ideal. In so doing, we first seek a vision of the ideal society that embeds the system that we wish to design. Our quest is guided by several questions:

- What kind of society do we wish to create? What is our vision of an ideal society that will give inspiration and meaning to our quest? What is that larger vision that will have the power to guide us in our design of the desired future system?
- Once our quest is rewarded with a vision of the society we wish to have, only then do we ask the question: What is our vision of the system (which we wish to design) through which we can make a contribution toward creating that society?

The two-pronged outcome of our vision-quest most likely will be a highly inspiration formulation, possibly some metaphor(s), that expresses at the most general level the ideals we seek.

In our present age, when we find that most of our social systems are out of sync with emerged new realities, we shall inform our vision-quest by an exploration of two additional questions. First, What are the new realities? and, second, What are the implications of the new realities for the society we wish to have and for the system we wish to create?

The questions introduced above should be considered "triggering" questions that guide the conversations of the designing community in its vision-quest. Well-conceived trigger questions and well-defined methods of a conversation provide a structured and disciplined approach to vision-quest and image creation. The conversation at this stage should lead to the formulation of a shared vision, a shared picture of the future that creates the first shared collective identity of the future system and a collective sense of destiny in the designing community.

A rich set of methods is applicable in creating a shared vision. Senge (1990) and Morgan (1993) developed compelling arguments for the significance of envisioning and imaging in the design of organizations. Moore (1987) describes several idea-building group methods that seem to be most useful in creating a shared vision.

I describe here a method, adapted from Moore, and develop it as one possible way of vision building. In collective design consensus building methods are used throughout the entire design process. A variety of methods is described in Section 4.7. An example of a vision-building scenario follows.

Small design groups of seven individuals or so initiate the vision-quest. Each group assigns the role of coordinator to one of its members. Then, the groups begin to formulate the kind of triggering questions introduced above. Next, members of the group have a contemplating period and begin to write down their visions that respond to the vision-generating triggering questions. Next, the items are read and clarifying questions are asked. The items are recorded on newsprint attached to the walls of the room. There is now an opportunity to integrate some of the visioning ideas and create systems of complementary and internally consistent ideas. Then, the ideas might be prioritized. If several groups take part, the groups might report back to the entire designing community. The newsprints, developed by the small groups, are attached to the wall and the designing community begins to work with the visioned ideas in order to create a synthesis of the visions. Findings are recorded. This process and others are introduced in some detail in Section 4.7.

Activity #24

To get a grasp of the process of envisioning, arrange for an activity that would enable you and others to experience vision-quest in a real-life context. The context can be your own family, an organization you belong to, or any other context. You should arrange an appropriate place and time that allows the group three to five hours to engage in the envisioning activity described above. Enter your findings in your workbook.

4.5.2. Initial Exploration: Learning about "New Realities"

In Section 2.7 of 2, I introduced the idea of the new realities of our changing world. These new realities are one of the main reasons for (re)designing our social systems so that they can become compatible with their changing environments and realize our aspirations. There is a very rich knowledge base that portrays the characteristics of major emerged societal changes. The designing community will engage in learning about emerged societal characteristics by studying various statements from relevant literature. An approach to such study might be to establish study groups that distribute among themselves the study of literature on change.

First, the groups could study the general markers of change by consulting the work of such authors as Bell (1976), Drucker (1989), Harman (1988), Harman and Horman (1990), Markley and Harman (1982), Maynard and Mehrtens (1993), Haisbitt and Aburdene (1990), Morgan (1986), Reich (1991), Theobold (1987), Toffler (1970, 1980), and others who characterize societal changes. The reading tasks could be distributed among members. Based on their study, the groups create a synthesis of their findings.

Next, the groups would study the various component dimensions of societal changes, such as the sociocultural, sociotechnical, economic, technological, organizational, scientific, and possibly some other dimensions they would identify. This exploration might lead to a synthesis of the overall dimensions of societal change.

The information in Table 4.1 stipulates some of the key features of the changed societal landscape in the post-industrial/information age. In order to highlight the change, I contrast the emerged features with the features of the industrial/machine age. The contrast shows sharp discontinuities between the two eras. The features of the postindustrial age cannot be extrapolated or predicted from those of the industrial age.

Even a quick review of the contrasting features helps us to see that most of our existing systems are still grounded in the thinking and practices of the industrial/machine age.

Having explored the emerged new realities, designers revisit the vision statement(s) they created and elaborate that statement in view of the findings of their exploration of societal changes. As stated earlier, the guiding (triggering) questions of this process are: What are the implications of the emerged societal realities for the societal vision we formulated? What are the implications of these realities for our vision of the system we wish to design?

We might ask: Why is it so important for designers of social systems to understand the larger context of societal changes and the emergent new realities? Ulrich (1993) provides us with a convincing answer to this question when he says that "perhaps the most fundamental concept of critical systems heuristics is the

TABLE 4.1
The Changed Societal Landscape

Industrial/machine age	Postindustrial/information age
Key marker: Extension of our physical powers by machine technology.	Key marker: Extension of our cognitive powers by cybernetics/systems technology.
Processes organized around energy for material production.	Processes organized around intellectual technology for information/knowledge development.
Paradigm: Newtonian classical science, deterministic, reductionist, linear causality, organized simplicity.	Paradigm: Cybernetic/systems science, emergence, expansionism, mutual/multiple causality, dynamic complexity, ecological orientation.
Technologies: Inventing, fabricating, manufacturing, heating, engineering.	Technologies: Gathering/organizing/storing information, communicating, networking, designing.
Principal commodities: Energy, raw and processed material, machines, manufactured goods.	Principal commodities: Knowledge used to support policy-making, planning, and design.
Focus of economy: High-volume production.	Focus of economy: High-value production.
Social consciousness: Based on national and racial identity.	Social consciousness: Also extends into global consciousness.

context of application" (p. 592). The context of application is that section of the natural and social world that is considered to be relevant to the design inquiry and that justifies the design's normative content. Our value judgments flow into the design, and we consider the practical consequences that a particular design may have for those who will be affected by the design. So the practical question is: What impacts are to be expected, and who will be affected by it? (p. 593).

Activity #25

Describe core ideas of design you discovered in this section. Continue working with the group you involved in the previous activity. If this is not possible, you might do it on your own. Members of the group (or you) should review at least three of the literature sources on societal changes. Based on the review, rewrite my statements in Table 4.1, or formulate other statements. Then, the group (or you) should generate a statement of features that characterize sociocultural, sociotechnical, economic, organizational, scientific, and technological changes. Next, consider the implications of these changes for the rewriting of the vision statements your group developed in completing Activity #24. Enter your findings in your workbook.

4.5.3. *Initial Exploration: Making Boundary Judgments*

This activity aims at exploring possible alternative boundaries of the design inquiry. This exploration establishes the boundaries of the design inquiry and defines the context of application. This context is never given objectively. "It needs to be delimited by judgments from the total universe of facts and value implications that might be considered" (Ulrich, 1993, p. 592).

> [The rightness of] boundary judgments depends on the subjective interest, values, and knowledge of those who judge, which is to say that boundary judgments (if recognized as such and laid open to everyone concerned) will tend to be disputed. A theoretically sufficient ("objective") justification will not be available; at best, "informed consent" of all those involved and affected can be attained. (p. 593)

The process by which such informed consent can be achieved is some type of consensus-building conversations. Thus, the two critical questions are: (1) How do we define the boundaries of the design inquiry of the system we wish to design? and (2) How do we define that section of the universe that we consider to be the "environment" that is relevant to our system? Therefore, we are to draw boundaries that set aside the design inquiry space of the future system from its environment and set the inquiry boundaries of the "system-relevant" environment. The boundaries of the design inquiry create the option field within which some initial alternative design configurations can be formulated and explored.

Having the vision statement in view, designers ask the question: What are some of the dimensions that could be selected and considered upon which the designers can draw alternative boundaries of their inquiry? The field defined by the boundaries should be broad enough to (1) accommodate the vision(s) we formulated, (2) allow unrestricted exploration, and (3) enable us to define the first comprehensive image of the future system.

In Section 3.3, Chapter 3, I introduced an example for creating a field of inquiry—called an option field—of four dimensions that are appropriate to making boundary judgments and exploring initial alternative design configurations in designing systems of learning and human development. Another example will assist us to contextualize the process of making boundary judgments in design inquiry. Figure 4.5 shows a field of inquiry for the exploration of possible options or patterns of interaction among a group of research and development (R & D) agencies. These agencies for years have been in competition. Upon the encouragement of their main founding source they initiated a consideration of some form of interaction, cooperation, coordination—and possibly even integration—to better serve their clients: the educational communities of the nation. Being a previous member of the community, at their invitation, I joined them in developing an option field that allowed the consideration of alternative design options in creating a "macro system" that would enable them to link up with each other.

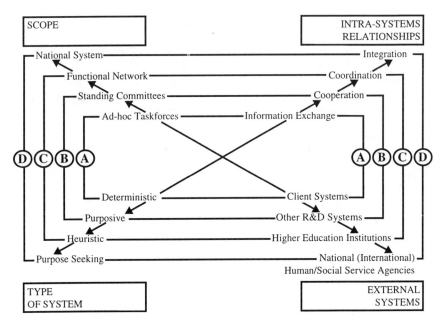

FIGURE 4.5. A framework for exploring design options.

In creating the option field, I proposed four dimensions and several options on each of the dimensions. The sequence of options from (A) to (D) shown in the figure implies gradual extension of the boundaries of the design inquiry.

- The *scope* dimension offers four options: (1) the establishment of ad-hoc (temporary) task forces, representing each of the agencies, that cooperate on special topics; (2) standing committees that work jointly on special issues for an extended period of time; (3) a functional/permanent network that establishes coordination of programs and projects; and (4) an integration of the various agencies as a "virtual" national R & D system.
- The dimension of *intrasystems relationships* offers four options: (1) information exchange about issues of common interest; (2) occasional cooperation on selected projects or programs, (3) coordination for a more permanent joint effort in selected programs, and (4) integration that creates a new "macro" system.
- *Relationship with external systems* might bring about linkage with (1) client systems of the agencies in order to develop selected projects, (2) other R & D systems of similar interests and programs, (3) higher educational agencies, (4) national/local human and social service agencies, and (5) possibly relevant international agencies.

- *Types of systems* offer such choices as (1) deterministic, (2) purposive, (3) heuristic, and (4) purpose seeking.

Using the inquiry space (pictured by Fig. 4.5), designers (1) consider the various options of the dimensions; (2) plot several possible alternative inquiry boundaries from selected options, within which they create various design configurations; (3) weigh the implications of the alternatives; (4) make a judgment of the most desirable alternative; and (5) display and describe it.

In Fig. 4.5, I drew some boundaries that indicate alternative, optional design configurations that designers might consider. Boundaries A, B, C, and D show markedly different inquiry configurations that eventually would lead to very different designs of the future system. The design strategy implied by the example above demonstrates dramatically the importance of making boundary judgments. But making boundary judgments and selecting a desired inquiry configuration—as any design decision we make in systems design—are tentative and always subject to reconsideration.

In the course of considering alternative configurations, designers will again apply the types of consensus-building group methods I introduced earlier or select some other methods, such as those presented in Section 4.7. They could ask "triggering" questions such as: What are particular design configurations that respond to the vision we created? What values do we want to realize in the system we want to design?

The collectively agreed upon values, called core values, will guide the designers to make design choices throughout the inquiry. For example, values, motivating beliefs, desires, and moral concerns may express aspirations of attaining higher inner quality of life; human dignity and human betterment; social and economic justice; and individual, social, and ecological ethics. These are examples of bases upon which decisions are made. In summary, designers ask: (1) What values would support/justify a particular option configuration? (2) What would be the implications of selecting a particular configuration? (3) What core values lead to the selection of a particular option configuration?

Activity #26

In Activity #25, you (and your design team) formulated a vision of a system you wished to design. Based on the vision, you (and your design team) should (1) Construct a framework, appropriate to your selected design issue, that would enable you to establish the boundaries of your design inquiry. (The framework would display the main dimensions of the inquiry and various options marked on those dimensions.) (2) Create potential alternative inquiry boundaries. (3) Display these alternatives on the framework. (4) Make boundary judgments by weighing the implications of the various alternative boundaries for the future system. (5) Formulate and describe collectively agreed on core values and core ideas that underlie particular boundary judgments. Enter in your workbook the design information generated in the course of the above activities.

4.5.4.   Initial Exploration: Creating the First Image of the System

The findings generated in the course of the design strategy, previously described, become the essential knowledge base that is used in the course of the entire design process. This "essential knowledge" emerges from the entwinement and integration of the vision that was formulated, the core values that the designers collectively articulated, and the core ideas that emerge from the boundary judgments made as the inquiry options were selected. This knowledge base is the main source of design decisions that lead to the creation of the first image of the future system. This image "provides us with a 'broad-stroked picture' or a 'macro-view' of the system" (Banathy, 1991a, p. 128). The design strategy applied here is driven by the following question: How can we organize the knowledge generated in the course of the previous activities into an internally consistent system of design ideas that becomes a "system of key markers" of the image and that reflects the stated core values?

The power of these markers, and the power and completeness of the image, depend on the thoroughness and completeness of the design work accomplished heretofore. The more complete and detailed the previous design work, the more core values and core ideas we agreed upon, the clearer and more powerful the "markers" around which we can create the image and the more confidence can we have in the power of the image.

The methodology for selecting image markers is one of the group consensus-building methods discussed earlier, or some other method selected by the designers. The designers will formulate several alternative marker configurations and evaluate those against the vision statement, the values, and the core ideas. An image example in Table 4.2 is organized around the following markers: (1) the overall scope of the integrated R & D system, (2) its main function, (3) the key organizing principle, (4) relationship with peer systems, (5) relationship with external systems, (6) the internal focus, and (7) designation of the system type.

Using these markers, the image of the desired future state is formulated and introduced in Table 4.2 by juxtaposing it with the image of the existing state of the system of R & D agencies. The statements in the left-hand column represent the perceived existing state, while the statements on the right-hand column project the image of the desired future state.

The image introduced above is speculative but it may serve the purpose of demonstrating an approach to the last task of the initial exploration stage of systems design. The image that designers create is the first collective systemic representation of what "should be." Components of the image should be internally consistent and mutually compatible.

TABLE 4.2
An Image of an Integrated R & D System

The existing state	The future state
The overall scope is defined by the sphere of influence of individual agencies.	The overall scope of the future state is the larger society, the nation, and beyond.
The main functional is to provide service by responding to the needs of state and local educational agencies within their geographic region.	The main function is the coordinated development of models and processes that the eductional communities of the nation can use to design their own learning and human development systems.
The key organizing principle is to respond to local needs that fall into the traditional means and methods of addressing specific problems.	The key organizing principle is to assume leadership and be at the cutting edge of theory formulation and methods development of disciplined inquiry.
Relationship with peer systems is limited to information exchange and occasional/limited and short-range cooperation.	Relationship with peer systems is full-fledged partnership and long-range integration of operations and services.
Relationships with other systems is occasional/self-serving cooperation.	Relationship with other systems aims at building alliances with local/national agencies that advance human development.
Internal focus is the prudent administration of specific projects and programs.	Internal focus is on developing organizational capacity and staff capability for organizational learning and educational design.
Type of system: deterministic, moving toward purposive.	Type of system: heuristic, moving toward purpose-seeking.

Activity #27

First, describe a set of core ideas of how to generate an image. Then, using the outcomes of Activity #26, you should now formulate some markers that will help you and your team to create an image of your selected system. For techniques to describe the image, consult Section 4.7. Introduce your image in your workbook.

Reflections

Today we recognize that most of our social systems are out of sync with the emerged new realities of the postindustrial information/knowledge age. We also recognize that this situation exists because most of our social systems are still grounded in the designs of the industrial/machine age. Our organizations are failing today precisely because they were so successful in the nineteenth and the early twentieth centuries. Attempting to improve or adjust them with the use of our traditional social planning methods is not working anymore. We have no other viable choice than reconceptualizing and redesigning our social systems. Confronted with this situation we now understand that (1) rather than trying to expand our existing systems we should transcend them, (2) rather than trying to revise them we should revision them and (3) rather than trying to reform our systems we should transform them.

In this section we have focused on the frontal part of the design journey, in the course of which we "revision and create a new image" of our social systems. We have reviewed approaches, strategies, and methods by which we can (1) engage in a vision-quest of the systems we wish to create, (2) establish the boundaries of our design inquiry and consider optional design configurations, and (3) formulate the first image of the system based on collectively articulated core values and core ideas.

The questions that have guided the three-staged initial exploration of the design journey are seldom asked in the design of social systems. And the approaches, strategies, and methods introduced here are rarely employed. But I am confident that, as you have worked with this text and completed the activities, you will recognize that exploring those questions and working with those approaches, strategies, and methods provide the foundation of a viable, authentic, and sustainable design. The proposition I set forth is that a social system design is viable, authentic, and sustainable only if it is carried out collectively by those who serve the system, and who are served and affected by it.

4.6. Approaches, Strategies, and Methods for Designing the Model of the Future System

In this section, I introduce various approaches, strategies, and methods that can be used in the process of designing the model of the future system. Constructing the model of the future system builds upon findings gained from the exploration space where we formulate the vision of the future system, from the selection of the design configuration, from the definition of core values and core ideas that guide our design, and from the creation of the image of the future systems.

The process of design is a continual elaboration of the image, the continuous specification and an ever more detailed description of the future system. This description unfolds as designers use strategies and methods for answering a cumulatively unfolding set of questions, such as:

- What is the system about? What is its definition or mission? What are its purposes and goals?
- What are specific functions that have to be carried out in order to attain the purposes/goals? How can we organize those functions into a system of functions?

- What is the system that enables the accomplishment of functions?
- What is the model that provides a prescriptive representation of the future system and what is the model of its systemic environment?

Chapter 3 described (1) the dynamics of interaction among the processes employed as designers address these questions, (2) the feedback/feed-forward interactions, (3) the dynamics of creating and evaluating alternatives, and (4) the continuous and expanding use of the knowledge base of the design inquiry. Here I review strategies and methods advocated by various design scholars that might be appropriate in responding to the questions introduced above. Whenever appropriate, I continue to suggest activities that help you learn to apply the reviewed material in a real-life context.

4.6.1. The Definition of the System: What Are Its Purposes?

We can speak of two major categories of purposes. Purposes that are "generic" to all social systems and purposes that are "specific" to a system of our interest.

4.6.1.1. General Purposes

Ackoff (1981) suggests that every system should attend to three purposes: the purpose of its parts, its own purposes, and the purposes of the system in which it is embedded. He proposes (Ackoff, 1995) that this three-way general purpose is adhered to only in social-systemically conceived systems, while organismic systems attend to purposes of their parts only partially and machine (autocratic) organizations do not pay attention to the purposes of their parts. Nadler and Hibino (1990) discuss four general purposes: greater effectiveness, higher quality of life, enhanced human dignity, and individual betterment. Vickers (1983) proposes that the overall purpose of social systems is maintaining relationships. The Pinchot's (1993) imperative is the engagement of everyone's intelligence, while Boulding's (1985) imperative is human betterment. Flood and Jackson (1991) suggest that the general aim is the emancipation of people (from oppression) and the liberation of human potential. Churchman's (1982) purpose is designing just systems for future generations. We could continue this list for a long while, but I turn now to the discussion of specific purposes.

4.6.1.2. Specific Purposes

A variety of strategies and methods exist in the design literature that address this question. Here I review the propositions of several design scholars and introduce examples that demonstrate possible outcomes.

Checkland and Scholes (1990) answer this question by proposing the formulation of a variety of "root definitions." A root definition defines (1) the customers, the users of the system; (2) the actors who serve the system and who accomplish the transformation process; (3) the transformation process, which processes and transforms the input into output; (4) a worldview, which makes the transformation meaningful in context; (5) the owners of the system who are also owners of the design process; and (6) the environment, the entities outside of the system that are a given or are to be defined. A requirement of formulating a root definition is that its components should be internally compatible.

An example of a root definition of the national R & D system follows:

1. The customers are communities across the nation that wish to initiate the design of their educational systems.
2. The actors are the national system and its component systems.
3. The system transforms the generic knowledge base of systems design into actionable design models, approaches, strategies, and methods that communities across the nation can learn to use to acquire organizational capacity and collective capability to design their own educational systems.
4. A worldview that makes the transformation meaningful in context. For example, individuals and communities have the right and responsibility to take charge of their future and design their own systems.
5. The owners are local communities who use the services and to a certain degree funding agencies who provide the support to the work.
6. The environment comprises supporting agencies, social service and human development systems, and the larger society.

For each component of a root definition several alternative formulations are possible. The following examples demonstrate alternative root definitions of health services in a community.

Example "A": (1) Customers are people who are ill; (2) actors are health professionals in doctors' offices, hospitals, and pharmacies; (3) transformation is treating illness and disease by medical and drug interventions; (4) worldview posits that health services are the responsibility of health professionals; (5) owners are the health professionals and their organizations; and (6) environment includes health facilities, government, and insurance agencies.

Example "B": (1) Customers are people and groups in a community; (2) actors are all people in the community, health professionals, educators, organizations; (3) transformation maintains and improves the physical, mental, emotional, and spiritual wellness of all people; (4) the worldview suggests that the primary responsibility for wellness lies with individuals and their groups and is shared with others who can provide help and support; (5) owners are all the actors mentioned above; and (6) the environment is the entire community and beyond.

Nadler and Hibino (1990) suggest that in design the word "purpose" has such connotations as utility, intent, and objective. These express a "whole range of motivations and results [that are] possible in applying change to an existing condition" (p. 107). An initially formulated purpose is "only the beginning and a great many more purposes emerge with scrutiny and creative thinking" (p. 108). In dealing with purposes they advocate expanded thinking, which will lead to many more purposes and many more possible solutions than initially stated. This approach leads to the formulation of a multilevel (hierarchical) system of purposes. The thrust is to seek ever broader (higher level) purposes. Purpose at one level gets its meaning from a purpose stated at a higher level. But eventually it is the task of the designers to select the primary purpose level. It is in relationship to the primary level to which they define the purpose at the next higher level and purposes at a level below, at the level of component systems of the primary system.

The purpose structure of the national R & D system (NR&DS) can be formulated at gradually expanded levels. For example, we can start formulating purposes of the NR&DS itself, then formulate purposes for individual agencies, and then, moving to the level higher than the NR&DS, formulate purposes at the level that supports the work of the NR&DS and other national/international social service and human development systems.

The context of the three-level purpose structure could be the entire society or it could be international educational and human and social development systems. In this example, the primary system level is the one at which the NR&DS operates.

Banathy (1991a) suggested that a core definition of a system comprises its statement of mission and its elaboration in the formulation of purposes. As a general approach, I suggested that a core definition should address the question of: What is our aspiration to serve humanity, the larger society, our community, people who serve the system, the clients and stakeholders of the system, and all those affected by the system—most importantly—future generations?

Based on the statement above, the purposes of an educational system might be as follows: (1) Become a societal institution that has a global vision of humanity as a whole, one that makes a purposeful contribution toward human betterment; (2) offer programs and arrangements by which individuals, families, and social systems in the community can attain their fullest potential, enrich their inner quality of life and the quality of their environment, and learn to give direction to their own life and the life of their systems; (2) develop the community as a learning society in which learning, becoming truly human, and ethical action and service to others are collective goals to live by; (4) create a coordinated linkage of all social and human development systems in the community to offer integrated arrangements and resources for learning and human development that are coherent, easy to use, and continuous; (5) ensure a high quality of working life for all those who serve the system, (6) nurture cooperation among individuals and groups, and support personal and professional development; and (7) become a self-organizing and self-directing system that can learn to give direction to its continuous development and manage itself by the widest possible participation of its members and members of the whole community.

Activity #28

Given the image you created for your system of interest, you should now formulate several possible root definitions of your selected system. Then, identify the particular root definition that best reflects the image, the core values and ideas, and the vision you created earlier. Enter your findings in your workbook.

Activity #29

Given the definition you formulated for your system, define the purposes for the system at the next (higher) level and for the subsystems of the system. Enter your findings in your workbook.

Reflections

I close the discussion on what the system is about by sharing with you the thoughts of Jones (1980). He suggests that answering this question is "a crucial step in the whole design process and needs to be done with the help of as many sources, wisdom and knowledge as can be enlisted." Furthermore, essential design objectives "should be stated with no more, and no less, accuracy than the current information permits, and should be restated more accurately as new information arrives at later stages." It should be further ensured that "statements of essential objectives are compatible with each other and with the information that becomes available while designing. The essential objectives will determine the area of search for a solution. The search will include all means of satisfying these objectives" (pp. 196–197).

4.6.2.   What Are the Specific Functions that Must Be Carried Out?

Answers to this question lead us to make the most central, the most essential design decisions. Checkland and Scholes (1990) call this activity the building of a conceptual model of the system. The model is a structured set of activities that are necessary to realize the root definition. The model is a systemic arrangement of "verbs," and nothing else but verbs. In my earlier work (Banathy, 1991a), I described a process by which we can identify functions and build them into a system of functions as we ask: What are key functions that have to be carried out?

How do these functions interact to constitute a system of functions? What are subfunctions of the key functions? How do these integrate into subsystems of key functions? What are the component functions of the subfunctions? How can we build them into a component system of subfunctions? Following this process of elaborating functions in more and more detail, we create ever higher-level resolutions of the systems complex of functions. The multilevel systems complex of activities/functions we design through the method described here represents the first model of the future system.

For example, given the root definition of a national R & D system identified earlier, key functions that might be carried out could include the following: (1) integrate member laboratories and centers into a national system; (2) promote the application of systems thinking and systems design among member agencies, clients, and stakeholders; (3) conduct joint R & D in order to develop a knowledge base relevant to systemic educational change; (4) establish programs for experimentation with systems design and systemic change; (5) collect information on ongoing systemic educational change programs; (6) share and publish findings relevant to (3), (4), and (5); (7) develop resources and conduct programs for systems and design learning; (8) initiate and guide systems building programs in all member organizations; (9) establish functional relationships with other R & D systems and national agencies that serve/promote human and social development; and (10) develop and guide arrangements for continuous systems evaluation and organizational learning. These key functions are to be relationally arranged to compose a system of functions. As functions are implemented, integration becomes ever more a reality and their interactive dynamics create the wholeness of the system.

Individual functions are evaluated by asking several questions (Nadler, 1981; Banathy, 1991a): (1) What are we trying to accomplish when we perform this function? (2) Why do we need to accomplish this function? (3) What other (alternative) functions should be considered? (4) What higher (level) functions call for the accomplishment of a specific function?

The system of functions is evaluated by asking the following questions (Banathy, 1991a): (1) Did we provide for all functions necessary to respond to the core definition (purposes/objectives) of the system? (2) Are there any redundant functions? (3) Are the functions mutually compatible?

Activity #30

(1) Describe core ideas of designing a system of functions. (2) Based on the core definition you or your design team developed, you should now design the system of functions for your selected system. Design this complex at a minimum of three levels: (1) at the system level of key functions, (2) at the level of subfunctions, and (3) at the level of the component functions of subfunctions. In

order to select methods/techniques appropriate in carrying out this activity review methods introduced in Section 4.7. Enter your findings in your notebook.

4.6.3. Designing the Enabling Systems and the Systemic Environment

The system complex of functions tells us what activities to carry out in order to attain the core definition, the mission, and purposes. The next—and crucial—task is to design what Ackoff (1981) calls the enabling systems and the systemic environment (Banathy, 1991a).

Enabling systems should have the organizational capacity and human capability to carry out the functions. Two such systems are to be designed: one that guides the organization and one that is the organization itself. As an outcome of this activity two models will emerge: the model of the management/guidance system and the model of the entire organization. These models, designed to carry out the "verbs" of the functions model, are constructed as a system of "nouns" (who does what). Inquiry that guides the design of these systems is described next, based on the work of Ackoff (1981) and Banathy (1991a).

4.6.3.1. The Management/Guidance System

This enabling system is created by asking what design will enable the system to (1) select the processes that "transform" the functions into ongoing actions; (2) conceive and plan the initiation of those actions; (3) motivate and energize the individual and collective action of those who carry out the processes; (4) work with the environment in order to collect and analyze information that is of value to the system and that enhances the accomplishment of functions; (5) work with the environment in order to acquire and manage the resources that are needed by the system; (6) identify actual and potential problems, threats and opportunities; and (7) engage the system in continuous organizational learning and nurture design capability.

4.6.3.2. The Organization

Next the task is to design the organization that carries out the systems functions. The challenge of designers is to design a system that has the organizational capacity and individual and collective capability to carry out the functions as specified in the functions model. Questions that drive the design inquiry include the following:

1. What organizational and personal capabilities are required to carry out the identified functions?
2. What potential systems components and what kinds of people have those capabilities and how can they be acquired and nurtured?

3. How can we distribute functions among the components of the system?
4. How should we organize components in relational (vertical/horizontal) arrangements?
5. What authority/responsibility should be assigned to whom?
6. What resources should be allocated to what component?

4.6.3.3. Designing the Systemic Environment

The third enabling system is the larger system that embeds the system we design. It is the obligation of the designing community to design a system that will have the commitment to provide support and guidance of the system the community designs.

Activity #31

(1) Describe the core ideas of designing enabling systems. (2) In Activity #30 you (and your design team) defined a functions model for your new system. Based on the model, work with the sets of items and questions introduced above and make an attempt at creating a tentative design of the management/guidance system and a design of the organization that will carry out the functions. (3) Speculate about the systemic environment that will be able to provide commitment and resources to support the system. Consider the use of methods and techniques described in Section 4.7. Enter your findings in your workbook.

4.6.4. Presenting the Outcome of Design: Modeling the System

The outcome of design is a presentation, a description of the future system and its systemic environment. This description will take the form of a set of systems models. The construction of these models was described in Section 3.6 of Chapter 3. Here I briefly review the three models in relationship to the design inquiry discussed above. A detailed description of the modeling process is presented in an earlier work (Banathy, 1992a).

4.6.4.1. The System-Environment Model

The systems-environment model describes the systemic environment of the new system. The systemic environment is that part of the general environment with which the system regularly interacts and from which the system receives the support and the resources it needs in order to become a viable system. The model describes systems-environment relationships, interactions, and the dimensions of mutual interdependence. A set of inquiries—built into the model-building process—guides designers in assessing the environmental adequacy and responsiveness of the new system and, conversely, the adequacy of the responsiveness of the environment toward the system we designed.

4.6.4.2.   The Functions/Structure Model

The functions/structure model describes the new system at a given moment in time. It guides designers in presenting much of what they accomplished in responding to the inquiries introduced in this section. Here we describe the final selection of (1) the mission/purposes of the system, (2) the systems model of functions that have to be carried out, (3) the management/guidance system, and (4) the organization that attends to the functions. A set of inquiries, coupled with the model, enables the designers to test the functions/structure adequacy of the new system.

4.6.4.3.   The Process/Behavioral Model

The process/behavioral model calls for a description of what the system does through time: how it transforms functions into processes, behaves as an open and dynamic social system, receives/screens/processes input, transforms input into output, assesses and processes output, makes adjustment in the system, or transforms the system in a new system state by redesigning it if indicated. Another set of inquiries, built into the model, helps designers to evaluate the system's process/behavioral adequacy.

The three models jointly provide a comprehensive description of the new system and its systemic environment and also provide for a comprehensive (conceptual) assessment of the outcome of design. In addition to the conceptual evaluation, designers will arrange for the real-life testing of the system. They will test the operational readiness of the design for the development and implementation of the system. This testing might indicate a need for a redesign. This process of conceptual and real-life testing and redesign will eventually reach the point when designers will have enough confidence in their design to proceed with systems development and implementation. This point can come quite early, due to the fact that designers build into the system the capability for continual organizational learning and (re)design. Learning never ends. Neither does design.

Activity #32

A full-scale modeling will require the use of knowledge base and resources for "modeling" a system. The process of how to model a system with the use of the three models identified above will require the use of my earlier work (Banathy, 1992a). You may not be in a position to engage in full-scale modeling. At the minimum, you should follow the modeling process described in Section 3.6 of Chapter 3. Enter your findings in your workbook.

Reflections

We have arrived at an important milestone in the course of our journey toward understanding systems design and applying it in order to take charge of our individual and collective futures. This milestone also marks an important achievement for you, namely, a comprehensive understanding of what design is, how it works, and how it can be used in social systems. This understanding developed as you worked with the text and, more importantly, as you completed the activities.

You had an opportunity to integrate the perspectives, propositions, and knowledge sources about design with your own knowledge of how to work with social systems. This working with the text—and integrating design ideas into your own scheme of thinking—has led to the development of an understanding and appreciation of systems design as a collective human activity. You have formulated your own view of design, constructed your own "meaning" of design, developed your own perspectives and propositions about design, and constructed your own understanding of what design is, why we need it and, how it works.

You have also been challenged by the activities to apply specific design activities in the context of systems of your own choosing. Such application experiences are the only way I know for grounding knowledge, skill, and insight into lasting capability.

Activity #33

I suggest that you take a break in your journey and reflect for a while. (1) Review your workbook and construct a set of statements that captures your most important findings about design, the most salient aspects of what design is, how it works, what it does, when should it be used, and why is it important for us— for all of us—to develop capability in engaging in design. Enter your notes in your workbook.

4.7. Group Techniques that Aid Design

An essential feature of the kind of design elaborated in Part I is that it is accomplished by groups of people who serve the system (to be designed), who are served by it, and who will be affected by it. This section introduces group techniques that are appropriate to accomplish various design tasks. I first describe a variety of group techniques that, in the course of the last two decades, have been used in organizational and social systems inquiry. At the end of the section, a software-aided design inquiry program is introduced, which makes use of the group techniques described here.

4.7.1. Group Techniques for Generating and Evaluating Design Ideas

A range of group techniques is reported here based on the works of Moore (1987) and Warfield (1990). These techniques are reviewed at a level of detail that you might find adequate enough to work with various design activities. Once you initiate a comprehensive design project, however, you will find it useful to use the sources defined above as design aids. In this section two of the most often used techniques are discussed and others are listed.

4.7.1.1. Nominal Group Technique (NGT)

NGT structures small-group conversations for developing ideas about issues of importance when faced with ill-structured design situations of uncertainty and ambiguity when we seek the emergence of ideas, encourage diversity, and eventually seek consensus among members of the design group. The activity includes (1) the formulation of triggering questions for the issue to be addressed, (2) the silent generation of ideas in writing, (3) their recording and display to the group, (4) a discussion/clarification of the ideas, and (5) the selection/designation of ideas for consideration as inputs to design. These five activities are now described.

1. The formulation of the triggering question (TQ) is a crucial task. The TQ drives the whole activity. The TQ should be simple and unambiguous. "It should elicit items at the desired level of specificity" (Moore, 1987, p. 25). For example, at the onset of design it is too early to ask: What should be the outcomes of our design? This issue is too complex to explore with the use of this technique. An appropriate question might be: What values should guide our design? Members of the design group should formulate and test the TQ in order to determine if, in fact, it will produce the desired response. Additional preparation includes the availability of a room in which the group can sit around a table, newsprint that can be posted on the wall, markers and sheets of paper for members of the group to write their ideas on (the preferred size of the group is between five and nine).

2. The silent generation of ideas proceeds in response to the triggering question. Ideas are recorded on sheets of paper distributed to group members. Instruct members to write phrases or short sentences and work independently in silence. Invite members to record as many items as they wish.

3. A round-robin reporting and recording of ideas map the thinking of the group. Members report one idea at a time without any discussion, until all the

ideas are reported and recorded and numbered on sheets of newsprint. Members may "hitchhike" on the ideas of others and/or add new ideas.

4. A discussion/clarification of listed items follows once all items are re-corded. Invite comments and questions but avoid by all means expressions of judgment. Keep in mind that a most outlandish or impossible-sounding item might become most valuable later. We can now "pull" or edit items with the consent of the authors. But resist combining items into larger categories.

5. Prioratizing/Selecting Items. This activity proceeds in silence. Group members work with the large list and rank-order items in sequence of their preference. Prepare a tally sheet on the newsprint on which next to a numbered item we record the preference number. We can now list the items according to preference and might proceed with further clarification and discussion. If we have more than one group working, the next task is to assemble all groups and proceed with the integration/consolidation and final tallying of items. The product of this activity becomes input to the design program.

Two decades of wide use of NGT suggests that the technique is easy to learn and apply and people enjoy using it. It is usually very productive and does not take much time. The quality of the items obviously vary, and the outcome is only suggestive.

The use of NGT is appropriate when collective idea generation is called for in the formulation of design issues. NGT is helpful in neutralizing dominant individuals. It is a starting place in the work of design groups and is usually followed by a technique of idea development/elaboration, described next.

4.7.1.2.   Idea Development

Idea development focuses on the further elaboration of a specific item that was an outcome of an NGT activity. For example, let us suppose that we produced a set of values that should guide our design. Now we might ask: What are the implications of a particular value to selecting the focus (or the scope) of the design inquiry? The ideas developed become a documented input to design. In general, this technique is helpful in exploring the meaning and implications of specific ideas to the content of design in the context of a specific design program. The activity involves (1) preparation, (2) individuals responding in writing to the item or the triggering question, (3) members commenting in writing to other members' written responses, (4) members reading others' comments to their responses, and (5) the group discussing the principal ideas that emerge from the written ideas and responses and the group developing and recording on newsprint the findings of the activity. These five activities are described next.

- The preparation is similar to the NGT activity. It includes the designation of a room, appropriate seating arrangement, means of recording and displaying the ideas, and a careful formulation of the triggering question. A briefing on the procedure is also part of preparation.

- The initial activity involves each member's written comment on the item, formulated by a triggering question. Members should take five to ten minutes to formulate their ideas on a sheet of paper, and should note their name next to the triggering question. Once completed, members place their sheet in the middle of the table. After all members complete their writing and place their sheets on the middle of the table the response activity commences.
- In the response activity, each person records his or her comments in response to the ideas of other members, using the idea writing sheets placed in the middle of the table. Moore (1987) presents a form that can be used for this activity.
- Members read to the group their own idea and the responses of others to their idea. Following each reading, conversation develops that aims at clarification and further development of ideas.
- The last activity is the development of an integrated/consolidated statement that is recorded on newsprint, reported to the larger group, and used as input to the design program.

The use of idea development is appropriate when the design group formulates design ideas, including: (1) core values and ideas, (2) the image markers of the future system, (3) guiding perspectives, (4) the purposes of the system, (5) systems specifications, (6) systems functions, etc. The ideas generated usually invite an exploration of relevance and impact assessment.

4.7.2. Assigning Roles and Responsibilities

To ensure a productive and satisfying experience to all members, I have found it useful to assign specific roles/responsibilities to all members of design groups when using the techniques described here. Such an arrangement is a way to share responsibility and accomplish the task of the group. Suggested roles include the following:

- The guardian of participation, ensures that all members have equal time and opportunity to make contributions and that no person shall dominate the group.
- The guardian of focus keeps the group's emphasis on the theme, on the specified issue, or on the triggering question.
- The guardian of the selected group technique ensures that orderly progress can be made toward the accomplishment of the task.
- The guardian of documentation ensures that whatever is developed by the

group is appropriately recorded and made available for successive design work.

- The guardian of accepting and honoring all contributions disallows the criticizing or "belittling" of ideas.
- The guardian of values implements principles that have been articulated individually and collectively as bases of making design choices and decisions.
- The guardian of the "burning fire" keeps the team spirit at the highest possible level. (Ideas: an inspirational quote, a logo for the team, a theme song, a team ceremony, envisioning an ideal image of the design group, an empty chair representing future generations, and literally keeping the fire burning in the fireplace.)
- The guardian of coordination ranks the work of the group, which includes an initial briefing on the task at hand, keeping time, and ensuring that the group has adequate opportunity and resources to accomplish the task.

Depending on the number of participants, a person might have more than one role to play. All the roles described above are open to all participants. Roles are assumed at the discretion of individuals and the group. Furthermore, the role designations can also change based on the experience of the group. Roles can be modified and new roles added.

4.7.3. Other Group Techniques

The first technique, heartstorming, has been developed recently by Frantz. The rest of the techniques are selected from a set of techniques used in the context of the interactive management program (Warfield and Cardenas, 1994).

4.7.3.1. Heartstorming

Heartstorming (Frantz and Miller, 1993) is a group technique that precedes the techniques described above and those that follow. In the front part of social systems design, as we commence with a vision-quest, designers discover and express the subjective, value-laden aspirations and idealistic dreams they hold deep in their hearts about the future they wish to bring about in their design. When these are expressed openly in an unconstrained way the designing community discovers shared aspirations, dreams, and values. This process creates the common vision that becomes the basis for formulating core ideas of the future design by using other group techniques. Designers may use three criteria to assess the result of heartstorming: (1) Is it attractive enough to draw the designing community into the inquiry and keep them inspired? (2) Is it powerful enough to sustain their commitment to the design effort? (3) Is it clear and focused enough to organize the design activities that follow?

It is expected that heartstorming creates an atmosphere conducive to (1) openly expressed aspirations, values, and ideas; (2) psychological safety, in which participants feel valued, accepted, and understood; (3) psychological freedom to think and to express feelings authentically and responsibly. These conditions are not established rapidly or on demand. They develop gradually, assisted by friendliness and kindness. They invite and nurture creativity, playfulness, and relaxed informality. They welcome complex, ambiguous, improbable, and contradictory ideas, and embrace willingness to appear naive and even foolish.

Heartstorming is an approach that promotes the practice of "generative dialogue," described in the "Design as Conversation" section of Chapter 5.

4.7.3.2. Delphi

Dephi is a means of generating, clarifying, and structuring ideas. It is used when groups cannot engage in face-to-face communication, thus the conversation goes on in exchanging ideas in writing. Its use can be accelerated with the employment of electronic communication, e.g., creating a "home page" for a specific Delphi application.

4.7.3.3. Interpretive Structural Modeling (ISM)

ISM (Warfield and Cardenas, 1994) "provides the means to enable groups to structure information with computer assistance, with simultaneously clarifying the component ideas. It allows for amendment of preliminary structures, again with computer assistance. It is self-documenting" (p. 91).

4.7.3.4. Option Profile (OP) and Option Field (OF)

An OP is a visual representation of a design alternative, consisting of a set of chosen options. Option profiles are organized into an option field, which is a means for the comprehensive development of potential design solutions.

4.7.3.5. Attributes Profile (AP) and Attributes Field (AF)

An attribute is an aspect that is deemed to be relevant to the design situation, and is assigned to a category, such as social, political, or technological. The AP is developed from and AF the same way that an OP is developed from an OF.

4.7.3.6. Trade-off Analysis (TA)

TA offers a means to select from sets of design solution alternatives based on developed evaluation criteria. Its use is described in Section 7.3 of Chapter 7.

4.7.4. The "Cogniscope" System of Social Systems Design

A salient "future" approach to social systems design has already arrived. The "CogniScope" system of design inquiry is the first, at least to my knowledge, that applies a thoroughly tested cognitive technology that integrates software as "groupware" in the work of a designing community. Based on the conceptual foundations created by John Warfield and his interactive management (IM) approach, the CogniScope system represents the third generation of IM. It was developed and applied in a large number of settings under the leadership of Alexander N. Christakis.

The CogniScope system is based on over twenty years of experience, involving over 200 applications, and it is documented by several reports and books and more than 400 scholarly papers. A summary description of CogniScope follows, based on Christakis's (Christakis and Conaway, 1995; Christakis *et al.*, 1995) work.

4.7.4.1. Definition

The CogniScope system is a disciplined inquiry process that enables a community of stakeholders to create a design and an action plan based on the design. "The 'action plan' is the outcome of the integration of the pluralities of realities of the stakeholders" (Christakis and Conaway, 1995, p. 10). Design conversations among members of the stakeholder community generate and clarify a large number of ideas, on which they base an interactively created design solution and action plan, which is co-owned by them because it has been co-created by them.

4.7.4.2 Principal Activities

A designing community employs the CogniScope system to perform three principal activities: (1) to generate and clarify ideas in response to carefully and properly framed triggering questions; (2) to produce "idea patterns" that result from an exploration of relationships among ideas in the context of carefully framed generic questions, and (3) to valuate idea patterns and action packages to agreed on criteria. In the course of these activities, the various group techniques elaborated in the first part of this section are used to generate, clarify, and structure ideas; explore and interpret influence patterns; and work out solution alternatives.

In the course of design the three principal activities are carried out in four interrelated stages, described as follows.

4.7.4.2a. Stage One: Define the Design Situation. At this stage, the designing community defines the design situation by responding to the question:

What should we do? Usually well over one hundred ideas are generated. Such a large number of ideas cannot be handled without the support of software technology offered by CogniScope. The technology enables the community to organize ideas into meaningful relational patterns very efficiently and enhance their insights and understanding of the desired situation. For example, one step at this stage is the structuring of the ideas for exploring "influence relationships" that reveal that the accomplishment of one idea will help in the accomplishment of others. By using an "inference logic system," a key element of the CogniScope system, designers discover these influence relationships in one-tenth of the time it would have taken without the support of the software.

4.7.4.2b. Stage Two: Designing Alternatives. Once the design situation is defined as the outcome of stage one, the designing community begins to explore design alternatives by responding to the question: How do we do what we want to do? Through group work and focused dialogue, the "hows" are superimposed on the "whats" and displayed on newsprint on the walls of the designing facility. The designing community will now see whether the "hows" are indeed addressing the fundamental "whats."

4.7.4.2c. Stage Three: Choosing the Preferred Alternative. Within the host of "how" ideas there is an almost limitless number of possible solution alternatives from which to choose the preferred alternative. With the use of the CogniScope instrument, the designing community can select a limited number of alternatives to be fully deliberated. The deliberation is followed by the formulation and consideration of criteria to be used to select the preferred alternative.

4.7.4.2d. Stage Four: Developing an Action Plan. Once the preferred design solution alternative is selected, the solution idea components are sequenced. This process responds to the question of "when" and "how" will we do the "whats."

4.7.4.3. Human, Technical, and Facility Support

The support components of the CogniScope system include a facilitation team, computer support, and a design facility.

4.7.4.3a. The Facilitation Team. The human support to the design inquiry is provided by several people who make up the facilitation team. One manages the conversation in a neutral way. Another operates the computer with the use of the CogniScope software, which generates the displayed patterns of ideas based on the deliberations of the designing community. A third member records and reports the proceedings of the design activities. If useful, a content specialist listens to the conversation and might provide advice regarding the thread of emerging meaning. One of the key purposes of the CogniScope process is to train and empower selected members of the stakeholder community to serve as their own facilitators.

4.7.4.3b. Computer and Display Support. Computer support (group decision and design support system) is used to record and organize the ideas, display influence patterns and solution alternatives. These are presented on a large screen to the designers for their consideration, exploration, and decision. In addition, as ideas are generated by the designers, these are recorded on large cards and displayed in the design room, as visual aids to the designers.

4.7.4.3c. Design Facility. An essential component of the CogniScope system is an appropriate environment for the design inquiry. The facility should ensure the availability of (1) a large design room with lots of wall space for the display of idea cards, (2) furniture that facilitates group work, and (3) space for the computer support system.

4.7.4.4.   A Relational Image of the CogniScope System

The system has five components: the stakeholder community of informed participants, a facilitation team, the computer support of the CogniScope system, consensus methods, and a designing facility. Fig. 4.6 displays my interpretation of a possible arrangement of the five components.

4.7.4.5   Added Benefits of the CogniScope Process

The CogniScope system literature notes that in addition to the inquiry power and proven results of the program, the system produces intangible gains that may be of even greater value than the product of the design effort. This added benefit is manifested in three forms: (1) The system nurtures organizational learning; (2) it fosters cooperation among stakeholders; and (3) it ensures the likelihood of effective implementation in that the design will be installed and operated by those who created it; that is, they own it.

Reflections

The first three items of this section displayed a variety of techniques as tools of social systems design. These are viewed as complementary, to be selected in the course of the design inquiry to best respond to the context of the inquiry. The best way of selecting particular techniques is to explore their use and see how they attain the desired outcome. The CogniScope system, described above, is

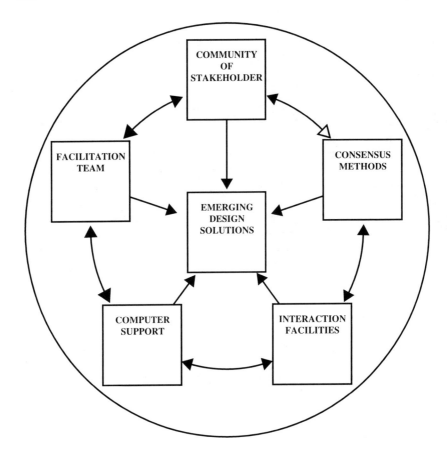

FIGURE 4.6. An image of the CogniScope system.

particularly appropriate to use in product development, system or process design or redesign, strategic management, or resource allocation.

Activity #34

In the context of what has been described in this chapter, it might be useful to think about the adaptation of the CogniScope system or the creation of additional software to be applied in two task contexts. Speculate about the use of Cogniscope or other software (1) to aid the design of a new system and assist in the comprehensive modeling of the new system and (2) to facilitate systems design, undertaken by a community whose members are dispersed geographically. Enter your thoughts in your workbook.

Adding Value to Systems Design and How Systems Design Adds Value to Society

In Part I we explored such issues as what design is, how it works, and why it is important for us today to develop design knowledge and design competence. You have worked with activities that guided you to construct your own meaning and understanding of social systems design and apply this understanding to systems of your interest.

In Part II we build on this understanding as we extend the horizon of our exploration into issues and knowledge that have not yet received comprehensive treatment in the design literature. In the course of this exploration, we shall visit various fields of disciplined inquiry and scholarship that seem to be relevant to social systems design and anticipate gaining new insights, knowledge, and understanding. We weave these into a tapestry and by so doing, we illuminate new color schemes and create a richer and more comprehensive image of social systems design. Overall, the thrust of Part II is to add value to social systems design and, in turn, understand ways that systems design can add value to society.

5

Design as a Multidimensional Inquiry

The topic of this chapter is social systems design as a multidimensional human activity of disciplined inquiry. First we inquire into the cognitive aspects of systems and design thinking, systems and design knowing and understanding. Then, we unfold multiple perspectives that extend the scope of design inquiry. We explore how values and ethics underlie our actions and our aspirations as we seek to envision and pursue the ideal. Then, we find that creativity and communication are two distinctive and unique properties of the design experience. These dimensions are explored to examine their special role in adding value to social systems design.

The sections of this chapter are organized as follows: (1) systems thinking as the enfolding context of design thinking; (2) design's own special way of thinking and knowing; (3) the use of multiple perspectives in social systems design; (4) the ethics of social systems design; (5) design in search of the ideal; (6) creativity and its central role in design; and (7) design communication and design as conversation.

From an integration of these domains we see design emerging as a system of activities that touches all domains of the human experience.

5.1. Systems Thinking

Our interest in this work is systems design and, in this chapter, design thinking. As we explore the conceptual realm of design thinking, our first task is to draw the boundaries of our exploration. Design is one of several disciplined inquiry modes we use in the domain of social systems. Thus, the boundaries of our inquiry into design thinking must be extended to include the ways of thinking we employ in viewing and working with social systems. We call this way of thinking systems thinking.

Laszlo (1972) views the history of science as an alternation between atomis-

155

tic and holistic thinking. He suggests that while early scientific thinking was holistic but speculative, modern scientific thinking is empirical but atomistic.

> Neither is free from error, the former because it replaces factual inquiry with faith and insight, and the latter because it sacrifices coherence at the altar of facticity. We witness today another shift in ways of thinking: the shift toward rigorous but holistic theories. This means thinking in terms of facts and events in the context of wholes, forming integrated sets with their own properties and relationships. (p. 19)

Flood and Jackson (1991) suggest that the concept of "system" does not refer to things in the real world but to a particular way of organizing our thoughts about the world. I always shock incoming students at their orientation when I say that there is no such thing as a system out there. Systems exist as mental pictures in our minds. Saying this another way, systems thinking structures thinking about whatever entity or phenomenon we become aware of and assign meaning to. Systems thinking is the conceptual environment of design thinking and design thinking is embedded in systems thinking. In this section we paint a broad-stroked picture of systems thinking as the conceptual parent of design thinking.

By studying and working with social systems in the course of the last several decades, we developed an increasing realization of the inquiry power we can gain from systems theory and systems philosophy and their application through systems methodologies. We have liberated ourselves from the constraints and limitations of the analytically oriented and reductionist inquiry mode of traditional science. Systems inquiry enables us to orchestrate the findings of various scientific disciplines within the framework of systems thinking and to develop and apply systems approaches, models, and methods in working with social systems. Systems thinking, and its relevance to design thinking, is explored here by (1) exploring systems thinking in social systems inquiry, (2) reviewing the evolution of systems thinking, and (3) considering how the ideas of systems philosophers have shaped systems thinking.

5.1.1. Systems Thinking in Social Systems Inquiry

Systems thinking is a property of the thinker, who organizes internalized systems ideas, systems concepts, and principles into an internally consistent arrangement, using a systems way of viewing and understanding, in order to establish a frame of thinking. As we observe what is "out there," this frame of thinking enables us to reflect upon what we experience; thus we construct our own meaning. We create our own cognitive map, which is our own interpretation of the out there. As we view and work with social systems, systems thinking enables us to create our own cognitive map of the systems of our interest. It enables us to explore and understand those systems, and describe:

- The characteristics of embeddedness of social systems as they enfold their component systems, and as they are nested in the community and in the larger society.

- The complex nature of social systems, operating at various interconnected levels, integrating their interdependent and interacting components.
- The purposes and boundaries of social systems as these emerge and are defined from a coevolution with their environment.
- The ongoing relationships, interactions, information, and energy exchanges between a social system and its environment.
- The dynamics of interaction, interdependence, and patterns that connect and integrate the functions and components of a social system.
- The characteristics of wholeness, meaning that the whole organizes and integrates the parts and, as a result of this organization and integration, properties emerge at the systems level that are not in evidence in, and are not the properties of, the parts.
- The dynamics of change that operates through time in social systems, and how these might affect (and change) the purposes, functions, and components of the system.

Even this brief scan of the characteristics of systems ideas shows their direct relevance to systems design. We design systems that should have those systemic characteristics. Thus, we readily recognize why we should consider systems thinking as the conceptual context in which design thinking is embedded.

Systems thinking generates insights in ways of knowing and reasoning that enable us to set forth and pursue a comprehensive system of disciplined inquiry in four complementary and interacting domains of social systems inquiry. These domains are (1) the description, modeling, and analysis of social systems; (2) their design; (3) the development and institutionalization of changes in social systems; and (4) systems management and the management of change. In addition to considering systems thinking as the conceptual environment of design thinking, we can designate domains (1), (3), and (4) as the peer inquiry domains of systems design.

5.1.2. The Evolution of Systems Thinking

The evolution of systems thinking during the last four decades or so is mirrored in the evolution of design thinking. The evolution of systems thinking is elaborated in the work of Jackson (1992). The various stages of this evolution are briefly outlined here, showing how they have been manifested in the evolution of design thinking.

5.1.2.1. Hard Systems Thinking

The first stage of this evolution is marked with the label of "hard systems thinking," which is a mode of thinking associated with operations research, systems analysis, and systems engineering. Hard systems thinking was mirrored in design thinking. We defined design as an orderly sequence of systematic activities, practiced by expert designers. Jones (1980) noted that systematic design keeps logic and imagination, as well as problem and solution, apart by an effort of will and by external rather than internal means. Systematic thinking, a key characteristic of engineering design, dominated the thinking of the design community during the sixties. Working in the arena where the objective of a system can be clearly stated up front, the systems engineer devises ways to improve output in the most cost-effective way. But this approach does not work in the context of ill-structured social systems.

The successes of hard systems thinking in its realm of applications led to attempts to transfer hard systems thinking into the social systems environment. This transfer was labeled social engineering. By the middle of the sixties we recognized that hard systems thinking and engineering applications were not only useless in the social systems arena, but they were dangerously counterproductive, resulting in some disastrous applications. Recognizing this state of affairs, we looked for a new orientation in systems thinking.

5.1.2.2. Organismic Systems Thinking

This orientation emerged from an open (organismic) systems orientation. It was grounded in a general theory of systems, represented in the works of the founders of the systems movement, including Bertalanffy, Boulding, and Rappoport. They recognized that organisms should be treated as wholes, that they have emergent properties that are unique to each, properties that are not manifested in the parts. This new systemic orientation was transferred into the social systems arena, and it led to a search for a systemic and holistic orientation in working with social systems. This orientation gave rise to a structuralist perspective that sought to define features of viability in dealing with sociotechnical systems. This trend was manifested in Forrester's (1969) "systems dynamics," Beer's (1979) "organizational cybernetics" and viable systems approach, and the "living systems process analysis" approach based on Miller's (1978) living systems theory.

5.1.2.3. Soft Systems Thinking

The search for systems thinking that would be more appropriate to social systems was rewarded by the emergence of soft systems thinking. Soft systems thinking brought about a sea change in design thinking in the social systems arena during the late seventies and eighties. Soft systems thinking established itself as clearly distinct and distinguishable from the two systems thinking types mentioned above. It was quickly embraced by scholars and practitioners of the design inquiry community through the works of such design scholars as Churchman, Ackoff, Checkland, Nadler, Cross, and Warfield.

Checkland (1981) suggests that "systems thinking implies thinking about the world outside ourselves, and doing so by means of the concept of system" (p. 3). Systems thinking orders our thoughts by making conscious the concept of wholeness inherent in the word "system." In Checkland's formulation the two pairs of core concepts of systems thinking are (1) hierarchy in systems and the emergent properties at the various systems levels and (2) communication and control in human systems. Based on soft systems thinking, Checkland developed the multistaged soft systems methodology, that was presented in Part I of this work.

Ackoff (1981) says that systems thinking reverses the analysis-focused machine-age thinking that aimed at understanding an entity by decomposing it, explaining its behavior by its parts, and aggregating these explanations as the explanation of the whole. In contrast, systems thinking identifies the whole that contains its parts, explains the behavior of the whole, and then explains the parts in terms of their role(s) and functions within their containing whole. Ackoff's ideal systems approach to systems design is based on the thinking described above, which he calls systems-age thinking. His design model was also introduced in Part I.

5.1.2.4. Critical Systems Thinking

A new trend in systems thinking has emerged in recent years. Pioneered by Ulrich, Jackson, Flood, and others, it is called critical systems thinking (CST). This orientation challenges some of the earlier aspects of systems thinking. Critical systems thinking is reflected in Ulrich's critical systems heuristic and Flood and Jackson's total systems intervention. (These also mark a new trend in design thinking.) CST embraces a set of core commitments, such as critical awareness, social awareness, human emancipation, and complementarity (Jackson, 1992). *Critical awareness* closely examines the values and assumptions that enter into systems inquiry, such as systems design. It provides tools that are useful for applying critical awareness, such as Ulrich's (1983) critical systems heuristics. *Social awareness* recognizes social and organizational pressures that guide systems interventions. It aims to guide users of various intervention approaches to contemplate the social consequences of their planned approach. This commitment also calls for an open and free debate on the justification of the use of a proposed approach. *Human emancipation* aims to ensure the well-being of all individuals and the full development of their human potential. It aims to prevent coercion and exercise of power that would prevent open and free discussion (for example, in design inquiry). *Complementarity and informed development* of all varieties of systems approaches is another commitment of CST. Various systems trends express various rationalities and theoretical positions. CST suggests that the positions and the methodologies that arise from these theories should be respected and their development should be encouraged. CST has a commitment to the complementary and informed use of all the various systems approaches whenever their use is appropriate to the context of various social conditions and situations. The relevance of these commitments of CST to systems design is more than obvious. Jackson (1995) sums up the aims of CST, saying that "it does not seek to recreate a unified systems theory—to overcome fragmentation through some totalizing vision. But it does want to take us beyond fragmentation by supplying means through which we can be critical in the use of various systems ideas and methods at our disposal" (p. 40).

5.1.3. Systems Philosophers and Their Systems Thinking

Contemplation of systems thinking is the primary domain of systems philosophers. A review of the contribution of several systems philosophers will enlighten our understanding and appreciation of systems thinking. Jackson (1995) notes that many of the ideas we associate today with systems thinking, such as rationality, comprehensiveness, human well-being, emancipation, and progress, are closely related with Kant's notion of "enlightenment." Kant's concern was with man's release from "self-incurred tutelage," with people freely thinking and deciding for themselves.

5.1.3.1. Churchman and Company

A most salient contribution to systems thinking in the social systems arena has been made by West Churchman. His contributions present an internally consistent set of core ideas of systems thinking and social systems inquiry. His major works (Churchman, 1968b, 1971, 1979, 1982) are grounded in the ideas of the philosophers of the Enlightenment (e.g., Kant). Churchman sets forth several core themes of systems thinking, such as whole systems judgment, the ethics of whole systems, unbounded systems approach, unfolding, the sweep-in process, and consideration of future generations. (These core ideas will be discussed later.) Churchman suggests that in social systems inquiry problems are unbounded and tightly connected. Every problem is an aspect of all others. This notion, he says, was already advanced in the fifth and sixth centuries by Greek scholars. One of them, Anaxagoras, said that no matter how far one goes in breaking an object down to parts and subparts, a resulting piece still contains everything—"in everything is everything." But the reductionist Western scientific community rejected this notion and promoted bounded inquiry. Churchman (1982), a student of the Greek scholars, believes that "we need an 'unbounded' systems approach which sweeps-in all that is relevant to our inquiry. The ethics of the whole system includes a study of the ethics of humanity, not within a problem area, but universally" (p. 8). All issues in the social domain are inherently ethical and not factual. They are first of all prescriptive and not descriptive. He calls for determining the ethics of the whole system. For example, someone concerned about education should first consider the nature of an ideal society and then ask how education can serve it. In our arrogance we often assume in our inquiries that we, the stakeholders of the present, are what counts. Churchman passionately asks: What are the implications of our inquiry for future generations? This forward-looking sweep-in is one of his primary imperatives. He also calls for a focus on the invariants of humanity, the aspirations, the values, and the hopes of our collective humanity. All the core ideas mentioned here are germane to social systems design. You will be asked to explore these later.

Many of us consider ourselves to be in Churchman's company. Here I refer to the recent work of Mitroff and Linstone (1993). They follow Churchman's thinking as they put forth the notion of unbounded systems thinking (UST) as the new thinking called for in the information/knowledge age. They found that in UST "all branches of inquiry depend fundamentally on one another, and the widest possible array of disciplines, professions, and branches of knowledge—capturing distinctively different paradigms of thought—must be consciously brought to bear on our problems" (p. 91). UST is not governed merely by conventional logic or rationality. It involves "considerations of justice and fairness as perceived by various social groups and consideration of personal ethics or morality" (p. 91). The idea of UST becomes one of the imperatives of design thinking.

5.1.3.2. Vickers

The systems philosopher Geoffrey Vickers (1981) presents a broad-based view of systems thinking. He suggests that although all systems have common characteristics, there are significant differences between them as there are between biological and social evolution. His main interest lies with the ecology of systems ideas as they relate to social systems—"the effect of systems thinking on our outlook on life and our philosophy of life" (p. 19). This focus tends to correct some of the "illicit" extensions of ideas (to the social realm) that have been derived from the natural sciences. In Newton's world, inert objects stayed put unless moved by some force. By contrast, our world is one of active and dynamic reactions "in which stability, not change, requires explanation" (p. 19). The interface between a social system and its environment distinguishes inner relations from outer. Inner relations hold social systems together and enable the system to act as a whole in the context of its environment. The scope of the system's "external relations depends on the coherence which its internal reactions secure. We are accustomed to regard 'relating' as something which entities do, rather than something which they are. Should we rather view all entities as systems, created by their relations which sustain them?" (p. 20.) Exploring identity, continuity, and change, Vickers asks: When does a system retain its identity and continuity through change and when does it itself vanish or become something new? These questions are of great practical concern in the context of systems design.

Human systems are the only kind that can succeed or fail. "A human system fails if it does not succeed in doing what it was designed to do; or if it succeeds but leaves everyone wishing it never tried" (p. 21). This observation is a lesson for systems design. So is the understanding that social systems are governed by the demands of stability on the one hand and by criteria of success on the other. Vickers says, "I find it ridiculous to try to reduce the second to the first as so many people try to do" (p. 21). Now, contemplate the implication of Vickers's ideas for the design of social systems.

5.1.3.3. Jantsch

Erich Jantsch (1980) made a unique and powerful contribution to the evolution of systems thinking. He is among those who shaped the intellectual process that has had a profound concern for self-determination, self-organization, and the openness and plasticity of systems and their freedom to evolve and coevolve. He says that the primacy is on process orientation, in contrast to an emphasis on solid components and structures. While "a solid structure determines the processes which it can accommodate, the interplay of processes may lead to the open evolution of structures. Emphasis is then on becoming—and even the being appears in dynamic systems as an aspect of becoming" (p. 6). Social systems are coherent, evolving, interactive processes that have nothing to do with the equilibrium and solidity of technological structures, which are geared to the output of specific products. Social systems are concerned primarily with renewing and evolving themselves, which are essentially learning and creating processes. "When a system, in its self-organization, reaches beyond the boundaries of its identity, it becomes creative. In self-organization, evolution is the result of self-transcendence. At each threshold of self-transcendence a new dimension of freedom is called into play for the shaping of the future" (pp. 183–184). Evolution is open through the self-organization of social systems, by the dynamic interconnectedness and the coevolution of the system and its environment. Evolution determines its own meaning. This meaning is the "meaning of life. We are not the helpless subjects of evolution—we are evolution" (p. 8). In the course of our unfolding discussion we often brought in the notions that Jantsch articulates so powerfully, such as self-determination, self-organization, openness, self-transcendence, the freedom to evolve and coevolve by design, and the notion that design is process and that the product of design is also a process. Design is becoming, it is learning, it is creating.

5.1.3.4. Laszlo

Laszlo (1995) suggests that rapid developments have been experienced in the systems field due to the combined effect of conceptual innovations in the systems sciences and advances in computational hardware and software, in sciences such as cybernetics and systems theory, in nonequilibrium thermodynamics, in general evolutionary theory and theories of self-organization and chaos. Systems approaches have become mainline methodology in the advanced branches of natural sciences.

> We may expect that they will gain increased application and importance also in the social sciences. This expectation is justified by the fact that not only nature, but also human society, exhibits growing complexity. . . . And for decoding complexity, no viable alternative has presented itself in the social sciences to the methods and models of contemporary systems sciences (p. 6).

Reflections

Even a quick review of our discussion on systems thinking will show us the unquestionable relevance of systems thinking to systems design, its power of shaping design thinking. It was appropriate to say up front that systems thinking is the parent of design thinking and systems inquiry embeds design inquiry. It appears that we could rewrite this section by replacing the term "systems thinking" with "design thinking." But there is more to design thinking than systems thinking. Design thinking carries the genes of its parent, but it has a DNA that is unique to it. We discuss this uniqueness in the next section.

Activity #35

Review the text and identify core ideas of systems thinking. Then stipulate how those core ideas apply to the design of social systems. For each core idea write a brief notation that explains its relevance to social systems design. Enter your findings in your workbook.

5.2. Design Thinking

In Section 5.1, I highlighted systems thinking as the conceptual environment or parent of design thinking. Design is one of several disciplined inquiry domains of social systems in which systems thinking is manifested. Others include the descriptive representations or modeling of social systems and their evaluation, systems development and institutionalization, and systems management. Each of these domains of social systems inquiry, as well as systems design, has its own specific ways of thinking and knowing.

In this section, design thinking is addressed by (1) introducing the design-relevant ideas of systems philosophers and scientists, (2) presenting statements of design scholars about design thinking and knowing, and (3) exploring the design implications of scientific insights we have gained from understanding the findings of what is called the "new physics."

5.2.1. Systems Scholars' Ideas on Design Thinking

Here we turn to a group of systems philosophers and scholars who have accomplished extensive work in systems design and revealed ideas that further enlighten our understanding of design thinking.

5.2.1.2. Churchman's Design Thinking

Churchman's ideas on design (1971, 1979) were summarized by Ulrich (1983). Churchman's thinking is reflected in his complex strategy for dealing with ill-structured design issues. The markers of his thinking include reflecting the viewpoints of the individuals and the collective of the designing community; the nonhierarchical nature of the system's complexity; the wholeness of the system and its irreducibility; the critical consideration of the purposefulness of the system; the nonseparability of a system's components; and the uniqueness of the description of the system. In Churchman's view, the designer's main tool is subjectivity, which includes social practice, community, interest and commitment, ideas and ideals, the ethics of the system and the moral idea, affectivity, faith, and self-reflection. For Churchman the issue is not whether we can design systems that are wholes and unique, but whether we can design systems that make us more whole, unique, and self-motivated. Churchman's strategy provides a powerful guide for a designing community.

5.2.1.3. Ackoff's (1981) Design Thinking

In Chapter 2 we explored Ackoff's (1981) four styles of how people think about and work with change. His choice is the interactivist or proactivist, who believes that the future is largely subject to our creation. Thus, design is the creation "of a desirable future and the invention of ways to bring it about" (p. 62). Ackoff's design thinking is reflected in three operating principles.

1. The *participative principle* holds that learning and using the design process (individually and collectively) is a more important outcome than the actual product of design. The other implication of the participative principle is that one cannot design effectively for someone else. In fact, no one should, because as we said earlier, it is unethical to design for someone else.
2. The *principle of continuity* challenges the well-known routine of making a plan or design and updating it periodically in an on-again, off-again cycle. There are two reasons for design continuity. First, continuous reflection on—and valuation of—the operating design and its systemic

environment enables us to be continuously in the design mode. Second, as we design something of value and pursue it, the value we place on it might change as time goes by, because our values might change. This, then, indicates changes in the design or implies redesign.

3. The *holistic principle* has two aspects. First is the principle of coordination. It tells us that no part of the system can be designed independently from others that operate at the same system level. The second aspect means that all system levels should be designed integratively.

Ackoff (1994, 1995) points to the antisystemic nature of such current popular panaceas as total quality management, continuous improvement, process reengineering, and right-sizing. First, most of these fail "to whole the parts." They manipulate the parts without focusing on how they effect the performance of the whole. Second, they fail "to right the wrongs." Focusing on efficiency, they do the wrong things better rather than doing the right things. Designers often say, as does Ackoff, that getting rid of what is not wanted does not give you what is desired.

5.2.1.4. Checkland's Soft Systems Methodology (SSM)

Soft systems methodology is a well-developed approach that applies soft systems thinking. Checkland (1995) defined SSM by contrasting it with hard systems engineering, which assumes systemicity in the "things world" of the system. SSM, on the other hand, assumes systemicity of the process used to inquire into the system. SSM allows users to articulate their different viewpoints, their worldviews, and explore the implications of those views in working with the system of their interest. This approach is basically a systemic process of learning about the target system, learning about each other's views, and jointly contemplating some possible changes and alternative design solutions. "In soft systems methodology models are not part of the world; they are only relevant to debate about the real world and are used in a cyclic learning process" (p. 47). (See Part I for a detailed review of SSM.)

5.2.1.5. Nadler's (1981) Design Thinking

Nadler's thinking is reflected in seven key ideas:

1. Specific human and organizational attitudes, desires, and behaviors are the starting points of systems design.
2. Individuals and systems are unique. There are no identical entities; therefore, the design approach adapts to the specific situation rather than the situation to an approach.
3. Design is a purposeful human activity to be carried out in the context of human values and objectives.

4. The design approach is descriptive and prescriptive, grounded in theory. It is based on some truths or axioms, supportable assertions, and empirical evidence about effective design approaches. For example, research shows that people fail to use solutions that were invented somewhere else (however, this negative reaction is usually interpreted as resistance to change).
5. "Design should operationalize the 'whats' and 'whys' of design into specific, day-by-day, week-by-week methods of interrelating with the human and organizational real world" (Nadler, 1981, p. vi).
6. In defining a design situation, the focus should be on what is desired rather than what is wrong.
7. Systems design is holistic and interdisciplinary; it is greater than the sum of its parts.

5.2.1.6. Warfield's (1990) Principles of Design

Warfield's set of principles further inform design thinking. These principles interpret a set of articulated laws of generic design, which in turn are linked with the foundations of the science of generic design. Here I present some of the principles that can guide our thinking about design inquiry. I am using language that is familiar to the users of this work.

1. Design inquiry should be carried out in groups whose members are selected in order to ensure that they articulate the variety in the design situation.
2. The design situation should be represented in dimensions that enable the consideration of a range of design solution options.
3. The dimensions of the design situation and the design solution are interdependent and should be explored integratively in order to avoid design errors.
4. The design process should be iterative in producing design solutions and in assessing solution alternatives.
5. In developing solution alternatives, the group should experience the process of considering the sequences in which choices should be made as they grade the choices. This enables them to make the most salient choices first and the least salient last.
6. The status of the unfolding design should be continuously made visible by displaying it. This enhances productivity and minimizes the need of reliance on memory.
7. The design environment should be carefully designed and equipped in order to provide maximum support to the inquiry.
8. The design program should include formal processes and specially

assigned roles by which we play continuous attention to the content, the context, and the process of design.

9. Participants in the design inquiry should develop their own self-reflective, self-assessing, and self-governing criteria that will guide and steer their participation.

10. To support the design program inherent in these principles, Warfield places special emphasis on the need to provide computer-aided support to the design program (the provision for computer support is inherent in his laws of design). His principles are manifested in a comprehensive program of systems design, called interactive management (Warfield & Cardenas 1994).

5.2.2.   Special Markers of Design Thinking

In Section 2.5, Chapter 2, design was differentiated from the cultures of science and the humanities. We can build on those differentiating markers of design as we explore ways of thinking and knowing that are unique characteristics of systems design. The phrase "designerly ways of thinking" pops up in the design literature around the late seventies and early eighties.

5.2.2.1.   Designerly Ways of Thinking and Knowing

Cross (1984) noted that there are ways of thinking and knowing that are very specific to design. Designerly ways of thinking are distinct from the scientific and other scholarly ways of thinking. The "world of design has been badly served by its intellectual leaders, who have failed to develop their subject on its own terms" (p. 223). These leaders have been seduced by the lure of *Wissenschaft*. They defected to the cultures of scientific and scholarly inquiry. Earlier, Archer (1984) noted that there exists "a designerly way of thinking and communicating that is both different from scientific and scholarly ways of thinking and communicating, and is as powerful as scientific and scholarly methods of inquiry, when applied to its own kinds of problem" (p. 348).

5.2.2.2.   Scientists and Designers Think Differently

Lawson (1984) compared the problem-solving strategies of scientists with those of designers. He found that scientists focus on studying the problem while designers concentrate on finding solutions. Designers learn about the nature of a problem as a result of exploring solutions. Designers are to produce practical results within a time limit, while scientists often suspend judgment until more is known. "Further study is required" is a justifiable conclusion for the scientist (Cross, 1990). A crucial distinction between scientific and design thinking is related to differences in their aims. Scientists aim to find out the nature of what exists. Designers create things that do not yet exist. Scientific thinking is analytic; design thinking is synthetic and creative. The scientist's interest is abstract forms, created by logic. The designer creates novel forms by exploring conjectured solutions. The line of reasoning pursued here emphasizes the creative, normative, and synthetic nature of design thinking.

Another critical distinction between the two ways of thinking is that the scientist is looking for patterns lying in the data, as in a perceptual puzzle. The designer actively constructs patterns from his own effort. Design is about pattern making, whereas science deals with pattern recognition (Levin, 1966). The pattern that the designer creates is perceived to be appropriate but cannot be proved to be right (Jones, 1980). This pattern making, says Jones, is the true creative act of the designer, which turns a complex problem into a solution by changing its form and by deciding what to emphasize and what to overlook.

Thinking, very particular to design, is characteristically constructive. Constructive thinking is distinct from inductive and deductive thinking, which are the modes of reasoning in science and the humanities. (C. S. Pierce called these modes of thinking and reasoning "abductive.") Another marker of the designer's thinking style is that it is dynamic and multilateral rather than linear or serialistic. Within the larger holistic context of solution seeking, the path of thinking may change directions; it moves back and forth. Often it backtracks or jumps forward.

5.2.2.3.   Complementary and Reflective Thinking

"When we are dealing not merely with formal-logical expressions, but real-world events, we can rarely be certain that we can see the totality of possibilities" (Ulrich, 1983, p. 274). Thus the dualistic thinking of "either a or b" has to be replaced by "a as well as b" thinking. Ulrich calls this way of thinking complementary thinking. Ulrich's characterization fits well with design thinking. Complementary thinking is in the very nature of design thinking. (Complementarity is one of the core ideas of total systems intervention [Flood and Jackson 1991].) In the design of social systems (Ulrich, 1983) as we make judgments about alternatives, we must think about their social consequences and must critically reflect on both our assumptions and the approaches we use in systems design.

5.2.3.   *Design Implications of the "New Physics"*

Going beyond what we usually consider the domain of sources of design thinking, we should be informed by new insights that have emerged in the thinking of scientists of the new physics, such as Bohm, Prigogine, and Davis. It would take a separate book to explore the relevance of emerging scientific ideas to design. And such a book should be written to further design thinking. Here we can have only a fleeting impression of how some of the emerging ideas of the new physics might shape design thinking.

5.2.3.1. Bohm's Wholeness and Order

Developed in the twenties by Bohr and his followers, quantum theory demolished then-prevailing concepts about reality. It blurred the distinctions between cause and effect and object and subject and brought forth a strong holistic element in our worldview (Davis, 1989). Quantum theory forms a pillar in new physics and it provides the most convincing scientific evidence about the essential role that consciousness plays in the nature of reality.

David Bohm, a quantum physicist and philosopher of science, has presented a sweeping theory of wholeness and order. In his *Wholeness and the Implicate order,* Bohm (1983) suggests that quantum theory dropped the notion of analyzing the world into autonomous parts and placed the emphasis on undivided wholeness. The world is not a collection of autonomous but coupled things but rather is a network of relations. A division into subjects and objects, inner and outer world, and body and soul is no longer acceptable. Bohm and Peat (1987) define "implicit order" as the enfolding of reality from which the explicit order of specific phenomena unfolds. His scheme of orders embraces "generative order." It is a deep and inward order out of which the manifest form of things can emerge creatively. Bohm suggests that generative order is relevant not only to science (e.g., the science of chaos) but to all areas of experience. I find the notion of generative order to be directly relevant to design thinking. Bohm sees the manifestation of generative order in the work of the artist. The artist, as the designer, begins with an overall vision, a general idea, a feeling that already contains the essence of the final work in an enfolded way. He captures the overall form in a sketch (the image of the future system in design) from which, as the painting progresses, details are created gradually, each time building on the whole. They unfold the whole. "The artist is always working from the generative source of the idea and allowing the work to unfold into ever more detailed form" (p. 158).

In the same way, the designer gradually creates the design solution from the original vision and the "generative" image of the future. The painting expresses a visual, outward perception, but it cannot be separated from the inner perception of the artist's values, beliefs, aspirations, his whole life, and knowledge; it is inseparable from his emotional and intellectual relationship to his theme. So it is with the designer whose inner perceptions, inner self, and social self are integrated with his outward perception of the future of the system of his interest. It seems that Bohm's example of the generative creation of art is a good metaphor for design. But there is much more to it. A thorough interpretation of his book *Science, Order, and Creativity* (1987) for design thinking would enable us to generate some new and powerful insight into what design is and how it works. We explore this later.

5.2.3.2. Prigogine's Theory of Change

The Nobel prize–winning ideas of Prigogine (Prigogine and Stengers, 1984) present us with a comprehensive theory of change that is directly relevant to the way we can think about design. He suggests that instead of being orderly and stable, reality is bubbling with constant change, process, and disorder. The subsystems of a system are continually "fluctuating" and may move far from equilibrium. Driven by positive feedback, at times these fluctuations may become so powerful that they "shatter" the whole system. At that moment the system is at a "bifurcation point," when it can either disintegrate into "chaos" or leap onto a higher-level "order" by self-organization, which Prigogine calls "dissipative structure." It is called dissipative because this higher-order structure requires more energy to sustain the emerged system. In open systems, such as social systems, this energy comes from the environment. In nonhuman systems the direction taken at the bifurcation point is up to chance; it is inherently impossible to determine. However, it is different in social systems. The various cultures of our societies are immensely complex systems. They are highly sensitive to fluctuation and potentially involve an enormous number of bifurcations. These, then, could lead our systems on an evolutionary path or reorganization at ever higher levels of complexity. "From a human point of view, all this is quite optimistic" (p. xx). For us this is truly hopeful, in that, even small fluctuations may grow and change the system. "As a result individual activity is not doomed to insignificance" (p. 313). Yes, we can give direction to the evolution of our systems by purposeful design. The concept of such "guided evolution" is one of the overall theme of this work. We carry the burden, the responsibility, and it is our privilege to guide our evolution and be responsible for it.

5.2.3.3. Paul Davis's Cosmic Blueprint

Paul Davis (1989), a theoretical physicist, rejects the ideas of the universe being at the mercy of randomness or a world slavishly conforming to mechanical forces. The universe is creative, progressive, and innovative in character. Its organizational aspects are collective and cooperative, and its perspective is synthetic and holistic. Our goal is not to understand what things are made of but how they function as integrated wholes. Human society, says Davis, "has evolved to the stage where it is shaped and directed by conscious decisions" (p. 194). Our self-organizing power creates ever greater organizational complexity. "As we consider systems of greater and greater complexity, the concept of a class of identical systems becomes less and less relevant because an important quality of a very complex system is their uniqueness. In their uniqueness social systems possess logical and structural relationships of their own that transcend the properties of individual human beings" (p. 194). They have their own dynamic behavior and generate their own consciousness and meaning that are their collective cultural attributes. In the evolutionary path of self-organization "the possibility arises that a new threshold of complexity may be crossed, unleashing still a higher organizational level, with new qualities and laws of its own" (p. 196). As a result collective activity might emerge that is now beyond our ability to conceptualize.

Creation is an ongoing process, constantly bringing forth new structures, processes, and potentialities. It is not deterministic. In its development it is intrinsicly unpredictable. It has "freedom of choice," which also means uncertainty. It is free to create itself as it goes along. The fact, says Davis, that "the universe has organized its own self-awareness is for me powerful evidence that there is 'something going on' behind it all. The impression of design is overwhelming" (p. 203). Bringing Davis's self-creating cosmic image down to the scale of self-creating coherence by design in social systems, we can be both humbled and inspired by the possibility of participating in the grand scheme of creation. Our question becomes: How can we capture the emerging cosmic blueprint and how can we mirror it in our designs?

It might be useful to consult additional scholars who have interpreted some of the social implications of the emerging new science. Among others, the works of Briggs and Peat (1984, 1990), Wheatley (1992), and Gleck (1987) might be useful.

Activity #36

Reflect on the text of this section and identify markers or core ideas that represent design thinking. Contemplate those and their use and construct your own understanding of design thinking as you explore the meaning of those core ideas. Then place each and every idea in the context of a design activity you have completed in working with this program. Test the ideas by asking: Does a particular idea characterize that particular design activity? Or can that design activity be made more "designerly" by infusing into it thinking represented by the core idea? Note in your workbook any insight you have gained in the course of testing or applying the core ideas.

Reflections

As I review this section, I cannot avoid thinking about the momentous task facing us in making it possible for our young and old to learn designerly ways of thinking. Our current educational system is geared to the development of knowledge, understanding, and capabilities that promote the cultures of science and the humanities. Clearly those two cultures are necessary parts of learning, but they are not sufficient. They must be complemented by designerly ways of thinking, knowing, and doing that capture and nurture the totality of human experience. If we wish to enable people and communities to determine their own future, if we want to bring about a truly participative and creating democracy, then the acquisition of design thinking and the development of design competence must be an essential part of the learning agenda.

Furthermore, I am truly inspired by the new horizons that open up for us as we begin to explore and understand the implications for social systems design of the emerging new scientific thinking. Such an exploration offers us a unique challenge and opportunity to enrich our conceptualization and understanding of what design might become and how it might work. We might be able to elevate design to a higher level of individual, social, and societal value.

5.3.   Multiple Perspectives

The design of social systems is a complex and dynamic process in the course of which designers are continually challenged to consider choices and make decisions about "what should be." These choices and decisions cumulatively shape the context, the content, and the form of the system they create. As design unfolds the concern and aspiration of designers, they should ensure that their choices and decisions are authentic, informed, and viable.

Design choices and decisions are authentic to the extent that they are made by all the people who constitute the designing community, namely, by all those affected by the future system. This requirement means that in the course of making design choices and decisions, the individual and collective values, aspirations, expectations, and ideas of the entire designing community should be taken into account.

Design choices and decisions are informed to the extent that they are based on all attainable information and knowledge that is relevant (1) to the overall context and content of the systems we design and its systemic environment and (2) to each and every choice and decision we are to make in the course of the design. Design choices and decision are viable to the extent that they are made in consideration of multiple perspectives and a variety of points of view that are relevant to the inquiry.

The requirements of being authentic and informed were discussed in the previous chapters. In the present section the requirement of viability of making design choices and decisions is explored. First the meaning and the need for multiple perspectives in disciplined inquiry is discussed. Then the constituent elements of multiple perspectives and the imperative of integrating multiple perspectives in social systems inquiry are considered. In conclusion, guidelines are proposed for the use of multiple perspectives in social systems design.

5.3.1. The Need for a Multiple-Perspectives Paradigm

In his *Design of Inquiring Systems,* West Churchman (1971) sets forth inquiring imperatives based on the work of Kant and Hegel and the pragmatist philosopher Singer. The common theme of these philosophers, as well as Churchman, is that in social systems inquiry one must "sweep in" the totality of relevant perspectives and points of view upon which judgments, choices, and decisions are made. For churchman, the perception of social situations and the viability of social systems design rest on the debate of many different perspectives and the maximum participation of different stakeholders. Churchman's position has been interpreted by several systems scientist, notably Ackoff, Checkland, Flood, Jackson, Linstone, Mitroff and Ulrich. His ideas have greatly influenced my own work over the years.

Linstone (1984) and Linstone and Mitroff (1994), recognizing the short-comings and pitfalls of underconceptualized and one-dimensional approaches to social systems inquiry, developed a multiperspectives paradigm that leads to unbounded systems thinking with an emphasis on ethically based social systems inquiry and social action. It is proposed that the use of a multiple-perspectives paradigm in social systems design is an essential condition to ensuring the viability of judgments and design choices and decisions.

In social systems scholarship and organizational science there has been a heavy reliance on the science-based technical perspective. Since the middle of this century, the technological revolution, while giving us earlier unimagined powers, has accelerated to the point where we have lost control over it (Banathy, 1993a). It seems that technological evolution runs on its own. It has been separated from sociocultural evolution. We have failed to match technological intelligence with an advancement in sociocultural intelligence and wisdom. As Peccei (1977) noted, the development of such wisdom is essential in giving direction to—and guiding—technological developments for the benefit of mankind. Thus, the development and nurturing of sociocultural intelligence is our central challenge today.

Ida Hoos notes (Linstone & Mitroff, 1994) that the most highly trained scientists and technologists have only a partial view of the world. Their analytic, data-and-model-based approach is only one element in really important decision questions. This approach is one-dimensional in a multidimensional world. Today we face complex systems where everything interacts with everything else, "where human and technical factors must both be fully appreciated and ethics means much more than logic and scientific rationality" (p. xx). The traditional "textbook approach," based on clear definition of the problem and a single set of assumptions-based definitions of the solution, is inadequate in a "messy world."

Thus, in order to assure the viability of the choices and decisions we make in social systems design, we must sweep in a range of other perspectives in addition to the technical perspectives.

5.3.2. Using a Set of Interacting Perspectives

In *Multiple Perspectives for Decision Making* (Linstone, 1984) and *The Challenge of the 21st Century* (Linstone and Mitroff 1994) the authors propose the use of three very different kinds of perspectives to illuminate complex systems: the technical, the organizational, and the personal or individual. "Each views a system through different lenses. Most importantly, each perspective provides insights not obtainable with the others. Together, they give us a deeper understanding of system complexity. Once again, the total is more than the sum of its parts" (Linstone, 1984, p. 12).

5.3.2.1. The Technical Perspective (T)

The technical perspective (T) has the following characteristics: (1) problems are simplified by abstraction and isolation from the real world, assuming that this approach permits solution of problems; (2) data and models are the focus of the inquiry; (3) logic, rationality, and objectivity are presupposed; (4) order, structure, and quantification are pursued; (5) experimentation and analysis aim to attain predictive validity; (6) validation of hypotheses and experimental replicability are expected. The power and success of the T perspective in science and engineering remain unchallenged. Thus, the attempts of economists and social scientists to adopt the T perspective is understandable. The development of operations research, systems analysis, management science, and econometrics illustrates the T perspective in action. These approaches work well in the context of tame and well-structured problems, such as factory or blood bank inventory, airline scheduling, and economic input–output analysis. The use of this "hard systems" approach in such fields as urban planning, criminal justice, and health care led to attempts to isolate the "hard" part of the problem from the human and organizational elements and to eliminate them from the analysis. The analysis-focused T perspective "is in danger of perishing between two options: addressing open, unsolvable, exceedingly complex problems or reducing them to closed, solvable, but irrelevant ones" (Linstone and Mitroff, 1994, p. 93). We now realize that in social systems we must simultaneously sweep in other perspectives that complement the T perspective.

5.3.2.2. The Organizational Perspective (O)

We human beings organize ourselves in social groups and societies, surrendering some of our rights and accepting responsibilities in exchange for gaining benefits. In our social groups we shape a common culture, and adhere to values and beliefs that promote bonding and loyalty. We are members of many groups and organizations, formal and informal, small and large, private and public, intimate and transnational. Each of these protects its identity and maintains collective perspectives that make up a unique "filter" that we call the O perspective. A given organizational decision involves diverse groups of organizational members as affected and affecting parties. Each may define the system in a unique way. Therefore, within any organization a variety of O perspectives should be taken into account. The O perspective focuses on process rather than product and actions and solutions rather than problem analysis. "The critical questions are: Does something need to be done? If so, What? Who needs to do it? How? rather than What is the optimal solution" (Linstone and Mitroff, 1994, p. 98). The O perspective is always intertwined with the T perspective. Changes in the latter must be accompanied by considering and rethinking the former. It also works the other way around.

5.3.2.3. The Personal Perspective (P)

The personal perspective (P) is the third filter that views systems through the eyes of the unique individual. The P perspective sweeps in aspects that are not captured by the T and the O perspectives. For much of our history the P perspective was the privilege of the elite. This changed during the Renaissance. In modern times it plays a particularly important role in America but is still submerged in such societies as China. Churchman observed that economic models aggregate a number of things, and one thing they aggregate is people. We can not face social problems without considering P perspectives. We are often driven to act on the basis of moral feelings, which have little to do with objectives or some measure of performance. Each individual has a unique set of values, beliefs, and perspectives that influence his or her choices and decisions in organizational context. By taking into account P perspectives we open up deeper mental and spiritual levels that can offer great potential value to social groups. The uniqueness of individuality is a basic and essential property of complex systems.

5.3.3. The Interaction and Integration of Multiple Perspectives

Interaction and integration are critical aspects of the multiperspectives paradigm. Today's complex systems are made up of many interacting elements and require many interacting perspectives. The integration of perspectives is crucial to thoughtful systems design. It can be formalized only to a modest extent and must draw on informal processes, such as consensus-building methods. Effective integration of perspectives is a hallmark of systems design. "Only uninhibited cross-cuing and feedback assure that important information is not overlooked.

Bringing differences among the perspectives to the surface facilitates constructive resolution and integration. The perspectives are dynamic, that is, they can change over time, therefore, the cross-cuing of perspectives may require iteration" (Linstone and Mitroff, 1994, p. 117). The authors quote Will Rogers's dictum: "it ain't what you don't know that hurts you, it's what you know that ain't so."

Social systems design must always draw on multiple perspectives. We know that people are not moved by reason, by the T perspective alone. Both the O and the P perspectives must be taken into account. The integration of diverse perspectives and the ability to use conflicting perspectives creatively are key requisites to ensure the viability of making decisions in social systems design.

Linstone and Mitroff (1994) conclude their exploration of multiple perspectives by suggesting that we can meet the challenge of the twenty-first century only if we are ready to manage ourselves and our social systems at higher ethical levels than we have done in the past. "Ethical management implies the ethical integration of T, O, and P. It means simultaneous, balanced action; individually, in a moral way; technically, in a rational way; and organizationally, in a just way" (p. 342). The issue of ethics in systems design is discussed in the next section.

5.3.4. Guidelines for the Use of Multiple Perspectives

There is a scarcity of treatment of the use of multiple perspectives in the design literature. It is therefore fortunate to have available the work of Linstone (1984, 1985, 1989) and Linstone and Mitroff (1994) as sources to use in thinking about the application of multiple perspectives in social systems design. In their book, Linstone and Mitroff (1994) introduce general guidelines for multiple perspectives users. Their guidelines are transformed here into the context of systems design.

In social systems design, the T perspective refers to the intellectual technology of systems design, findings of research on systems design, and case studies of design. The O perspective stands for the specific system (organization) of interest, its culture, and its systemic environment. The P perspective represents the perspectives of the individual members of the designing community. It seems important to add a fourth perspective—the sociocultural, or C perspective—when dealing with social systems, particularly systems in the public and nonprofit sectors, the design of which is our primary interest in this book. In the work of Linstone and Mitroff this perspective is folded into the O perspective. In our rapidly changing world we often find a significant—and widening—evolutionary gap between the larger sociocultural environment and specific social systems embedded in it. For example, the design of our current educational and human development institutions is still based on the thinking and practices of the nineteenth century design of the industrial/machine age (Banathy, 1991a). We turn now to a consideration of the guidelines in social systems design.

5.3.4.1. Each Perspective Offers Specific Insights

Each perspective offers insights not obtainable with the others. Together they provide a far more meaningful basis for decision than reliance on any one perspective. In a design inquiry program that involves stakeholders, a good deal of time should be devoted up front to understand the implications of using the four perspectives. The designing community should be empowered to (1) use the T perspective, to apply the strategies and methods of the intellectual technology of systems design; (2) learn approaches and methods of how to "give voice" individually and collectively to the O and P perspectives; and (3) to explore the C perspective and use the findings of this exploration as an input to design.

5.3.4.2. The Choice of Perspectives Requires Judgment

There is no "correct" or "complete" set of perspectives (the proposed inclusion of the C perspective reflects this advice). Other perspectives might be evoked, whenever appropriate, as bases of making design decisions. In the next chapter we bring into focus the ethical perspective. Taking into account the aesthetic perspective may focus on the aspirational quality-of-life-enhancement aspects of the system we design. The economic perspectives often consider constraints or lead to stating requirements of support to the systemic environment. The consideration of political perspectives often influences the O perspective. In designing social systems, we should always keep the door open to the consideration of new perspectives.

5.3.4.3. The O and P Perspectives Are Usually Case Specific

We cannot deal with generalized O and P perspectives. The C perspective is also changeable, as are the O and P perspectives. In each and every case, the content and context (the environment) of design are unique. The design issue addressed, as well as the designing community itself, is also unique. In social systems design, this uniqueness is also an important organizing concept of the T perspective. There is no single design model that can be "transferred" for use in a specific design program. It is not only the "complementary" concept of systems inquiry, but most specifically, it is the uniqueness of the design situation that invites the consideration of several design models, approaches, and methods. The understanding of the influence of "uniqueness" on the T perspective is the reason why we must engage in the design of design inquiry.

5.3.4.4. A Balance of Effort among the Perspectives Is Desirable

Interaction or cross-cuing of perspectives is vital at all stages of the decision process. The design teams should represent a collective balance of the perspectives. The "stakeholder/participative imperative" of social systems design will require that we place special emphasis on the selection and use of T methods that enable the generation of collective and shared O, C, and P perspectives in arriving at design decisions. As part of the design process, a special "guarantor of balancing" function should be established. In order to ensure the cross-cuing of perspectives, the perspectives should be considered in a parallel fashion.

5.3.4.5. There Is No "Correct" Weighting Formula

There is no weighting formula for the integration of perspectives. This recommendation speaks to the uniqueness of specific design contexts. It also suggests the need to generate alternative representations of weighting formulas. From a presentation of alternative representations a conversation will unfold that enables the design community to arrive at a judgment about weighting formulas that will lead to the substantive integration of the various perspectives. An important by-product of this approach is that it enables the design community to learn which perspectives to apply and how, and to attain increasingly refined "perspective literacy."

5.3.4.6. Perspectives Are Dynamic and Change Over Time

Changes in perspectives are the product of two intertwined aspects of social systems design. One aspect is a dynamic change in the perceptions and positions of members of the designing community over time. This change is due primarily to the designers involvement in the process of design as well as to their understanding and appreciation of the unfolding design solution. They learn as they design. They learn about the system, they learn about themselves and each other, they learn increasingly more about how to design, and they learn about the function of perspectives in design. Another aspect is grounded in the nature of design as an intellectual technology. Systems design is not linear. It is a multi-level, dynamic, feedback/feed-forward, unfolding process, in the course of which we have changes in perspectives, their interaction and integration, and their influence on generating a design solution.

Activity #37

(1) Develop a set of ideas based on the multiple-perspective orientations. (2) Select from the various previous activities a set in which you completed designed tasks. (A possible set might be activities 26, 27, 28, and 29 in Chapter 4.) With

the description of your findings of those activities in view and considering the core ideas of applying multiple perspectives in social systems design as presented in this section, work out an illustrative example of how you might use multiple perspectives in the design of your selected system. Pay special attention to the guidelines presented in the section. Record your findings in your workbook.

Reflections

This chapter is designed to bring forth information and knowledge you can use to construct your own meaning of design as a distinctively human experience. The chapter is designed to broaden and deepen your understanding and appreciation of the significance of design as a human activity that affects our lives and enables us to affect changes in our lives and in the systems in which we live. Sections 5.1 and 5.2 focused on portraying systems and design thinking as a cognitive experience distinctively different from other modes of thinking. The special designerly way of thinking and knowing was further amplified in this section, where we discussed the need for using a range of different perspectives in social systems inquiry and specifically in the design of social systems.

The first three sections of this chapter also helped us to grasp the nature of complexity, the complexity of design thinking, the complexity of using multiple perspectives in design, and, in general, the complexity of social systems. Furthermore, ethics is perceived to be a key aspect that should guide the use of multiple perspectives. The key role of ethics in systems design is explored next.

5.4. The Ethics of Social Systems Design

In the previous section it was proposed that ethics is the key organizing idea that integrates multiple perspectives in the design of social systems. This section explores this integrative role. Various sources that discuss ethics in societal contexts are consulted. We return to Churchman's core idea of ethics in social systems design and explore ethics in systems inquiry and the design of ethical systems. In conclusion, the notion of ethical accounting is explored.

5.4.1. *Ethics Integrates Multiple Perspectives in Social Systems Design*

Linstone and Mitroff (1994) suggest that our individual and collective accomplishments are determined by our readiness to manage ourselves and our systems at higher ethical levels than has been done in the past. No single perspective can attain ethical systemic behavior. Ethical social systems inquiry implies balance among personal and moral action, rational and technical inquiry, and just organizational behavior. The integration among the perspectives means that (1) norms of morality of the individual are to be linked to the larger enfolding sociocultural system; (2) connection has to be created between the technical solution and organizational/institutional justice; and (3) connection should be made between the technical inquiry and personal ethics.

At any given moment in time, there is usually agreement in the scientific/technological community about the *T* perspective. But we cannot expect such agreement to be in evidence in social systems in terms of the *P* and the *O* perspectives. Furthermore, even in the *T* perspective of systems design there is a variety of possible approaches, depending on the nature of the system, the kind of systems we wish to design, and, most significantly, the degree of involvement of stakeholders.

In designing social systems in a genuinely participative, user-designer mode there will be always a variation in the *P* perspectives in the designing community as there are different perceptions in the *O* perspective. In top-down (coercive) organizational settings, the authority's *P* and *O* perspective dominates, and members of the organization are expected to "buy into it" and to be guided by those perspectives. By now we know that this approach cannot evoke the full potential of people. The organizational/management literature is filled with discussion of this topic, advocating genuine involvement. Unfortunately, in most cases the issue does not go beyond rhetoric.

Therefore, a key task in social systems design is to enable the designing community to use the appropriate *T* perspective, to understand and respect the *P* perspectives of its members, integrate those perspectives, and forge an *O* perspective that is acceptable to all. A conversation on approaching this task might unfold as follows. The *P* perspective: we have the right and responsibility to shape our lives and contribute to the design of the system in which we live. The *O* perspective: we jointly and collectively have the right and responsibility to shape and design the systems in which we live. Therefore, from the *T* perspective we should learn to work with a participative (user-designers) design approach, which will enable us to design our own system. Because, as we said earlier: No one has the right to design social systems for someone else.

5.4.1.1. Ethics as a Mirror in Making Design Choices

At the Annual International Conversations on the Design of Social Systems (of the International Systems Institute), a recurring topic has been the ethics of systems design. In one of its summary reports, the ethics research group (Pruzan, 1994a) explored the issue of the role of morality and ethics in arriving at a consensus on the *P* and *O* perspectives. They asked the question: How should we deal with the "blind spot" that each of us has as we see the world from our own moral perspective. The group found the answer in a metaphor: I can see the other person (and vice versa) but cannot see myself seeing the other person, unless I hold a mirror before me. Ethics provides such a mirror they said. When people in groups listen to each other, they can discover their "blind spots" via a conversation that focuses on the values of each that underlie the group's morals. Morality employs norms, rules, and guidelines for members of the group to engage in design inquiry. Ethics employs principles to solve conflicts between individuals/groups having different moral perspectives. Thus, ethics can be viewed as second-order morality.

5.4.1.2.   Ethical Conversations

Values, morals, and ethics are key topics of conversation among the stakeholders. Their explicit discussion, aimed at finding common ground and developing consensus, is at the very heart of making viable choices and decisions in systems design. From this perspective, a design decision is ethical if the stakeholders (the designing community) give their informed approval. This presumes that certain conditions are met, for example, that the participants are not coerced to agree and that they have access to all relevant information and are capable of using the information in presenting their position in the course of making design decisions. Ethical conversation replaces aggressive conflict with an informed and value-based exchange of ideas and perspectives. It is a powerful learning experience for all involved. Each member of the designing community is a member of many social systems. We each bring with us to the ethical discourse a wide variety of values and moral attitudes. Although this creates a more complex discourse, it also empowers the conversation with the capacity to deal with increased complexity.

5.4.1.3.   Ethical Process and Ethical Product

What is the relationship between an ethical process of design and an ethical product of design? The ethics research group, mentioned above, proposed that the outcome (product) of the design process is ethical insofar as the process used to generate it is ethical. Thus we can speak of (1) the ethical design of a system, (2) an ethical system that produces the design (an ethical designing system), and (3) an ethical system that is the product of both. Furthermore, the product of the design should include explicitly stated ethical standards as guidelines of system behavior.

In an ethical conversation about the design of social systems, it is more appropriate to focus on giving voice to the legitimate interest of those who serve the system and who are served and affected by the system, including future generations, than to focus on listening to the voice of any single individual or a few who are part of the conversation. This type of conversation can develop if the various stakeholders, mentioned above, agree on the ethical principles that guide selection of a criteria by which to judge the ethics of the ensuing decision/conversation process.

5.4.2.   Ethical Perspectives in Social Systems Inquiry

We evolve a meaningful wholeness of an ethical stance by the integration of a three-pronged ethical perspective—the self-realization ethics, social ethics, and ecological ethics. Keeping these three criteria constantly in view in establishing and judging the ethical quality of social systems design is of primary importance. These three ethical perspectives are complementary and they serve as guiding principles of social systems inquiry.

5.4.2.1.   The Self-Realization Ethic

Markely and Harman (1982) propose that "the proper end of all individual experience is the evolutionary and harmonious development of the self (both as a person and as a part of wide collectives), and that the appropriate function of social institutions is to create an environment which fosters that process" (p. 115). This ethic will supersede the man-over-nature ethic and the material-growth-and-consumption ethic that have given rise to much of our current problems by focusing solely on material aspects and the exploitation of nature. An emphasis on this ethic is of paramount importance in the design (redesign) of our social systems if we expect that our systems will truly serve the individual's full and valued participation in and contribution to the society. This is the way to empower people!

As corollaries to the self-realization ethic, self-determination of individuals and groups would be fostered, diversity of choice would be honored, and social/political decision making could be largely decentralized. Applying these perspectives in social systems inquiry would lead to the creation of a truly participative "designing" democracy as a preferred choice over representation or bureaucracy for the accomplishment of most social tasks.

The perspectives developed here reinforce the idea of building a design culture in our communities and in the larger society. We are proposing nothing less than the empowerment of people to design their own lives and shape their own future by design.

5.4.2.2.   Social Ethics

Geoffrey Vickers (1982) complements the self-realization ethic discussed above by setting forth the idea of social ethics. This perspective stresses "the specifically social nature of man, humanized by membership in a specific society" (p. 225). Self-actualization ethics has to be related and integrated with social ethics. This integration will enhance the individual's contribution to the larger good and the good of the social system, while the social system is expected to be the channel through which individuals actualize themselves. "A more human world will be a more socially responsible world and this responsibility will have costs and benefits, limitations as well as enlargements in terms of self-actualization" (p. 226). On the other hand, a self-actualized person who fully developed his or her potential can make far more contribution to the social group than one who has not. It is the balanced, mutually supportive integration of the two ethics that we seek in social systems inquiry. This balance ensures the design of systems in which the individual serves the common good (the social system) and the social system is designed to nurture the development of the individual.

5.4.2.3. Ecological Ethic

This ethic views man as an integral part of the natural world. It implies the movement toward a balance between the economic, social, and ecological systems. Humans and social groups act in partnership with nature and harmonize ecological and social relationships (Markley and Harmann, 1982). "Such an ethic is necessary to achieve a synergism of heterogeneous individual and organizational micro-decisions such that the resultant macro-decisions are satisfactory to those who made the component decisions and to society" (p. 114). There is another dimension of ecological ethics. It is a concern for the coordinated and balanced well-being of social groups (social ecology) and cultures (cultural ecology), as well as among various types of activities such as the arts, the humanities, the sciences, politics, etc. In systems design both the ethics of social ecology and cultural ecology are primary bases for making design choices and decisions.

5.4.3. *Ethics and Morality: An Evolutionary Perspective*

In our earlier explorations we became aware of the relationship between the design of social systems and societal evolution. In the course of this century we have developed an evolutionary consciousness and more recently we have grasped the potential of conscious evolution: the potential of giving direction to our own evolution and the evolution of the systems we inhabit, the evolution of our communities and our society by purposeful and deliberate design. Here we explore the role of ethics and morality in evolution and in evolution guided by design.

5.4.3.1. Evolutionary Ethics

Jantsch (1980) defines ethical behavior as that which enhances evolution. Ethics emerges with evolution and it takes on a regulatory function. This regulatory function in the human world consists of rules of behavior but also of morality as a distinct inner experience. As an integral aspect of evolution, ethics is experienced directly by way of the dynamics of self-organization and creative process. For example, ethics is directly experienced in the creative processes of systems design.

> The direct living experience of morality becomes expressed in the form of ethics—it becomes form in the same way in which biological experience becomes form in the genetic code. The stored ethical information is then selectively retrieved and applied in the moral process in actual life situations. (p. 264)

In the life of social systems, in which possibilities for action are available in such a rich spectrum, it is primarily our intentions, desires, and preferences that are guided by morality and determine our behavior. This statement speaks directly to social systems design, in the context of which we designers explore a wide range of alternative solutions, representing our intentions, desires, and preferences, which in turn are guided by our morality and shared ethical convictions.

5.4.3.2. Multilevel Ethics

Human ethics is multilevel ethics, including personal and transpersonal ethics, the ethics of our social systems, and ultimately, what Jantsch calls evolutionary ethics. He notes, however, that we are still far from formulating and implementing evolutionary ethics. "What in the Western world we call ethics is a behavioral code at the social level which is primarily geared to ensure the free unfoldment of the individual" (Jantsch, 1980, p. 265). For this reason we talk almost exclusively of the rights of individuals and particular groups but almost never of responsibilities. As Vickers (1982) pointed out, "rights" are static and defensive, while responsibilities imply creative participation in the design of the human world.

The ethics, says Jantsch (1980),

> that dominates the Western world is therefore an individual ethic in the disguise of a socially committing behavioral code. It is not a multi-level ethics in the true sense. Morality, in contrast, is the direct experience of ethics inherent in the dynamics of evolution. The higher the number of levels and the intensity at which we live, the higher the number of levels and intensity at which our morality becomes effective. (p. 265)

Ethics is a manifestation of consciousness, and a major aim in our design of social systems is the increasingly mounting application of ethics at multilevels, such as the personal, transpersonal, group, and whole systems, as well as at the level of societal evolution. Designing and managing our own lives, the systems we live in and our communities, we have to satisfy and live by "the ethics we ourselves have established as the guardian of our actions" (p. 266). Reviewing the concepts Jantsch developed on the essence and role of ethics, we can only conclude that what he says is directly relevant to an understanding of the essential role of ethics in the design of social systems. Rowland's (1994) translation of Asimov's laws of robotics is an appropriate guide to ethical principles in systems design. He says that, "systems design must benefit (do no harm to) the world: it must insure ecological harmony. Systems design must benefit society (except where social benefits harm the world). It must ensure social justice. Systems design must benefit the individual (as long as such benefits do not harm society or the world); it must insure individual freedom" (p. 286).

5.4.3.3.   Morality and Evolution

Csikszentmihalyi (1993), discussing morality and evolution, says that "in every human group ever known, notions about what is right and what is wrong have been among the central defining concerns" (p. 139). Moral imperatives have become necessary because

> evolution, in liberating humankind from complete dependence on instinct, has also made it possible for us to act with a malice that no organism ruled by instinct alone can possess. Therefore, every social system must develop "memes" to keep the intergroup harmony that genes no longer can provide. These memes constitute the moral system, and generally they have been the most successful attempts humans have developed to give desirable direction to evolution. (pp. 159–160)

For over a century, it was fashionable in the social sciences to suggest that different cultures develop entirely relative and arbitrary moral systems.

> In fact what is so remarkable is how similar the world's major moral systems are in considering "good" to be the achievement of the kind of harmony within consciousness and between people that we have called negentropy, and which in turn leads to higher levels of complexity. (p. 160)

The great moral systems of the world are congruent, despite differences in emphasis and variations in the metaphors used to explain why some things are right and others wrong.

All ethical systems propose to direct evolution by channeling thought and action from the past to the future.

> The past—represented by the determinism of instinct, the weight of tradition, the desires of the self—is always stronger. The future—represented by the ideals of a life which is freer, more compassionate, more in tune with the reality that transcends our needs—is by necessity weaker, for it is an abstraction, a vision of what might be. (Csikszentmihalyi, 1993, p. 162)

The new, the hopeful, and the creative appear to be more ephemeral than what was tried and worked. The realist, who deals with the here and now belittles the "impractical" idealist who invests energy in the stuff of a "blue-sky world." Without the realist we could not survive. But without the idealist we could not evolve. The choices made in our continuing evolution are to be guided by a moral system that takes into account the wisdom of tradition, yet it is inspired by our vision of the future. "It should specify right as being the unfolding of the maximum individual potential joined with the achievement of the greatest social and environmental harmony" (p. 162).

The above discussion on morality and evolution brings into focus our individual and collective challenge to give direction to the evolution of our lives and the systems we inhabit by purposeful and ideal-seeking design. Our design inquiry is to be guided by ideals of a life that is freer and more compassionate, that is guided by the desire to create conditions that lead to the unfolding of the maximum individual and collective potentials, coupled with the achievement of the greatest social and environmental harmony.

5.4.4. Churchman: The Ethics of the Whole System

The ethics of the whole system was advanced by West Churchman (1968b), and it has become a major source of inspiration for many of us in design thinking and practice. The ethical value of design, he said, can be determined only in terms of the whole system. "The problem of systems improvement is the problem of the ethics of the whole system" (p. 4). For Churchman the whole system includes all those who are affected by the system we design. Churchman envisions ethics as the guide to the life of the whole system. The boundary between the individual and whole system ethics is not easy to draw. These integrate into the life and behavior of the individual as well as into the life of the collective. But what is important to understand is that the primary level at which we measure the value of the design is that of the whole system. Thus, the key task of designers of social systems is to determine the "ethics of the whole system" and apply it in making choices and design decisions in the course of their inquiry.

In designing various social service systems, first we are to envision an image of the ideal society, and then ask the question: What should be the design of the educational, human development, health, social services system that would enable these systems to contribute to the creation of that ideal society? When we measure the outcome of our design, for example, the design of an educational system, we are not going to valuate the outcome of our design or justify the resources invested, based on test scores, attendance records, success in future careers, etc. The question we should ask is: To what extent is the system instrumental in moving the larger system toward the ideal?

For Churchman, the whole system idea includes the consideration of future generations. His ethics of the whole system most pointedly addresses the issue: What are the implications of our design for future generations? Concern for future generations is the key imperative for Churchman (1982). We are obliged, he says, to consider the impact of our design on those who come after us. Our design should be such that it enhances and expands their options. For Churchman the idea of "humanity within" is not only an invariant idea but a sacred one.

Thus, the moral law with respect to future generations is: We should undertake to design our societies and their environments so that people of the future will be able to design their lives in ways that express their own humanity (p. 21).

Ethics, says Churchman (1982), is an ever ongoing aesthetic conversation. This conversation should never stop.

> Ethics is an eternal conversation. The reason that ethical relativism ("different strokes for different people") is so bad is that it stops conversation. Relativists are only sure of one thing, their relativism. They actually think that ethics is a search for absolute values, and since it is, it is a hopeless enterprise. Since ethics is an eternal conversation, its conversation retains its aesthetic quality if human values are regarded as neither relative nor absolute. (p. 57)

In systems design the ethical conversation never ends. In each and every choice we make, at each decision point, we ask questions such as: What is the value base of this choice? What does the ethics of the whole system tell us about the choice we are to make? To what extent is the ideal image realized by the design decision we have made? These questions can be answered only from the perspective of the ethics of the whole system. But the ethics of the whole system does more than guide our design inquiry. Ultimately, it is the guarantor, the conscience, and the soul of social systems design.

5.4.5. Ethical Accounting

Ethical accounting (Pruzan, 1994b) provides criteria as it probes into how well the design we are creating lives up to (1) the shared values to which the designers committed themselves and (2) the ethics of the whole system that designers have articulated as a basis of making design decisions. The threshold requirements that an ethical accounting system should satisfy include the following:

- The ethical accounting statement (EAS) should employ concrete specifications, developed by the designing community. These "specs" should operationalize their shared values and the ethics of the whole system.
- The specifications of the EAS are to be used in the course of the design inquiry. They should probe into the issue of how well the design responds to the agreed upon (stated) value system and to the ethics of the whole system.
- The EAS is to be designed so that it is impossible to identify individual respondents. It should not give a particular person monopoly on interpreting the results. The interpretation takes place in the course of an ongoing conversation among the designers and it should lead to a shared judgment.
- Findings of the ethical accounting inquiry should lead to a continuing conversation that would ensure an ever-increasing match between the accounting specifications and the design solution.

An EAS is not a static measure. It is subject to continuous redesign and refinement. It is used not only in the course of the design inquiry, but throughout the implementation of the design. Earlier, we conceived of design as continuous and ever ongoing. So is the EAS. It is continually reinterpreted as it is used. It can be instrumental in the continual re-creation of the system.

Reflections

It is appropriate to reflect on the core ideas embedded in the discourse developed above in order to appreciate their collective impact on our understanding of the ethics of social systems.

In connecting multiple perspectives with the ethics of design we can say that ethical design balances personal moral actions, rational technical inquiry, ethical organizational behavior, and the ethical action of the designers. This notion is consistent with what we said earlier: no one has the right to design social systems for someone else.

Ethics provides us with a mirror in which we can see if our design is guided by ethical imperatives. If we have such a mirror, then ethical conversation replaces aggressive conflict among the stakeholders with informed and value-based exchange of ideas and perspectives. Thus, we can say that the outcome of design is ethical insofar as the process used to generate the outcome is ethical.

The three-pronged ethical perspectives—the self-realization ethics, social ethics, and ecological ethics—should also be in constant view in the course of the design inquiry. Consideration of self-realization ethics is of paramount importance if we expect that our system will truly serve the individual's full and valued participation in and contribution to society. The consideration of social ethics is equally paramount since it ensures the design of just systems. Furthermore, the ethics of social ecology and sociocultural ecology are also bases for making design choices and decisions.

Ethical behavior enhances evolution. In the course of evolution of every human group, notions of what is right and wrong have been among the central defining concerns. It has become fashionable to claim that moral systems are arbitrary products of various cultures. In fact, what is remarkable is how similar the world's major moral systems are. Ethics involves an eternal conversation. Ethical relativism stops this conversation. Relativists are sure about only one thing: their relativism.

A key task of designers is to determine the ethics of the whole system and apply it in making design choices. The idea of whole systems ethics includes consideration of the impact of our design on future generations. Our design should be such that it enhances and expands their options.

A key issue implied by our discourse in this section is to determine how well the design lives up to (1) the shared values to which the designers committed themselves and (2) the ethics of the whole system that designers articulated as the basis of design decisions.

Activity #38

(1) Generate core ideas that represent the ethics of systems design. (2) Select a previous activity in which you developed a scenario for systems design. In view of the discourse on ethics presented in this section, develop a "parallel" scenario that highlights (1) at what points of the inquiry and (2) in what manner would you given considerations to values, morals, and ethics in the course of a design inquiry. Describe your scenario in your workbook.

5.5.   The Design of the Ideal System

In this chapter we extend and broaden our exploration of the design of social systems, as well as sweep in new understanding and new perspectives that help us to establish a more informed view of not only what we know about what design inquiry is, but more importantly, what design inquiry should be. In this section we introduce the imperative of the ideal: design inquiry in the context of social systems should always be guided by the ideal. Here we tie in ethics in design with the ideal in design. Then the notion of ideal-seeking systems is explored, which includes defining the concepts of ideal-seeking and idealized design. The definitions lead us to discuss the design of the ideal systems model. As you work through the text, note concepts or phrases that you consider to be core ideas of the notion of ideal systems design.

5.5.1.   Ethics and the Ideal System

The sections of this chapter are developed with the intent to integrate them to create a seamless exploration of the "shoulds" of design inquiry. The ethics of design will now be integrated with the ideals in design. In his *Search of a Way of Life,* Singer (1948) developed the notion of ideal in the philosophy of ethics. He proposed that ethics should always be discussed in the context of human ideals. Singer, however, recognized that there is a dialectic between the idealist and the realist. Commenting on Singer's view, Churchman (1982) says that

> the realist is a down-to-earth, practical person who tries to solve te practical, hard problems of everyday life in a practical, coherent fashion. The realist goes to management development programs and expects to find out what to do next Monday to become a better realist. The idealist tries to understand the saga in terms of human ideals and their meaning in the very long run, and he sees that there is a constant struggle towards an ideal society. He tries, as best he can, to explain what that ideal might be. (p. 133)

It was said earlier that designers of social systems should paint the largest possible picture in the largest possible context and that the context for their design should always be their vision of the ideal society (Banathy, 1991a). Churchman further admonishes that we should not only seek ideals, but we should also find and define what that ideal should be. We must also understand that ideals keep changing. "What the utopians of the nineteenth century thought was the ideal community, we would no longer regard as the ideal. That does not mean that we are wrong to try to design utopias" (p. 135).

Concluding the tie-in between ethics and the ideal, we believe that ethics is aimed at searching for the "ultimate good," which is the ideal. We recognize the interaction between the inspiration—coming from the beauty of this ultimate ideal and the search for it—and the aspiration that aims to define it. Design inspired by ethics leads us to create visions and images of an ideal future we aspire to attain. Inspiration and aspiration jointly give us the courage to pursue the ideal.

5.5.2. Ideal-Pursuing Systems

> We humans are unique in our ability to formulate and pursue ideals, desired states that we can never attain but to which we can always come closer. If we are to pursue the ideal continuously, we must never be willing to settle for anything less: that is, we must never be either permanantly discouraged or completely satisfied. . . . We must always be able to generate visions of something more desirable than what we have and must pursue these visions. (Ackoff, 1981, p. 40.)

"Humans are more than ends-seeking animals; we are ideal-seeking." We seek "those ends that are believed to be unattainable but towards which we believe progress is possible" (p. 63). In an earlier work, Ackoff and Emery (1972) explore the nature of ideal-seeking systems. In our social systems we collectively pursue states we know we cannot attain. We still draw satisfaction from approaching such states. "The approach is called progress and the end-state is called an ideal" (p. 137). As many wise men observed, there is more satisfaction in pursuing an end than in attaining it. People who play games well often say that the play in itself gives tham more satisfaction than winning. "The continuous pursuit of more desirable ends is an end in itself, and hence the attainment of a specific end can be conceptualized as a means to such pursuit" (p. 137).

There is an ancient Hungarian legend of hunters of antiquity who pursued the "miraculous white stag." They followed it through endless obstacles, but they never hunted it down. It was the joy of the chase that inspired them. They had the vision that the miraculous stag would eventually lead them to the promised land. This story is a metaphor for ideal-pursuing systems.

Ackoff (1982) says that only the pursuit of the ideal can provide cohesiveness and continuity to life. By formulating and pursuing ideals we put meaning and significance into our lives and by so doing we derive satisfaction. An ideal-pursuing system must be able to derive at least as much satisfaction from moving toward the ideal as it does from attaining short-range goals.

> Only if this is so would we be willing to sacrifice the present for the future, because progress toward an ideal cannot take place along a straight line. The ability to see the long-range consequences of current activity, the ability to do the right thing for the right long-range reason, is the essence of wisdom. (p. 246).

5.5.3. The Ideal Systems Concept

In his early work, Nadler (1967) suggests that the ideal system refers to the perfect, the best and flawless, "prime" system that achieves the ideal we seek to attain. Ideal conditions refer to a state in which systems can be designed in an ideal way. An ideal systems model is a conceptual representation of a system that can attain the ideal. The ideal systems model can guide us throughout the design inquiry. Thus, the conception and articulation of the ideal is a most practical approach to the design of social systems. Nadler (1981) identified three levels of the formulation of the ideal systems: the ultimate ideal system, the contemplative ideal system, and the feasible ideal system.

5.5.3.1. The Ultimate Ideal System

The ultimate ideal system is formulated at the conceptual level. It occupies the widest possible design space with no limitation on the thought process of the inquiry. It has a quality of the infinite. It focuses our attention on the ever-forward-moving process of progress in pursuing the ideal. The ultimate ideal system serves as the guide to the design of the contemplative ideal system.

5.5.3.2. The Contemplative Ideal System

The contemplative ideal system is a visionary formulation of the future system that in fact we design, but that cannot be implemented until further developments render it feasible to implement. It presents the challenge of the "can become if . . ." (My story for this is the design of the *Duomo* in Florence, Italy. The architect envisioned and designed the cupola, which was eventually built 300 years later, once the technology for its construction was developed.) The contemplative "if" operates in the possibility space of the design inquiry. It becomes the guide for the design of the feasible ideal system.

5.5.3.3. The Feasible Ideal System

The feasible ideal system is a workable, technologically feasible and doable system that may have several alternative representations from which we select the most desirable and obtainable. In the process of selection, the feasible ideal

system is our guide in choosing the finally recommended system that we shall
fully design and implement.

5.5.4. *Idealized Systems Design*

Three topics are discussed as an introduction to a discussion in idealized
systems inquiry: (1) the properties of idealized design, (2) reasons for—and the
advantages that are inherent in—the process of participative design, and (3)
reasons that are inherent in the current dynamics of societal evolution.

5.5.4.1. Properties of Idealized Design

For Ackoff (1981) the selection of ideals lies at the very core of social
systems design. "It takes place through the idealized design of the system that
does not yet exist or the idealized design of the system that does" (p. 105).
Designers conceptualize a system they want to have now, not at a future date.
Thus, the environment in which the system will operate exists now. It does not
have to be forecast. Assumptions about the environment in which the system will
operate necessarily enter into the design. Ackoff states three properties of the
idealized system. "It should be: technologically feasible, operationally viable,
and capapble of rapid learning and adaptation" (p. 105).

5.5.4.1a. Technological Feasibility. Technological feasibility means that
the design should not incorporate technology that is unknown or unusable. In the course
of design, we can make innovative use of technologies such as "CogniScope," a
software/groupware-supported design program (Christakis & Conaway, 1995)

5.5.4.1b. Operational Viability. Operational Viability means that the
system we design should be capable of functioning and sustaining itself once
installed. The issue of implementability becomes clear when we consider the effects
of the design on people in the system. For example, is the design really implementable with someone other than the stakeholders who designed it?

5.5.4.1c. Capability of Rapid Learning and Adaptation. Capability of
rapid learning and adaptation can be properly met by several conditions: (1)
empowerment of the stakeholders to modify the design whenever they wish; (2)
introduction of design experimentation in order to resolve issues that emerge in the
course of the design; (3) use of processes that enable the system to learn from its own
experience and improve its design over time; and (4) the establishment of evalua-
tive/monitoring processes in the course of the design inquiry that introduce
corrections in the design whenever indicated.

For Ackoff (1981), the product of an idealized design is not an ideal system, because it is capable of improving itself. "Rather it is the most effective idealseeking system of which its designers can conceive" (p. 107.)

5.5.5. Benefits of Engaging in Idealized Design

Engaging in idealized design (Ackoff, 1981) includes such benefits as participation, aesthetic values, consensus, commitment, creativity, and feasibility.

- Idealized design facilities participation in the design inquiry. It provides opportunity to the designing community to work with others in the system, to think and learn about their system, to contribute their ideas to the design, and thus, to affect the future of the system.
- Participation in idealized design enables stakeholders to incorporate their aesthetic values into the design and thus improve the quality of their individual and collective lives. Furthermore, participating in the design of an idealized system is a most rewarding aethetic experience in itself.
- The idealized design process enhances the generation of consensus among those who participate in the inquiry. Consensus arises in idealized design because it focuses on ultimate values. When agreement is reached on ideals and ultimate values, differences over means often can be easily resolved.
- Participation in idealized design, and the consensus that emerges from it, generates commitment to bringing the design to life. Those who have a hand in developing the idealized design will have stronger commitment to the design and its implementation than those who do not participate.
- The idealized design inquiry engenders and stimulates creativity and focuses it on collective and individual development. The ideal-seeking nature of idealized design invites and releases creativity, removes many of the constraints that would inhibit it, and resolves concerns about implementation of the design.
- Idealized design inquiry enlarges the designer's conception of what can be implemented. In traditional design, a major obstruction to implementation is design developed by someone else. Participative idealized design not only reduces such concern but it also affects the participants' confidence of what is implementable.

5.5.6. Evolutionary Imperative of the Ideal Systems Design Approach

In times of relative stability and slow change, characteristics of the past, piecemeal adjustments, and incremental changes were adequate to bring social systems in line with gradual changes in the societal environmental. Under those conditions, the task of social planning was one of making ad hoc improvements in existing systems. However, those so-called good old days are passé.

Around the middle of this century, Banathy (1991a), we entered an era of rapidly accelerating dynamic changes, discontinuities, and transformations. Today, incremental changes and fixing the existing systems no longer work. We have also found out that extrapolating social planning and design based on experts' predictions of the future do not work either. This situation left us with no other serious option than to develop disciplined inquiry that enables people in our social systems to define and design their own future as they best see it for themselves, in the context of the larger societal characteristics. This disciplined inquiry in fact has emerged during the last two decades as the ideal systems design approach.

5.5.7. Implications and Consequences of Ideal Systems Design

Systems design intentionally creates social systems that fulfill the purposes stated by their designers. It is a process by which visions, ideals, ideas, values, and aspirations are shared, collectively agreed upon, and articulated by the stakeholders of the system. The stakeholders are people who serve the system, who are served by it, and who are affected by it. Stakeholders engage in design to create a system that will manifest their shared vision, one that represents their shared ideas, aspirations, values, and ideals. The kind of design described here is not a top-down, design-by-directive. It is not the kind that is designed by experts for other people—not even the kind an expert designs with people in the system. It is the kind that is designed collectively by the stakeholders. In the ideal systems design approach the target is always the ideal. The target cannot ever be less than the ideal. Design is a journey toward the ideal. Only the ideal is worth the effort that is required to undertake the journey of social systems design. The value of the often demanding design journey lies in the progress we make toward the ideal.

5.5.7.1. Implications of Commitment to the Ideal

There is the story of a prince, born with ugly deformities, who commisioned the sculpturing of a statue of an ideal image of a young athlete. The statue was placed in the royal garden. The young prince sat in front of the statue all day long, wanting to transform himself into the image of the statue. Years of persistance rewarded him. He became a living image of the statue.

Commitment to the ideal means a determination to create the most inspiring ideal system, one that will act as a magnet and pull us toward its realization. We conceive and design an ideal representation—an ideal model—of our system. The ideal system is not some science fiction speculation. It is a system that we want to pursue and attain. Once the ideal design of the future system is created, then—and only then—shall we consider constraints and enabling resources in order to attain a feasible, and now implementable, design. But the ideal model—like the statue in the royal garden—will be always in front of us. As we focus our eyes on the ideal, it will guide our continuous movement toward it. It is the ideal—like the beautiful statue created for the prince in the fable—that will inspire us to become ideal-like. The ideal, and our aspiration to attain it, give the inspiration and the courage to pursue our design. Inspiration and aspiration shape each other as we shape our ideal system.

An implication of having the ideal design "out there" on the horizon emerges from the realization that as we move toward the horizon the horizon moves ahead of us. The landscape changes. This "law of the moving horizon" applies in social systems design. As time goes by in our journey toward the future, the ideal model will most likely change, as we might "remake" it. As we move toward the realization of the ideal, the environment in which our system lives and the situational context in which our system operates will also change. We will be led to continuously reexamine and possibly reshape our ideal model, based on our commitment to coevolve with our continuously changing and transforming societal environment.

It is not only what is "out there" that leads to changes in the ideal but also what is within us. Our perceptions, insights, and ideas also change as time goes by, changing our aspirations of the ideal. Design is a journey that never ends. Designers are like the hunters of antiquity, who pursued the miraculous stag with the hope that it would lead them to the promised land. The design journey does not end with us. We now design systems that offer learning opportunities, arrangements, and resources for learning and human development by which future generations will be enabled and empowered to attain their full potential and become competent so that they can envision their ideals, shape their own future, and continue their journey toward their promised land.

5.5.7.2. The Consequences of the Ideal Systems Design

Social systems design is authentic only if it is created by those who serve the system, those who are served by it, others who have a vested interest in it, and all those who are affected by it. It is from their dreams, ideas, aspirations, and preferences that the "ideal should" emerges as a collective definition of the system to be brought to life by their design. They engage in design because they genuinely and deeply care about the future state of their system. Thus, ideal systems design by definition is participative. It is design that is carried out by, what I have called, "user-designers." Beyond attaining a design that truly represents what these user-designers believe in and one that they will consider to be their own, a design by user-designers has other significant consequences and benefits.

Ideal systems design engages the creative potential and focused imagination of all who participate in it. Such engagement of the user-designers makes possible their meaningful contribution to the definition and design of the system of their choice. At the same time, it provides them with unique learning experiences about how to engage in design, what are the characteristics of their emerging system, and what their individual and collective role in it will be. User-designers will also learn how to generate consensus among themselves. They will continuously apply consensus-building methods in making design decisions. Because the design is their own creation, user-designers will take part more effectively and with a greater level of commitment in the implementation of the design. Participation is empowering and design is always empowered by it.

Furthermore, by learning to design and engaging in it, people learn as individuals and collectively as an organization. As individuals, they will gain a genuine understanding of what their system is about and how it works. They will also realize how their performance affects the performance of the whole. As an organization, they will learn how to examine and continuously reexamine their purposes, perspectives, values, and modes of operation. Thus, they can collectively develop new insight and knowledge, from which, if called for, they can redesign their system and make continuous contributions to its life and to its future development.

A most demanding consequence of electing to use the ideal systems design approach is that user-designers or stakeholders have to learn how to carry out the ideal systems approach. Unfortunately, ideal systems design is not on the agenda of our social science departments or our professional schools. As a rule, it is not used by our social systems R & D agencies. And it is not practiced by our communities. It is this underdeveloped state of competence in systems design that confronts us with a challenge of utmost significance. The challenge to higher educational and relevant R & D institutions is defined as follows: (1) to become experts in the intellectual technology of ideal systems design (ISD); (2) to develop professional preparation programs in ISD; (3) to prepare resources and programs that people in our communities can use to "get ready" and to engage in ISD; and (4) to assist our communities in their design learning and in carrying out their design program. The challenge to our social systems and our communities is to seek out and use resources, programs, and learning opportunities that enable and empower them to engage in systems design.

Reflections

Design brings forth novelty, something that does not yet exist, something that has to be envisioned and then created. Designers who engage in this creation formulate ideas about what they aspire to attain. They integrate these ideas into a system of ideas, the ideal systems image of the future. The ideas emerge as they explore the largest possible context, as they ask: What kind of society do we wish to attain? They place an ideal image of the desired society on the "outer horizon" as their vision of the future. This image becomes an inspirational force as they ask: What contribution should our system make to create that ideal society? They formulate the ideal image of this system and place that image on their "inner horizon." The images placed on the two horizons interact dynamically, recursively, in a coevolutionary way.

Activity #39

(1) Review the discourse of this section and list core ideas (of the ideal systems notion) that you consider salient and relevant in the design of your selected system. (2) Select a few core ideas of an ideal image of a desired society. Now propose core ideas of your system of interest that would enhance the attainment of the selected societal core ideas. Enter your findings in your workbook.

5.6. Creativity in Systems Design

As a point of departure to our continuing journey on the landscape of social systems design, we consider general definitions of creativity and connect creativity with the ideal systems approach to design. In the course of this journey, I explore the realms of creativity and address such aspects as: (1) its nature and characteristics, (2) its process, (3) its internal and external conditions, and (4) barriers and ways of removing them. In closing, I explore the societal significance and imperative of creativity.

As you work with the various parts of the text you are challenged to search for and make note of the core ideas of creativity and ask yourself: What do these core ideas mean to me? What are their implications for the design of social systems? The activity at the end of the section further elaborates these tasks.

5.6.1. Creativity: Definitions

As an overall context of a definition of creativity, Whitehead (1968) suggests that creativity is the actualization of potentiality, and the process of actualization is an occasion of experiencing creativity. Viewed in conjunction, they carry the creative act, which drives the world. "Creativity expresses the notion that each event is a process issuing in novelty" (p. 236). Whitehead's notion is tied in here with Jantsch's (1980) notions of the self-organization paradigm of systems inquiry, and the conception of evolutionary progress as a movement through time in a manner and direction that is guided by our values and visions of the good and the ideal. Creativity, therefore, connects the process orientation of systems inquiry with the progress of guided societal evolution (Montiori, 1989).

Creativity (Barron, 1988, p. 80) is "essentially the ability to bring something new into existence purposefully. [It] is seen to be in the service of increased flexibility and increased power to grow and/or survive." There are no limits to the areas in which creative responses can manifest themselves, says Barron (1969). The defining properties of creative products and processes are their originality, validity, adequacy to meet needs/aspirations, and their aesthetic quality or elegance. "The emphasis is on whatever is fresh, novel, unusual, ingenious, divergent, clever, and apt" (p. 20).

Creativity implies novelty and innovation (Harman, 1984). Studies in the fields of motivation and learning have disclosed the power of novelty as an inducement to action. In human experience there is tension between maintaining equilibrium, security, and stability and seeking and creating new possibilities. This tension is manifested in such dualities as rationality and intuition, conformity and nonconformity, complexity and simplicity, certainty and uncertainty, and convergence and divergence. Convergent thinking tends to use rationality to focus down to a single goal. Divergent thinking, drawing on a richness of creative ideas and original thinking, is characterized by moving away from set patters and goals. "While both convergent and divergent thinking are involved in creative activity, it is divergent thinking that especially characterizes that which is most widely recognized as creative" (p. 51).

Rogers's (1961) definition of the creative process is as follows:

> It is the emergence in action of a novel relational product, growing out of the uniqueness of the individual on the one hand, and materials, events, people, or circumstances of his life on the other. The mainspring of creativity appears to be man's tendency to actualize himself, to become his potentialities. (pp. 350–351)

In human life and in society there is a tendency toward an urge to expand, extend, develop, mature, and to express and activate all our capacities. This tendency may be deeply buried under encrusted defenses of the familiar. Still, it exists in every person and awaits the proper conditions for its release and expression. The primary motivation of the release of our creativity is in forming new relationships with the environment and in aspiring to become fully ourselves.

The experience of creativity (Csikszentmihalyi, 1993, p. 175) "stretches our skills in new directions as we recognize and master new challenges. Every human being has this creative urge as his or her birthright. It can be squelched and corrupted, but it cannot be completely extinguished." The enjoyment of creativity comes from such experiences as surpassing ourselves and mastering new obstacles. These experiences involve concentration, absorption, deep involvement, joy, a sense of discovery, and accomplishment. They involve "the excitement of finding out something new about ourselves, or about the possibilities of interacting with the many opportunities for action that the environment offers" (p. 177).

5.6.2. Connecting Creativity with the Ideal Systems Design Approach

Design creates novelty. It creates new forms and processes. Design is a manifestation of creativity. Design is creation. These phrases underline the central and dominant role of creativity in systems design. We cannot understand design unless we understand creativity. Yet, with few exceptions, the design literature does not say much about creativity. I will now connect the ideal systems design notion, discussed in the preceding section, with creativity. Figure 5.1 provides the image that helps to make this connection. The arrow on the top stands for the act of creating/designing the ideal system. The ideal system can be conceived only by a collective, creative envisioning of the future by all stake-

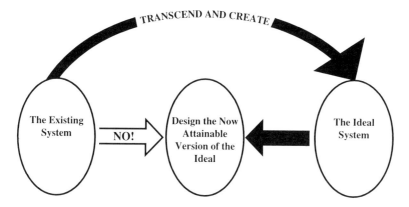

FIGURE 5.1. "Turning the arrow around."

holders. We cannot design the ideal system by staying within the boundaries of the existing system or working out from it. We have to transcend, to leap out from what exists and create a description of an ideal system. Then we turn the arrow around, and work out from the ideal as we define the now attainable version of the ideal.

5.6.3. *Exploring the Realms of Creativity*

The various realms of creativity are explored next, including the meaning and characteristics of creativity, the creative process, the conditions of creativity and ways of fostering it, the barriers and blocks to creativity, and misconceptions about creativity. Then, we reflect upon these realms and explore their relevance to design.

5.6.3.1. The Nature and Characteristics of Creativity

New creative ideas emerge from us as a play of the mind (Bohm and Peat, 1987). Failure to appreciate the creating role of this play is a major block to releasing creativity.

> Within the act of creative play, fresh perceptions occur which enable the person to propose a new idea that can be put forward for exploration. As the implications of this idea are unfolded, they are composed or put together with other ideas. Eventually, the person supposes that these ideas are correct. (p. 48)

From this process of propose, compose, and suppose, new perceptions emerge through the creative play of the mind.

> It is the very nature of this play that nothing is taken for granted as being absolute or

unalterable. The whole activity is not regarded as a problem that must be solved but simply as a play itself. To sustain this creative activity of the mind, it is necessary to remain sensitive to the ways in which similarities and differences are developing, and not to oversimplify the situation by ignoring them or minimizing their importance. (pp. 49–50)

Bohm and Peat (1987) further suggest that

communication is as essential to the creative act as is perception through the mind. Indeed, within this context, perception and communication are inseparably related, so that creation arises as much from the flow of ideas between people as in the understanding of the individual alone. When insight occurs, it emerges out of this overall structure of communication and must be unfolded so that it obtains its full meaning within it. Creativity flows out of a free and open communication. Indeed, it is not possible to consider any fundamental separation between the mind's perception and communication; they are an indivisible whole. (pp. 63, 65, 70)

Discussing creativity in the whole of life, Bohm and Peat challenge the assumption that creativity is necessary only in some specialized fields. Assuming a restricted nature of creativity has very serious consequences for the society as a whole. A free and open creative communication is the most effective way of addressing the crises that face society. A free exchange of ideas is of fundamental relevance for transforming culture by liberating creativity. If we restrict the spirit of free play of ideas, all the problems that have plagued civilization will surface again and overwhelm us. Thus, Bohm and Peat call for a creative surge that will involve all phases of human life. A radical transformation must take place that embraces all fields of inquiry and develops a "new view of humanity, culture, and society. What is needed today is a new surge that is similar to the energy generated during the Renaissance but even deeper and more extensive" (p. 265).

Rogers (1961), in exploring the nature of creativity, says that "the very essence of the creative is its novelty, and hence we have no standard by which to judge it. Indeed, history points up the fact that the more original the product, the more far-reaching its implications" (p. 351). The genuinely significant creation is most likely to be seen first as erroneous. Later it may be seen as obvious and self-evident. Still later it receives an evaluation as a creative contribution. The individual creates primarily because the creative act is satisfying, it is felt to be self-actualizing. "To the degree that the individual is open to all aspects of his experience, and has available to his awareness all the varied sensing and perceiving which are going on in his organism, then the novel products of his interaction with his environment will tend to be constructive both for himself and others" (p. 354).

We cannot expect to provide an accurate description of the creative act until it occurs (Rogers, 1961). The creative act is a natural behavior that has a tendency to arise when we are open to all of our inner and outer experiencing and when we are free to try out in a flexible fashion all possibilities and all manner of relationships. From these possibilities we select those that most effectively meet our inner need, or those that form a more effective relationship with the environment. There is one quality of the creative act, says Rogers, that may be described. "In almost all products of creation we note a selectivity, or emphasis, an evidence of discipline, an attempt to bring out the essence" (p. 355). Through his creative act the person says that this is "my way of perceiving reality, and it is this disciplined personal selectivity which gives to creative products their aesthetic quality" (p. 356).

Rogers also emphasizes the desire to communicate the creative act.

> It is doubtful whether the human being can create without wishing to share his creation. It is the only way he can assuage the anxiety of separateness and assure himself that he belongs to the group. He does not create in order to communicate, but once having created, he desires to share this new aspect of himself-in-relationship-to-his-environment with others. (p. 356)

Creativity is a function of interaction of knowledge and imagination (Parness, 1972). "Without knowledge, there can obviously be no creativity" (p. 194). What this statement means is that creativity is the innovative, novelty-creating use of knowledge. The author uses the kaleidoscope as a way of analogy. The more pieces we have in it the more patterns will show. The more knowledge we use in creativity the more novel ideas we can produce. But in the kaleidoscope, without revolving the drum, we merely have lots of pieces with no patterns emerging. Without working with, juxtaposing, combining, and synthesizing the available knowledge, no creative ideas will emerge. Therefore, without knowledge, and without its dynamic use in the course of the creative act, our effort will not be productive.

5.6.3.2. The Process of Creativity

The creative process is at the very core of transformation (Markely, 1976). Transformation is a process of metamorphosis, moving from what exists toward a desired novel state. This move is fueled by the creative act. The creative act involves preparation, incubation, illumination, and verification (Wallace, 1926).

5.6.3.2a. Preparation. After trying to use prevailing conventional approaches in addressing issues of interest and recognizing that they do not work, we become aware that we have to set aside our existing assumptions, which seem to block rather than help our inquiry. (This is a confirmation of what Einstein said: "We cannot solve a problem from the same consciousness that created it.")

5.6.3.2b. Incubation. We open up ourselves for new ways of thinking, new perceptions, and new assumptions. The term "incubation" "suggests the cessation of deliberate attempts to force insights" (Markley, 1976, p. 220). "Cease striving: then there will be self-transformation" (Chuang Tzu, Book XI).

5.6.3.2c. Illumination. The moment of creative insight occurs with vivid clarity as a period of realization of new relationships with new creative ideas that emerge following the incubation period. It enables us to see the issue of our interest in a completely new way. Such a moment of insight is common to creative discovery as well as to transformation.

5.6.3.2d. Verification. Verification is the process that validates the product or the discovery of creativity and brings it to fruition in its environmental context. We have to anticipate that the novelty or discovery we created will upset established ways and patterns. It will face the resistance of established authority.

Henri Poincaré provided powerful insights into the way nonlinear chaos operates inside the creative mind (Briggs and Peat, 1990). He showed "that in our creative activity, the ancient tension between chaos and order is forever renewed" (p. 191). In the chaotic period, ideas rise in crowds and collide until pairs interlock and create a new stable combination: order emerges from chaos. But in Poincaré's case, when he further pursued this first breakthrough insight of interlocked ideas, a new scale of chaos emerged and from its confusion sprang another conversation that led to another perception of a new order.

Koestler (1964) used the term "bisociation" to refer to the kind of flashes that Poincaré called "order out of chaos." By bisociation he meant the conjunction and integration of two distinct frames of reference or "matrices," which he thought to be the central processes of creativity. Wrestling with a problem, the mind usually keeps to the habitual patterns and a single frame of thought, which delimits thinking. "The term 'bisociation' is meant to point to the independent, autonomous character of the matrices which are brought into context in the creative act, whereas associative thought operates among members of a single preexisting matrix" (p. 656). The solution does not lie in the same frame of reference as the problem, as Einstein also said. Koestler says the solution is not found in the familiar context of previous solutions to related problems. The creator's frustration mounts and the search for a solution becomes increasingly more erratic. Limit cycles break down and produce a far-from-equilibrium flux. At a critical point, in this bubbling of thoughts, a bifurcation is reached where a small piece of information becomes amplified, causing the emergence of a new frame of reference, and within it, we create new insights and new order.

5.6.4. Inner Conditions of Creativity

The conditions reviewed here are considered to be the conditions of the design act. Rogers (1961) proposes a set of inner conditions that he associated with the creative act. One of these is openness of experience as opposed to defensiveness, in which creation experiences are prevented from coming into awareness. If we are open we are "alive to many experiences which fall outside the usual categories" (p. 525). It means lack of rigidity and the permeability of boundaries in concepts, beliefs, and perceptions. We are able to receive conflicting information, we have an "extensional orientation." Complete openness is an essential condition of constructive creativity.

The internal locus of evaluation is another fundamental condition. The value of what is created is established not by outsiders but by the one who creates. "Have I created something satisfying to me? Does it express part of me? These are the only questions that really matter to the creative person" (p. 354). If we create something that has not existed before that is satisfying to us, no outside evaluation can change that fundamental fact.

The ability to play with elements and concepts is the third condition of creativity. This condition is associated with openness and lack of rigidity. It means the ability to play spontaneously with ideas and relationships, make impossible juxtapositions, make the given problematic, translate one form to another, and transform into improbable equivalents. "It is from this spontaneous toying and exploration that there arises the hunch, the creative seeing of life in a new significant way" (p. 355). It is from such a seemingly wasteful spawning of a host of possibilities that eventually will emerge one or two creative forms or processes with unique and novel qualities that give them true value.

5.6.5. Conditions of Fostering Creativity

By the very nature of the inner conditions described above, it should be clear to us that those conditions cannot be forced. To highlight this, Rogers (1961) uses the analogy of the farmer who cultivates the soil, plants the seed, and supplies the nurturing conditions that enable the seed to develop its potentialities. He proposes two categories of external conditions that might foster and nurture creativity: safety and freedom. The psychological fostering/nurturing is to be offered by those who are in the social environment of the person (e.g., in a designing community these are the individual's fellow members of the design teams).

5.6.5.1. Psychological Safety

Psychological safety has three associated processes:

1. Unconditional acceptance. Acceptance conveys the feeling that individuals have worth in their own right. They can be whatever they are without sham or facade. They can actualize themselves in new and spontaneous ways. They are, in other words, moving toward creativity.

2. The absense of external evaluation. When we cease to form judgments of other individuals from our own locus of evaluation, we are fostering creativity. For individuals to find themselves in an atmosphere where they are is not being evaluated, not being measured by some external standard, is enormously freeing. Evaluation is always a threat and always creates a need for defensiveness.

3. "I accept you empathetically, see you and what you are feeling and doing from your point of you. . . . This is safety indeed" (Rogers, 1961, p. 357). In this climate, individuals feel free to express themselves in varied and novel ways. We feel free to create.

5.6.5.2. Psychological Freedom

Psychological freedom fosters the individual's freedom or symbolic expression and creativity. It nurtures freedom to think and to feel (Rogers, 1961). "It fosters the openness, and the playful and spontaneous juggling of perceptions, concepts, and meaning, which are part of creativity" (p. 358). Permission to be free also means that one is responsible. This type of freedom, coupled with responsibility, "fosters the development of a secure locus of evaluation within oneself, and hence tends to bring about the inner conditions of constructive creativity" (p. 359).

Now that we have explored some of the internal and external conditions of creativity, we should also examine conditions and forces that hinder or block creativity as well as ways that we might remove those blocks.

5.6.6. *Conditions and Forces that Block Creativity*

In *Science, Order, and Creativity,* Bohm and Peat (1987) identify a range of conditions that block creativity. One is the common tendency toward the unconscious defense of ideas "which are assumed to be necessary to the mind's habitual state of comfortable equilibrium" (p. 50). There is a tendency to impose and cling to familiar ideas, even when there is evidence that they may be false. This then creates the illusion that no fundamental change is needed. To cling to familiar ideas maintains a habitual sense of security and comfort and blocks the mind from engaging in creative play and the release of vibrant tension and passionate energy that is needed to free the mind from rigidity and engage it in the creative act. In taking familiar ideas and concepts for granted, the mind defends itself against the disturbance of novelty and what is different and maintains its "fixed position in situations that call for fundamental change" (p. 53).

Creativity is a natural and powerful potential of human beings. It is clearly manifested in the playful and imaginative actions of young children before they start school. But as they enter school, they are forced to accept "a single right answer," and they are told what they are supposed to do. This is indicated to them by praise and disapproval, and by expectation of conformity "to what other children around them are doing" (p. 231). Bohm and Peat explored in depth research findings that suggest that creativity is incompatible with external rewards and punishments. "The reason is clear. In order to do something for reward, the whole order of activity, and the energy required for it, are determined by arbitrary requirements that are extraneous to the creative activity itself. This activity then turns into something mechanical and repetitious" (p. 231). As a result, the intense passion and vibrant tension that goes with creativity dies away. In essence, the reward of creativity cannot be anything else but the creative act. When creativity is subservient to external rewards, the whole act withers and degenerates.

5.6.7. Misconceptions about Creativity

Various misconceptions also hinder and block creativity. One is the widely touted assumption that creativity is necessary only in specialized fields (Bohm and Peat, 1987), e.g., art, literature, music, etc. This restricted assumption about "the nature of creativity is obviously of serious consequence for it clearly predetermines any program that is designed to clear up the misinformation within society" (p. 239). This in fact suggests that ordinary people, groups or organizations collectively, and society in general, cannot be creative. A related misconception about creativity is two widely shared personal beliefs. One is the belief that ordinary people just don't have the talent, the necessary passion, and the courage to act in a truly creative way. Creativity is the privilege of the genius. The second commonly held belief is that creativity is "marked for emergency only" (Harman, 1988). In much of the literature on creativity the assertion is made that "one must strain to try to solve the problem with the conscious mind first, and absorb a great deal of information about it before the behind-the-scene creative mind should go to work" (p. 79). One should struggle to find solutions with the rational, analytical mind, and turn to creativity only in desperation. Otherwise, we are removing the ego-mind from its position as a gatekeeper, which, of course, will threaten its domination.

The preceding statements about blocks and hindrances of creativity, and misconceptions about them, imply the remedies that we should take to remove those blocks and correct the misconceptions. A further mitigation is the promotion of the internal and external conditions of creativity, discussed earlier.

Activity #40

The tasks in this activity invite the activation of your imagination and creativity. (1) Review each and every paragraph in the text and search for and identify "core ideas" of creativity. (2) Ask yourself: What is the meaning of the identified core ideas in terms of my own understanding? What do they mean to me? (A key condition of working with this book is that you are challenged to construct your own meaning of the various design ideas.) (3) Transform the various core ideas about creativity that you noted into a functional context of your choice. Here you ask the question: What are the implications of a few selected core ideas for the design of my systems of interest? Answering these questions will enable you to synthesize core ideas into sets of organizing perspectives that can guide your thinking and actions in using the power of creativity in carrying out systems design. (4) Think about and formulate means and methods that you would use to remove blocks and hindrances of creativity and overcome misconceptions about it. Enter your findings in your workbook.

5.6.8. The Significance and Power of Creativity

In reflecting on the significance and power of creativity, I yield to scholars who have expressed their views with singular power and conviction on the crucial role and function of creativity in society. Arnold Toynbee (1964) describes the value of creativity to a society. He declare that to give a fair chance to potential creativity is a matter of life and death for any society. Creative ability is mankind's ultimate asset. He warns that potential creative ability can be stifled, stunted, and stultified by the prevelance in the society of adverse attitudes of minds and habits of behavior. Therefore, society has a moral duty to ensure that the individual's potential ability is given free play.

In the same vein, Harman (1988) says that we have to learn to harvest creativity and enhance it in our culture. We can increase the number of creative persons about fourfold and put ourselves beyond the point of critical mass. When this level is reached, as it was in Periclean Athens, there is an escalation of creativity, resulting in a great leap forward. We can have a golden age such as the world has never seen. Creativity harvest would enable us individually and collectively to achieve fundamental insights. Profound inspiration will become a meaningful dimension of our lives. The power of deep intuition and creativity will be an accessible source for us individually and collectively, and each of us will attain the capacity to become much more than we think we can be.

Jantsch (1980) celebrated the role of creativity in societal evolution. He suggested that creativity enables self-transcendence, the reaching out beyond the boundaries of our existence. "When a system, in its self-organization, reaches beyond the boundaries of its identity, it becomes creative" (p. 183). Evolution is the result of self-transcendence. It is the creative reaching-out into the future.

5.6.9. Implications for Systems Design

I open the role of creativity in design by reflecting on Bohm's core idea (Bohm and Peat, 1987, p. 207). "The challenge that faces humanity is unique, for it has never occurred before. Clearly a new kind of creative surge is needed to meet it. This has to include not just a new way of thinking but a new approach to society, and even more, a new kind of consciousness." This new type of consciousness emerges from the dynamic interaction of self-reflection and creating consciousness. It is from such interaction that a new surge of creativity emerges, manifested in "designing consciousness," that can fuel the development of design culture in the society. This will lead to a new approach to society and to the creation and re-creation of social systems.

Cognitive mapping (Banathy, 1993a) is a process by which individuals, groups, social systems, and societies make individual and collective representations about their perceptions of the world and their understanding of their place in the world. These representations are implicit (but observable by our behavior) and can be made explicit in a variety of forms of mediation and expression. We draw cognitive maps on the basis of the values we hold and the ideas we have about how the world works. Cognitive maps are alive; they are created, confirmed, disconfirmed, elaborated, changed, and redrawn. They dynamically affect each other and the environment they represent. World views of the mapmakers affect their environment, while the world around them affects them and the maps they draw. This mutually affecting dynamic is constantly ongoing; it represents coevolution.

This creating, confirming, changing process is an interacting function of self-reflection and creating consciousness. This function produces two kinds of cognitive maps. One is the primary creation of self-reflection, which contemplates the "here and now" and "maps" what is. The other is the primary product of creating consciousness, as we transcend the here and now and aspire to change our lives and our systems by creating new substance and form. Its product is a representation or mapping of what should be. Self-reflection without contemplating creation is dormant and barren. On the other hand, evoking creating consciousness without reflecting on its value and meaning is idle speculation.

The act of purposeful design springs forth from creating consciousness. The act leads to the creation of an image, and, based on it, to the cognitive/prescriptive mapping of the desired future. This creation is based on the belief that, although the future is influenced by the past and present, it is not determined by them. The future remains open to conscious and purposeful creation, accomplished by systems design. This future-creating process and its outcome are constantly benefited from the process of reflection that looks at what is being created, reflects on it, and makes judgments about it. This reflection then becomes the springboard for further creations. This interplay harmonizes creation and reflection in systems design.

In the context of specific social systems, this future-creating process is manifested in the evolution of our systems, guided by purposeful design. The genesis of this creative process emerges when a new ww stage in societal evolution is recognized. At the midpoint of this century, with the emergence of the postindustrial information/knowledge age, continued use of the old cognitive map, grounded in the by-gone industrial/machine age, become increasingly more problematic. The old map cannot be used any longer to guide our actions. This recognition may create a great deal of anxiety and uncertainty. However, if this "negative" energy is channeled in a positive direction, and if it is guided by creating consciousness, it leads us to reconstruct our map or create a new cognitive map by design and to use that map to guide the transformation of our system. This systems creation and transformation involves (1) the sighting of a vision of a desired future, (2) the elaboration of the vision and the creation of an ideal image of the future system, and; (3) based on the image, we design a prescriptive/cognitive map or a conceptual model of the future system. This new cognitive map will exert a "magnetic pull," will bring about harmony with the new realities of the current era, and will guide our social systems into the twenty-first century.

5.6.10. Creativity in the Design Literature

It is generally recognized that creativity is a central force in systems design. Design is about creating novelty. It is about envisioning, imaging, and bringing forth something that does not yet exist. It seems, however, that in the design literature this essential role of creativity is taken for granted. One would expect more substantial statements about creativity (in design) than what one can find. This sparsity prompted me to devote a good deal of attention to creativity in this work by learning about creativity as a human experience. I have found attending to this issue most rewarding. It surely has given me new understanding, new insights, and new perspectives on the role of creativity in design. To summarize this metaphorically: "I believe that creativity is the soul of design."

Ackoff (1981) assigns several functions to creativity.

1. He suggests that we can treat a problem situation in three ways. We can (1) resolve it by producing a satisfactory outcome, which requires experience and judgment; (2) solve it by producing an optimum outcome, which requires experiment and science; (3) dissolve and remove it by redesigning the system—"dissolution requires experience and experiment, judgement and science, and, particularly, creativity" (p. 193).
2. Idealized design releases creativity as it removes many of the constraints that inhibit it. It "tends to liberate the imagination and stimulate the desire to innovate and invent" (p. 121).
3. The design process itself can be a rich and satisfying aesthetic experience. "It gives free reign to creative imagination of those who participate in it, and, because it is fun, it also has great recreational value" (p. 118).
4. Ackoff assigns special role to creativity in finding means by which to close the gap between the reference scenario and the idealized design. (As discussed earlier, reference scenario stipulates what would happen

to us if we did not change our system.) The search for those means is an inventive effort. Its success "depends greatly on how creatively it is carried out" (p. 173).

The contributors of the design methodology compendium of Cross make references to the role of creativity in design. Jones (1984) suggests "the method is primarily a means of resolving the conflict that exists between logical analysis and creative thought" (p. 10). The difficulty is that creativity does not work well unless it can freely roam among all aspects, in any order, and at any time, while logical analysis is a step-by-step sequence. The two kinds of thought should coexist and proceed together if any progress is to be made. (This reminds me of the idea of harmonizing reflection and creation in design, discussed above.) Jones argues that the expression of creativity in design needs a free atmosphere in which any idea can be posited at any time without regard to practicality. An uninhibited expression of ideas nurtures the releasing of creativity and the obtaining of large number of ideas from many areas of experience. (This reflects Churchman's notion of sweeping in. Archer (1984) suggests that arriving at solutions by strict calculation and from the interaction of data is noncreative; it is nondesign. On the other hand, "It is characteristic of creative solutions that they are seen to be apt solutions—but after completion and not before" (p. 58).

Nadler and Hibino (1990) suggest that an open mind-set that (1) nurtures reasoning at a higher level of abstraction than the limited perspectives of conventional reasoning and (2) fosters maximum creativity at the individual, group, and organizational levels is the most critical factor in social systems design. "Creative ideas are born in the human brain when two thoughts, two models, or two abstractions intersect. The various creative, purposeful, alternative solutions that emerge are potential components of the ideal target solutions" (pp. 148–149). One of the great advantages of the group process in developing innovative solutions lies in the potential intersection and expansion of the individual's creative ideas. We can maximize the development of creative solutions by setting aside all constraints that limit our vision. A creative environment prevails if people move cooperatively toward the ideal vision of the future. Such long-range perspective makes people more tolerant of ambiguity, "more willing to consider every idea, acknowledging the possibility that any idea may have some merit. These attitudes maximize the likelihood of developing creative, innovative solutions" (p. 147).

Flood and Jackson (1991) and Jackson (1992) developed the methodology of total systems intervention (TSI), which uses a range of systems metaphors to encourage creative thinking about finding solutions. The three phases of TSI are creativity, choice, and implementation. The task during the creativity phase is to use systems metaphors to help stakeholders think creatively about their enterprise. For example, what metaphors might better capture what we want to achieve? The tools provided by TSI assist in the use of systems metaphors, which focus attention on different aspects of organizational behavior and functioning. Examples of systems metaphor are the organization as (1) a machine, (2) an organism, (3) a brain, (4) a culture, etc. "What is expected to emerge from the creative use of metaphor is the selection of a 'dominant' metaphor that highlights the main interest and concerns and can become the basis for a choice of appropriate intervention methodology" (Jackson, 1992, p. 273).

5.6.11.   Exploring the Role of Creativity in the Course of Design

Banathy (1991a, 1994) projects a major role of creativity in the design of social systems. The three steps of transcending the existing state by envisioning the ideal future, creating the ideal image, and transforming by design integrate design strategies with creativity. The infusion of creativity in these strategies is accomplished by discovery, imaging, and bisociation. These systems-creating acts are aided by the use of metaphors, stories, imaging, and visualizations.

5.6.11.1.   Envisioning the Desired Future

Envisioning the desired future provides the main motivating force that reduces the anxiety and uncertainty associated with transcending and leaping out from the existing system. Vision is defined (Webster, 1979) as the act or power or seeing, the act or power of imagination, a revelation, and an unusual discernment of foresight. All these are appropriate definitions and markers of envisioning a desired future we wish to create. In design we envision a grand idea, rooted in its underlying value system, that creates the excitement and inspiration we need to transcend the existing state and leave the "here and now" behind.

The vision of the designers can take a variety of forms. It can be presented as a metaphor, an event foreseen, a visual representation, or a description of our aspiration. The Greeks had a grand vision of "padeia," the learning society in which learning, fulfillment, and becoming truly human were the desired qualities. In education we can contrast two metaphoric visions. A metaphor for the old vision is a visual description of an assembly line using outdated machinery, producing more rejects than good products. This visual metaphor could stand for the current state of education. The metaphor for a vision of future systems of learning and human development could be a picture of a creative artist's studio in which everyone is excited by learning to create and express their own uniqueness. Another contrast of an old and new vision of education can be expressed by two phrases: "learning to make a living" versus "learning to make a life." A vision-quest will engage imagination and creativity and generate excitement and inspiration in the designing community as the community collectively formulates grand ideas that represent their aspirations as they begin to shape the future of their system.

5.6.11.2.   Creating the Image of the Future System

A statement of a vision of a desired future is a brief but powerful idea or a set of ideas. The vision guides the formulation of an image that contains a set of key markers of the desired future system. The image not only makes society, society continuously remakes the image, says Boulding (1956). Boulding's first proposition is that behavior depends on the image. His second proposition is that our experiences provide us with messages that produce changes in the image. There are messages that confirm the image. Others call for adjustments in it or call for clarifications. However, when a message hits the nucleus of the image "the whole thing changes in a quite radical way" (p. 8). Such a message "overthrows the previous image and we revise it completely" (p. 9). The recently emerged images of the postindustrial information/knowledge age call for the radical revision and reconceptualization of the images of our social systems. Based on the new images, we are to redesign and transform our social systems. The basic bond of any society, culture, or organization is a public image, that is, "an image the essential characteristics of which are shared by the individuals participating in the group" (p. 64).

In designing social systems, it is the larger public image that designers seek to capture and understand so that the characteristics of that larger image can be reflected in and faithfully interpreted and recreated (at an appropriate scale) as the public image of the system they design. The kind of creative effort projected here will ensure that the system we design will not only be internally consistent with its embedding larger system but that it will be "on the same wavelength" with it and, thus, will be able to coevolve with it.

5.6.11.3.   Creating the Design of the Future System

It is this "transformation by design" strategy that invites the most intensive creative effort. In the course of this process the designing community engages in the transformation of the image into a detailed and comprehensive prescriptive representation of the future system. Saying this metaphorically, the image is the "seed" of the future system; it is the "nucleus" around which to build the system; it is the DNA of the system that is to be faithfully represented in all domains and aspects of the new system. An understanding of the use and the combination (juxtapositions) of the "planes" of these metaphors with the "planes" of the various realms of the design strategy represent the kind of fertile and creative "bisociation" described by Koestler (1964).

The various realms of the design strategy are the formulation of the core definition, the purposes and specifications of the system, the creation of the functions that respond to the purposes and specifications, the design of the organization, and the synthesis of all the above into systems models. All these call for the creation of a large number of alternatives from which to select the most desirable solutions that are also most compatible with the ideal image. The overall context of this strategy comprises two components: extensive (prodigious) divergence of creating large numbers of alternatives in all the various realms and, then, focused convergence. Divergence calls for a great deal of imagination and intense creativity. The selection from alternatives through the convergence process calls for the creative (e)valuation of alternatives from a variety of perspectives (discussed earlier), as well as the creative combination and synthesis of often incompatible alternatives.

Throughout this process creativity is evoked by a variety of means. For example, we might use a "what if" story: "What happens if we commit ourselves to this or that purpose? We might develop scenarios that extrapolate the consequences of selecting specific function alternatives. We might use various "lenses" that project the relevance of different perspectives (e.g., organizational, cultural, personal, technical) in the context of a proposed design solution. We apply bisociations by combining metaphors with potential alternatives.

An example that highlights the potential application of bisociative/metaphoric use of creativity is offered by Morgan's (1986) organizational images. Using these images as metaphors, we can evoke powerful new ways of seeing and understanding organizations. Morgan's metaphors describe alternative ways of looking at organizational types, such as the organization as a machine (e.g., bureaucracies), as an organism, as a brain, as a culture, as a hologram. The same metaphoric approach can be used to ask the question: What type of social system do we wish to design? The interpretation of the implications and the meaning of the machine, organism, brain, culture, or hologram metaphors to various potential alternative organizational characteristics would provide an information-rich context in which design choices can be made.

In representing the future system we synthesize the design choices made in the form of system models of the future system. Banathy (1992a) proposed the use of three "lenses" in modeling social systems. The three lenses produce three complementary images. One lens projects a bird's-eye view of the system (as it is embedded in its larger environment), called the systems-environment model. The second lens produces a still picture, or snapshot image, called the functions/structure model, of the system. The third projects a motion picture image of the system, called the process/behavioral model of the system. The significance of this three-dimensional modeling is that it calls for the collapsing of the three images, such as using different lenses in a telescope—as Galileo did—in order to get a clear image, a comprehensive representation of the system.

Reflections

Design creates novelty. Thus, creativity is central in design. Novelty cannot be produced by analysis of what is known or by associating aspects within the same frame of reference. The literature of social/organizational design does not question the role of creativity in design, but very little of the literature does dwell on the significance of this role. It seems that design scholars just take this role for granted and focus on the technical aspects of approaches and methods. The sources used in this section begin to hint at the potential powerful role and contribution that creativity might offer. However, it comes through that whatever has been written about the function of creativity and its function in the design of social system, it is far from being an adequate explanation of creativity in design. We must devote far more attention to creativity if we want to advance the power of social systems design. My intention here was to contribute to such advancement.

Activity #41

First identify and note the core ideas you find in this section that stand for creativity in systems design. Next, select from the various previous activities a set in which you completed design activities. (A possible set might be activities 26, 27, 28, and 29 in Chapter 4 and/or Activity 37 in this chapter.) With such an activity at hand, and in view of the core ideas you identified, work out an illustrative example of how you might use creativity in the design of the system you selected. Enter your findings in your workbook.

5.7. Design as Conversation

The focus of this chapter is on developing new insights that enrich our understanding of what social systems design is and how it works. In the previous sections we have explored the nature of design thinking in the context of systems thinking, brought into the picture multiple perspectives that give increased content viability to design, gained an appreciation of the ethics of design, broadened our view of the ideal systems approach to design, discussed the relationship between design and creativity, and explored the role of creativity in design. We weaved these topics into a tapestry and by so doing we illuminated new color schemes and created a richer and more comprehensive image of social systems design. In this last section we add another color to our tapestry: the color of conversation, which blends seamlessly into creativity.

Social systems design is a process that carries a stream of shared meaning by a free flow of discourse among the stakeholders who seek to create a new system. In order to understand the critical nature of this communication function, the various modes of social discourse are explored to search for the mode that is the most appropriate to systems design.

In recent literature the notion of "dialogue" has gained prominence as the most viable form of collective social discourse. An exploration of knowledge based on the use of dialogue indicates that it indeed does have the power to generate collective meaning and collective consciousness, the attainment of which is critical in design. However, as defined in the current social discourse literature, dialogue is not aimed to pursue a specific task, such as the design of a new system. It aims to create a common frame of reference, a shared worldview among the parties of the dialogue. Thus, for use in the design inquiry context, an extended form of dialogue is proposed here. Designated as "conversation," it combines two specific modes of dialogue, namely, "generative" and "strategic," as the most appropriate modes of social discourse in design inquiry. In closing, an example of the application of conversation will demonstrate design conversation.

In order to tie in creativity with conversation in design, I turn to Bohm and Peat (1987), who proposed that a free flow of ideas, beliefs, and meaning among members of groups

> may well be the most effective way of investigating the crisis which faces society, and indeed the whole of human nature and consciousness today. . . . Such a form of free exchange of ideas and information is of fundamental relevance for transforming culture and freeing it of destructive misinformation, so that creativity can be liberated. (p. 240)

5.7.1. Dialogue as Social Discourse

Following a brief review of historical roots and background, the concept of dialogue is defined and explored, and the culture-creating role of dialogue is described.

5.7.1.1. Historical Roots

Zeldin (1994) traced back in history the invention of dialogue and suggested that Socrates was the first known conversationalist, who replaced the war of worlds by dialogue and introduced the idea that "individuals could not be intelligent on their own, that they needed someone else to stimulate them" (p. 33). He proposed that if two unsure people engaged in discourse, they could achieve what they could not do separately.

> By questioning each other and examining their prejudices, dividing each one of those into many parts, finding the flaws, never attacking or insulting, but always seeking what they could agree between, moving in small steps from one agreement to another, they would gradually learn what the purpose of life was. . . . He [Socrates] argued that it was inadequate to simply repeat what others said, or borrow ideas. One has to work them out for oneself. (p. 34)

Socrates is the father of creative social discourse and the most advanced theory of learning.

5.7.1.2.   Background

In design scholarship and practice, we have evidence of the designation and use of various group methodologies. These include "conversation" that explores the ideas and values of designers that are relevant to the subject of design, and "consensus-building" tools that seek to establish collective judgment about solution alternatives. Conversation at times is noted as "strategic dialogue," which implies communication among designers that focuses on specific tasks of seeking solutions.

In the course of the last several years we have seen a surge of attention devoted to another mode of social discourse, called "dialogue" or more recently "generative dialogue." This mode is applied to generate a common frame of thinking, shared meaning, and a collective worldview in a group. The primary architect of generative dialogue is the physicist and systems philosopher David Bohm. He has devoted much attention to exploring the idea and use of dialogue as social discourse. His theoretical and philosophical perspectives (1983, 1985, 1990; Bohm and Peat, 1987; Bohm and Edwards, 1991) have been the main sources of discussions and interpretations in the organizational/social discourse scholarship community. Several of these interpreters developed Bohm's ideas in organizational settings, including Senge (1990), Isaacs (1992), Beck (1994), and Schein (1994). Bohm and his followers consider dialogue to be the most enabling form of free discourse in organizational/social/cultural settings.

Senge (1990) suggests that Bohm has developed a theory and method of "dialogue" as a vehicle by which a group of people becomes open to a flow of collective intelligence, suggesting that collective learning through dialogue is vital to realizing the potential of human intelligence. Bohm's work, Senge says, presents a synthesis of the two major intellectual currents: the systemic/holistic view and the interaction between our internal/mental models and our perceptions and actions.

5.7.1.3.   Definition and Exploration of Dialogue

Bohm and Peat (1987) make a sharp distinction between dialogue and ordinary discussion. In discussions people hold relatively fixed positions and argue their views in trying to convince each other. At best, Bohm says, this form of discourse may produce some agreement or compromise, "but it does not give rise to anything creative" (p. 241). The word "discussion" has the same root as "concussion" and "percussion." It is a process of shaking apart and hitting. Bohm uses the metaphor of a game of Ping-Pong, in which we pass the ball back and forth with the sole purpose of winning the game. Furthermore, in the course of a discussion when something of fundamental importance is involved, positions often tend to be nonnegotiable and confrontational. This leads to either a situation where there is no solution or a polite avoidance of the issue.

The term dialogue is derived from the Greek *dia,* meaning "through," and *logos,* standing for the "meaning of a word." So, Bohm considers the meaning of dialogue to be a free flow of meaning between people in a communication situation. In dialogue, people may prefer a certain position, but they are willing to suspend it; they are willing to listen to others in order to understand the meaning of their position. They are ready to change their point of view and blend it with others. In dialogue, people are able to face disagreement without confrontation and are willing to explore points of view to which they do not subscribe personally. "They will find that no fixed position is so important that it is worth holding at the expense of destroying the dialogue itself" (p. 242).

Pattakos (1995) adds a deeper meaning to generative dialogue. He also suggests that the term comes from the Greek "dialogos," made up of the root words dia ("through") and logos ("the meaning"). Pattakos, addressing the word "logos," says that various interpretations of it reveal that it has deep spiritual roots. "Interpreting logos this way, that is viewing it as a manifestation of spirit or soul, carries with it significant implications, both conceptual and practical. Dialogue, as a concept, takes on a new and deeper meaning when it is perceived as a group's accessing a larger pool of common spirit through a distinctively spiritual connection between the members. This suggests more than 'collective thinking' although dialogue certainly is a determinant of such a holistic process. Spirit flowing through the participants in dialogue leads to collective thinking which, in turn, facilitates common understanding thereby resulting in 'common education' or to use today's jargon, 'collective learning'" (p. 322–23).

5.7.1.4. The Process

The kind of social discourse described here as dialogue (Bohm and Edwards, 1991) leads to an exploration of shared meanings. It opens up into honesty and clarity. Everything that happens in the course of dialogue "is 'grist for the mill' and serves as an opportunity for learning how thoughts and feelings are weaved together, both collectively and individually" (p. 186). If members of the group are able to hold all their assumptions in suspension, they can generate shared consciousness. (The root meaning of "consciousness" is "knowing it all together.") In a dialogue the individual's and the group's "knowing it all together" form a subtle higher unity and come together in a harmonious way. In the dialogue event people are able to be honest and straight with each other, they level with each other, and they share content freely. They develop a common mind, a shared mind, that can think together in a new and creative way. They awaken their collective intelligence and feelings of genuine participation, mutual trust, fellowship, and friendship. They can think and talk together. Shared meaning and understanding flow freely in the group. However, they can do none of this if there is hierarchy or authority represented in the group.

5.7.1.5. Dialogue Is Culture Creating

The organizing principle of dialogue implies a change of how the mind works. In true dialogue a new form of consensual mind emerges, generating a rich, creative order between the individual and the community as a more powerful force than the individual mind is alone. This creative order "arises from a spirit of friendship dedicated to clarity and the ultimate perception of what is true" (Bohm and Peat, 1987, p. 247). People who learn the potential power of such a dialogue will be able to transfer the spirit of dialogue into their activities and social relationships and into the systems and communities in which they live. Dialogue, therefore, may create a new culture in the dialogue community and furthermore "members of the community can explore the possibility of extending the transformation of the mind into a broader sociocultural context" (p. 247).

Considering the contribution of dialogue to organizational learning and culture building, Schein (1994) suggests that dialogue is a vehicle for understanding cultures, and subcultures in organizations and organizational learning depend upon such cultural understanding. It facilitates the development of a common language and collective mental models. Thus, the ability to engage in dialogue becomes one of the most fundamental and most needed human capabilities. Dialogue becomes a central component of any model of organizational transformation.

5.7.1.6. Dimensions of Dialogue

Isaacs (1993) proposes that the development of the act of dialogue is composed of the following: collective learning and coordinated action, paradigm exploration, cultural healing, and collective creation.

1. Dialogue is a disciplined inquiry of collective learning and action. Most learning is individually based and moves from parts to whole. Dialogue presumes that the whole organizes the parts. Isaacs (1993) believes its premise is "that there is an underlying implicate wholeness that can be made explicate" (p. 8). The level of learning that can take place in a collective setting cannot ever be mastered individually. In dialogue the group creates a pool of common meaning and new levels of coordinated action.
2. Paradigm exploration in dialogue enables people to step back from the context of specific problems, reflect upon what lies beneath them, and learn a new way of seeing and attaining a new kind of consciousness. The thought that created the problem cannot be the thought we use to solve it. We have to shift our way of thinking. (As Einstein said, "we cannot address a problem from the same consciousness that created it.")
3. Dialogue enables us to create a bridge between diverse cultural differ-

ences, reach back into our shared cultural background, and create a common flow of meaning.

4. Dialogue fosters the power of collective creation. As we suspend our assumptions and begin to listen to each other in a deep way, new creative insights and new levels of wisdom emerge. We not only transform existing patterns of thought but transmute them and create new levels of consciousness.

5.7.2. Generative and Strategic Dialogue

In the dialogue scholarship community there is a shared understanding that dialogue is not a new tool for addressing specific issues or problems. Rather, it is a means to help people to think together. Dialogue offers an environment where people can create shared meaning. The type of dialogue discussed heretofore is often called "generative," meaning that it generates a collective worldview. Beck (1994) contrasts generative dialogue with strategic dialogue. According to Beck, strategic dialogue focuses on specific issues and tasks and it is applied in finding specific solutions in organizational and social systems settings. The dialogue scholarship community has not substantively addressed strategic dialogue. Still, Schein suggests (1993) that the test of the importance of generative dialogue will be "whether or not difficult, conflict-ridden problems can be handled better in groups that have learned to function in a generative dialogue mode. Because severe conflicts are the result of cultural or subcultural differences, I would assume that initial (generative) dialogue will be always necessary" (p. 4). Clearly, Schein describes here design problem situations.

5.7.3. Design Conversation

In what follows, the two types of dialogue are connected into a communication mode, most appropriate in pursuing the disciplined inquiry of social systems design in groups. Then, an ongoing experience with conversation is described, including a report on a conversation aimed at designing conversations.

5.7.3.1. Design Conversation = Generative + Strategic Dialogue

The statements of Beck and Schein create the opening needed to connect the discussion on dialogue with the communication mode of systems design. We can now enter the realm of social systems design and explore the method of discourse or the type of communication that is most appropriate to apply in social systems settings. It is proposed that the combination of generative dialogue and strategic dialogue composes a comprehensive method of social communication that is the most viable to use in a designing community. We call this method *design conver-*

sation. The root meaning of "conversation" is "to turn to one another." Members of a group turn to one another without reserve and in truth and openness, excepting and honoring each other. Shein (1933) aptly characterized the conflict-laden nature of social systems. For this reason it is important that before the design group engages in the substantive task of design, it involves itself in generative dialogue. This involvement will lead to the creation of collective consciousness, collective inquiry that focuses on the thoughts, values, and worldviews of the group and creates a flow of shared meaning, shared perceptions, a shared worldview, and a social milieu of friendship and fellowship. Generative dialogue becomes the core process of transforming the group into a designing community. Once the group feels that it has reached the stage where it has created a collective cognitive map for itself, generated a shared worldview, and attained shared consciousness, then, and only then, should the group turn to the tasks of systems design by engaging in a strategic dialogue.

5.7.3.2.   The Evolution of a Conversation Program

In the late seventies, a group of us in the international systems science community became increasingly disillusioned with the practice of traditional scientific meetings where papers are presented (often read) but rarely discussed in depth. But occasionally, creative conversations happened away from scheduled sessions. Whenever we had those rare occasions, we always had a high level of learning and satisfaction. So we decided "to make the rare the norm" in our meetings. We created an opportunity for the "rare" when we organized a one-week conversation in 1982 at Fuschl Lake in Austria. Arranged into groups, we asked the question: How can we use the insights gained from systems science for the improvement of the human condition? By the end of the week our conversation groups came up with some 80 action items, which we clustered and designated as an agenda for our international conversation program. By now we have had over 21 conversations in seven different countries. For quite a while, these conversations were the strategic dialogue type. During the last few years, however, we became aware of how an up-front, generative type of dialogue could enhance the potential of our conversations. In fact, in the course of the last three conversations some of our groups focused on the design of such generative and strategic conversations.

5.7.3.3.   Findings of a Design Conversation Group

The following presents some of the key ideas of the "Designing Conversation" group's report (Dieterle *et al.*, 1994) of the Fifth Annual International Conversation on Systems Design. As the group met, the essential ingredients for a conversation were already in place: (a) passion for the theme of experientially and intentionally exploring the attributes of the group's common experiences with conversations and an unstructured space for five days of intensive involvement. Out of this structureless group process an elemental idea seed leaped into existence. Rather than turning to task and beginning to design, the group felt that "we first have to become friends." Operating in a formless and open conversation zone, the group placed in the container of the conversation respect for each other's process and respect for the common experiences and meaning emerging within the group flow. They viewed conversation as collective learning, an inquiry into the assumptions that structure common experiences. "The practical pay-off was people participating in a shared community of meaning which leads them to aligned action and inspired performance" (p. 31).

Since conversations happen in an open space of evolving relationships, group facilitators temporarily emerge, but none are anointed as leaders. The process moves the group from an unstructured to a structured, task-oriented conversation (from generative to strategic dialogue). "Heartstorming" is freedom from the requirement to fit into the normal modes of scientific inquiry. The heart issue is related to the cultural perspective. Culture limits us to keep within paradigms of thinking and behaving as it defines the way we perceive. Heartstorming and honest and open exchanges enable the group to create collective cultural perspectives. This type of communication enhances the making of shared judgments and attaining wisdom. (I call this the "wisdom paradigm" of inquiry.) "A proposed elixir to access wisdom is to listen to the soul-voices each of us have within ourselves . . . the voices that we do not listen to often." (p. 31). But what unifies a diverse group of complexity of cultures, operating without the familiar hierarchy? It seems that the "scientific paradigm" of either/or, rational, closure, and order attributes is replaced by the "wisdom paradigm" of both/and, paradox, ambiguous, tentative, tolerance, and contingency attributes.

The group concluded that some other attributes of genuine communication include the following: the recognition of (1) the essential nature of authentic participation; (2) emotional commitment, which becomes the groups's source of energy toward change; (3) a stance of curiosity; and (4) conversation as a process "to be given into," that is, surrendered to. The evolving functions/structure allows the group to become self-renewing, as members pay attention to the unfolding process and engage authentically in it. The group seeks what is important and maintains compassion for conversation partners. There is no changing others, no right or wrong, just being witness to others' words.

The process described above is the process of creating the container of the conversation. A container can be understood as a system of assumptions, intentions, and beliefs shared by members of a group. These create a collective atmosphere or climate (Isaacs, 1993). Now that the container is formed the process to create collectively can start. This requires intentionality. The reaching of this state is required in order to initiate task-oriented conversation, or strategic dialogue.

During the last day of the conversation several issues were explored. Does conversation need a goal? Is the conversation journey directed to a specific result or is it an adventure that has as its goal the adventure itself? So what is the purpose of conversation? Does conversation aim at discovering attributes/manifestations of the group's shared view of the world? Or should it create a practical, action-oriented plan? Reflecting on these questions, a collective judgment was made that in a design-oriented conversation instead of an either/or choice, both aims should be chosen. The report ended by saying that "questions continued emerging, enriching the flow of meaning pooled in our evolving group container" (p. 33). Members of the group decided to keep the conversation flowing in an at-distance (e-mail) mode and to continue it on-site in the future.

In the long-range program of our international conversation community, the core ideas of the report introduced above mark true progress in our journey toward ever more meaningful, genuine, authentic, ethical, and sustainable conversation experiences. The work of this group—and all others in our ongoing program—demonstrates the power of conversation to tap into the collective intelligence of groups, create communities with shared meaning and a shared view of the world, and generate collective wisdom and capacity to engage in purposeful design.

Activity #42

(1) Review the text of this section and mark the core ideas that represent the intent and the process of conversation. Map the core ideas in a relational arrangement by asking: What experiences should the group have that would enable it to attain shared meaning, a feeling of a caring community, and a collective view of the world. Based on this exploration, write a "briefing paper" (and enter it in your workbook) that introduces to a novice group the meaning, the purpose and the process of conversations. (2) Use the briefing with a small group of your choice and have a conversation with them about the purpose and nature of conversations. Describe the experience in your workbook.

Reflections

Conversation as described in this section appears to be appropriate to use as a mode of communication in groups that engage in the design of social systems. In design groups there is an understandable tendency to "jump into" a design-task-focused discourse right away. However, such rushing into functional tasks does not allow us to explore the assumptions, beliefs, values, and implicit ideas that underlie the cognitive and affective beings of group members, even though these aspects are the bases for forming judgments and making design decisions. Unless we provide opportunity by design for surfacing and openly addressing these up-front issues in a generative dialogue, they will in some form bubble up later in the course of our strategic dialogue. At that time, their earlier neglect will cause the discourse to bog down and lead to disagreements. At that point, most likely we shall dismiss or suppress them as unwanted, as interfering with the carrying out design tasks. We shall wind up having debates and arguments. We might be able to coerce agreements, take votes, and we may even call the outcome a "shared decision." But such decision stands on very shaky ground and a shallow collective base. That which we acceded to, we have very little care for and we feel very little ownership or responsibility. Thus, we shall pay a high penalty for neglecting the use of the generative dialogue experience of our conversation program by jumping into a strategic dialogue.

6

Getting Ready for Design

As you worked with the first four chapters, you developed an understanding of what social systems design is, why it is important for us today to be able to engage in design, and how systems design works. In the last chapter we went beyond the technical and contextual dimensions of systems design and explored the various human-experiential dimensions that have relevance to design. Furthermore, working with the activities, you constructed your own meaning of design and applied what you learned in functional contexts of your interest.

The first three chapters provided the rich knowledge and experiential base that is needed to explore the issue of: How can a designing community get ready for design? This exploration invites the consideration of two component issues: the creation and empowerment of a community of designers and the development of capacity to engage in design. These two issues have their own component themes as described below.

The creation and empowerment of a community of designers is approached as follows: In Section 6.1 we ask: Who should be the designers? In Section 6.2 we address the issue of building a true community of designers. In Section 6.3 we discuss empowering the designers individually and collectively by building a design culture.

The development of capacity to engage in design is explored next. Section 6.4 describes the systems complex of design. The design of the designing system is the topic of Section 6.5. In Section 6.6, the question of the type of social system stakeholders should design is taken up as an issue of selecting the type most appropriate to the future system. In the last section, the design of design inquiry is discussed.

6.1. Who Should Be the Designers?

In *The Conference of the Birds* the Persian poet Fariduddin Attar tells the story of the assembly of the birds. The assembly was called together by the wise Hoopoe bird, who convinced the assembly that they have to find the King of the

223

Birds and ask his advice if they wish to live their lives to the fullest. But the wise Hoopoe warned them of the many dangers of the journey that leads to the castle of the King. Still, a large delegation of birds embarked on the search. During the journey many fell victim to deathly perils. At the end, only a few reached the castle. At the gate, the Chamberlain first refused to hear their plea, but they persisted and finally they were allowed into the throne room. To their amazement, no one was sitting on the throne. After great hesitation the birds approached the throne one by one and sat on it. First, they became confused. But after a while they were astonished as they looked into each other's eyes and realized that they themselves were, as a collective, the King. The throne was theirs and they together shared the responsibility for the kingdom.

In this section we explore the issue: Who is—or who should be—the designer of social systems? This question should be addressed before we explore the issue of how to get ready for design. The story of the birds represents a philosophical metaphor for answering the question. In this section, first the evolution of several generations of design approaches is discussed, followed by an exploration of the viability of the third-generation "user-designer" approach.

6.1.1.   The Evolution of Design Approaches

For years, the question of who should design wasn't even asked. It was the "natural order" of things that design was handed down from the top either by legislation or by the top echelon of the organization. The approach was "design by decree or design by dictate." People in the system or organization were expected and coerced to carry out the dictate. As the story goes, a worker went to the supervisor saying, "I think I have a design idea," to which the supervisor's answer was, "We didn't hire you to think." This top-down approach could be called the predesign-age approach in social systems design.

6.1.1.1.   The First-Generation Approach

Half a century ago, with the emergence of systems science, we entered the design age. In the span of five decades we have witnessed the emergence of three generations of design approaches. The first generation, the "designing for others" approach, was dominated by the expert designer. He was brought in by top management and was tasked to find and define a design solution. He conducted a problem analysis, engineered a solution, presented it to the decision maker, and was paid (at times to never return). If the solution was implemented—and often it was not—it was by applying coercion. And it worked as much as coercion works in social systems. The first-generation approach was heavily influenced by systems engineering, operations research, and systems analysis methods applied in large-scale military and space programs. The phenomenal success of these methods in hard systems settings led to their direct transfer to social systems. The expert, who called himself a social engineer, prescribed solutions to social problems, often resulting in disasters. We have a myriad of examples of the failure of this approach. In fact, these failures had much to do with discrediting systems science for a long time. The first-generation design approach can be characterized by what John Warfield often called "throwing the blueprint over the wall and having someone else build it." Its use was based on the belief that human systems can be manipulated, the expert knows best, and people better do what the expert says. Expert-driven design dominated the scene during the fifties and sixties. It is unfortunate that at certain places the expert designer still reigns.

Rittel (1984) characterized the first-generation approach this way: the designer is invited in by a client and studies and analyzes the problem. Then he withdraws and, following a step-by-step approach, works out a solution. Then he comes back to the client and offers the solution to him. But often he runs into an implementation problem because the client does not believe him. And the client is well advised not to believe the expert, because at every step in developing the solution the expert made ought-to-be judgments that the client may or may not share, and cannot read from the finished product, offered as the solution. It is for this reason that "the proponents of the first-generation methods, like operations researchers, tend to withdraw from attacking wicked (design) problems and concentrate on the art of linear programming and queuing theory as objects for their own sake" (p. 323).

6.1.1.2. The Second Generation

The next two generations are discontinuous with the first-generation approach. Their emergence was guided by a gradual and increasing recognition of the open, complex, indeterminate, and self-organizing nature of human systems and an understanding of their value-laden, purposeful, and even purpose-seeking characteristics. Their emergence was influenced by the rejection of the use of hard systems thinking and analysis/engineering practices in social systems and the coming to the scene of soft systems and, later, the critical systems thinking approaches and methods of the seventies and eighties. The initial stage of this emerging new thinking guided the development of the second-generation approach. Its advanced state brought about the third generation.

The second-generation approach is labeled as "the expert designing with users." In this approach the decision makers bring the design expert as a consultant into the system, where he or she stays for a while and works from time to time with selected groups of people who represent a cross-section of the system. They engage in a practical, task-oriented discourse, led by the expert. Depending on the designer's and the decision makers' inclination, the "designing with" approach can be regarded as more or less participative. If it is more, then it may generate a certain degree of commitment to the design solution. If it is less, it calls for compliance and coercion.

6.1.1.3. The Third Generation

Summarizing the characteristics of the third-generation, participative design approach, Rittel (1984) suggests that the first characteristic

> is the assumption that the expertise is distributed as well as the ignorance about the problem; that both are distributed over all participants, and nobody has any justification in claiming his knowledge to be superior to anybody else's. We call this "the symmetry of ignorance." The consequence of this assumption is the attempt to develop maximum participation in order to activate as much knowledge as possible. This is a nonsentimental argument for participation. It is a logical argument. There are many sentimental and political arguments in favor of participation, but this is a logical one. Whenever you want to make a sentimental or political case, it's good to use a logical argument. (pp. 324–325)

Churchman's (1971) argument for sweeping in as much information, knowledge, and point of view as possible in design is in concert with the notion of maximum participation. Rittel's quote ties in well with the third-generation design method, and it is an appropriate introduction to it.

The third generation of design approach is standing on the shoulder of the second generation. It is continuous with it. The story of the birds represents the insight that generated this approach. The "designing within the system" or the "user-designer" approach is based on the belief (Banathy, 1993a) that although the future is influenced by the past and the present, it is not determined by what has been or what is. It remains open to our individual and collective purposeful intervention, accomplished by design. Human activity systems, organized at various levels of society, from the family to the global system of humanity, can give direction to their own evolution and can shape their own future by engaging in purposeful design. Even more, this approach asserts that designing our future is our responsibility; we can and we should take charge of shaping it. This line of argument supplements Rittel's position that maximum participation is supported by both rational and emotional/political arguments.

We find much support for the third-generation "designing within the system" approach in the design literature. Nadler and Hibino (1990) present a detailed reasoning for a third-generation approach in setting forth their "people design principle." Their principle is based on the premise that the concerns and ideas of people in the system should be the basic fabric of design. In design, people should work "from the center (themselves) out rather than from the outside (others) in" (p. 220). The great need for creative and innovative design solutions is matched by the need to install the solution, which requires the active involvement of those who operate the system. "Their commitment is built by understanding what the solution is and how it was developed" (p. 223). Implementing the solution starts at the beginning of design, by getting people involved. Bringing people actively into design is a need, not only a desirable social value. Conventional approaches assume that we can separate the technical aspects of design from the human aspects; experts should design the technological solution and those in the system should accept it. We should demolish any lingering belief in this "presumption so at odds with both common sense and human nature" (p. 224). "We all want to be involved in decisions affecting our lives. And we accept and feel good about implementing a solution we help to devise. We should 'maximize individuals' participation and secure their commitment to the solution even before it is fully known" (pp. 225–226). The authors suggest that in some instances the benefits of group participation in creating the design solution can be more important to the system than the solution itself.

An early promoter of people's participation in design, Ackoff (1981) suggests that "when it comes to considering what a system 'ought to be' no one is an expert in preparing an idealized design of it" (p. 116). Every stakeholder can make an important contribution. Their ideas, aspirations, dreams, and preferences are all relevant. The fullest participation provides people with an opportunity to think deeply about the system and to share their ideas with others. This encourages the exploration and development of new ideas and facilitates personal and collective development. Weisboard (1992) sets forth a set of core values about participation in design. He suggests that ordinary people are extraordinary sources in design. They aspire to create their own future and want opportunities to engage their heads and hearts in the design of their future. He proposes egalitarian participation; everyone is equal. Baburoglu and Garr (1992) suggest that "designing a future collectively unleashes a creative way of producing organizational philosophy, mission, goals and objectives; enriched by shared values and beliefs of the participants" (p. 74). Lenford and Mohrman (1993) use a "self-design" approach, in which the members of the organization are the architects of the design. They propose that "central to the self-design strategy is the notion that organizational members are redesigning the organization. That is, the design is not being imposed by outside experts or reproduced in full from a model that existed somewhere else" (p. 147).

Jackson (1992) brings Vickers's (1983) notion of the "appreciative system" into design, which is an interconnected set of standards of judgment by which we order and value our experiences. If human systems are to achieve stability and effectiveness, "then the appreciative system of their participants need to be sufficiently shared to allow mutual expectations to be met" (Jackson 1992, p. 135).

The legitimacy of systems design (Churchman, 1971) rests on the consideration of the many different perspectives that stakeholders have and set forth in the course of the design conversation. Therefore, the design process receives its legitimacy from ensuring the maximum participation of stakeholders. The "designing within" approach is based on the assumption that human activity systems must be designed by those who are served by the system, who serve the system, and who are affected by it.

When it comes to the design of social and societal systems of all kinds, it is the users, the people in the system, who are the experts. Nobody has the right to design social systems for someone else. It is unethical to design social systems for someone else. Design cannot be legislated, it should not be bought from the expert, and it should not be copied from the design of others. If the privilege of and responsibility for design is "given away," others will take charge of designing our lives and our systems. They will shape our future.

6.1.2. The Key Markers of the "Designing Within" Approach

During the seventies I guided an R & D program aimed at developing an educational system that enhanced and nurtured the authentic culture of Native American youth. Our staff was primarily Native American. Going on site to a reservation, one of the tribal representatives told us kindly that if we came to tell them what to do, we better go home. If we work with them for a while and advise them what to do we might as well go. But if we stay with them, live with them, and learn with them we are welcome. This episode represents well the three generations of design approaches. The story has two ramifications. At a personal level, from that moment on I did not accept any consultancy role. Then, later on, one of our senior researchers went to a small fishing village in Costa Rica, where he lived with the native people, learned from them, and worked with them for nine months on a project called "Participatory Design for Social Empowerment: A Journey of Trust" (Kavanaugh, 1989).

In responding to the question Who should be the designers of social systems? three answers have been offered in the course of the last several decades. They are the three generations of design approaches: "designing for," "designing with," and "designing within." As justification for the third-generation "designing within" approach, the position was taken that the right to design rests with the stakeholders of social systems. This position meets the test of the criteria of authenticity, sustainability, responsiveness, uniqueness, personal development and organizational learning, and the ethics of design.

6.1.2.1. Authenticity

The design of a social system is authentic only if it is carried out by the stakeholders of the system. An authentic design has to build on the individual and collective values, aspirations, and ideas of those who serve the system and who are served and affected by it. The design should reflect their collective vision and it should make use of their collective intelligence. A design cannot produce an authentic and intelligent organization if it reflects only the intelligence of an expert (first generation) or if it is limited to use of the intelligence of only the top echelon or some selected groups in the organization (second generation). It is authentic only if it makes use of the individual and collective intelligence of all the stakeholders.

6.1.2.2. Sustainability

A system is sustainable only if its design is accomplished and put in place by the creative, collective, and unconstrained participation and contribution by all people in the system. Such contribution ensures sustainability several ways:

1. Participation enables people to understand their system thoroughly, since they designed it.
2. Participants know first hand what their role is in the system (since they shaped the roles) and know what they have to learn to play their role.
3. Participation enables the creation of consensus among those who work together.
4. Participation creates genuine respect for each other and develops fellowship.
5. It ensures that people will take part more effectively and at a deeper level of commitment in the implementation of the design since the design represents their individual and collective values, ideas and decisions. Participation is empowering and the design is empowered by it.

6.1.2.3. Uniqueness

The uniqueness of the individual, the uniqueness of cultures, and the uniqueness of situations have been generally acknowledged. These types of uniqueness have been considered within the contexts of such disciplined inquiries as psychology, biology, and antropology. But the uniqueness of human activity systems always frustrates those who approach human systems and organizations from a systematic (not systemic) perspective. Conventional statistical approaches—standards, averaging, aggregating, and other systematic methods—are unable to consider such characteristics of the uniqueness of social systems. We were not able to address the uniqueness of social systems until we have learned to use soft systems and critical systems thinking and their methods of inquiry.

If we aspire to create viable systems, we are to take into account a whole range of distinctions and differences embedded in the uniqueness of the system, such as the uniqueness of (1) the systemic context, (2) the nature of the system to be designed, (3) the individual and collective readiness and capability of the people involved, (4) the resources available, (5) the design situation, (6) the values and worldviews of the designing community, and (7) time, space, and complexity factors. The consideration of all these aspects of uniqueness will lead us to understand that a viable design can be attained only if it is accomplished by people who embody the uniqueness of the design task and design situation.

6.1.2.4. Opportunity for Learning

In design research and scholarship we often remark that the most important product of participation in design is the unique opportunity for learning and personal development. Some of us are prepared to say that such learning may be more important than the product of design. Few people would question that the ability to design is one of the highest personal and collective capabilities in an age of ever ongoing change. The second and equally important benefit derived from the collective involvement of all stakeholders is that it provides the best possible opportunity for organizational learning. By engaging in design, an organization engages in "double-loop learning" (Argyris and Schon, 1978), as people in the system learn to envision or "re-vision" the purposes, perspectives, values, functions and modes of operation of their organization and develop insights on which they can base change or (re) design of their system and continue contributions to its life.

6.1.2.5. Ethics

The fifth criteria is the ethics of social systems design. It is suggested that the design can be termed to be ethical only if it enables the self-determination of the stakeholders and respects their autonomy and uniqueness. Design should be self-guided and self-directed by the users of the system. Their genuine and unrestricted involvement would "liberate" them from a second-class status of being just employees or members and would ensure equity for all. The ethical and liberating involvement set forth here is based on the understanding that we have the right and the responsibility for the design of our lives and for the design of the systems in which we live. User-designers, like the birds in the Persian poet's story, understand that "we collectively are the king," and we collectively have responsibility for our lives and for the design and well-being of our systems. Our salvation comes from within: "the kingdom is within us."

At the onset of this section the question was asked: Who should be the designers? A review of the evolution of design approaches and an exploration of their justification have helped us to understand that the only viable approach to the design of social systems is the one that empowers the stakeholders of the system to become competent user-designers. In the next two sections the concept of the community of designers is developed, followed by the issue of developing a design culture in the designing community.

Reflections

It has taken us over 40 years to reach the understanding that the design of social and societal systems is the right and responsibility of those who serve the system, and who are served and affected by it. Like the perilous journey of the delegation of the birds, our quest for an authentic, sustainable, and ethical approach to design has been marked with many failures and disappointments. Disheartened by these, we often left the shaping of our future and the design of our systems in the hands of "kings"—the authorities, futurists, and experts. But now we know that the age of social engineering is over. We can reclaim the "throne." It is rightfully ours.

Activity #43

First, review the text and select and organize sets of core ideas that you deem to be relevant to the question: Who should be the designers of social systems? Then in view of the core ideas and the generations of design approaches presented in this section, speculate about means and methods by which you would enable the community of designers in the system of your interest to get ready to engage in design. Enter your findings in your workbook.

6.2. The Designing Community

The conclusion of the previous section was that the design of social systems is the right and responsibility of people who serve the system and those who are served and affected by it. Collectively, these three groups constitute the designing community. The term "community" has a wide scope of meaning. In the dictionary (Webster, 1979) we find two main relevant categories: (1) a unified body of individuals and (2) society at large. Under (1), eight subcategories are denoted, including a scientific community, an educational community, and a geopolitical community. Of the eight, the one that best describes the designing community is "the people with common interest, living in a particular area." The people of the designing community have not only a common interest but a common purpose, which is to design their system. Furthermore, the degree to which they are a community is marked by the degree of effort they devote to attain the purpose, the degree of their commitment to it, and the degree of their commitment to each other.

In recent literature, we have seen an ever-widening interest in exploring the implications of how to become the kind of community we describe here. There is an awakening of aspiration and a maturing resolve in the society at large and in many segments of the human community toward self-determination, toward claiming the right and assuming the responsibility for shaping our own future, and toward empowering ourselves to design our own lives and our own systems. Consequently, there is general disillusionment with the way representative democracy works, when others are making decisions for us. And the way participative democracy works today is limited and highly constrained. It is more label than substance.

I first explore the meaning of community in general and then reflect on the meaning of community as it applies to a community engaged in the design of social systems. It is important to note that in this book our main interest is exploring the whats, hows, and whys of designing certain types of social systems. Systems in business and industry, and bureaucracies in government, can also be called social systems. And the material in this work could be useful in those contexts. But our primary interest here is the design of various social service and human development systems, education, health service, volunteer agencies, and community service systems. In the second part, design teams, the core units of a designing community, are discussed from a variety of perspectives that demonstrate their value.

6.2.1. Authentic Communities

First, characteristics of designing communities are explored. Then, the notions of equity and diversity are addressed, followed by the idea of stewardship as the leadership mode in communities. Next, the health of designing communities is explored. In conclusion, a manifesto of an authentic community is presented.

6.2.1.1. General Characteristics

A review of community and organizational literature assists us to propose some general characteristics of communities that are deemed to be appropriate to a characterization of a designing community. MacCallum (1970) describes the community as an association of people having a feeling of solidarity within their group who achieve coordinated action "by virtue of shared enthusiasm and common dedication to a purpose which transcends immediate self-interest" (p. 80). During recent years we have seen the development of the Communitarian Movement (Etzioni, 1993) as a new moral, social, and public venture that reaffirms our shared values and aims to build a responsive and responsible civic society in which we are each other's keepers. The movement aims "to bring about the changes in values, habits and public policies that will allow us to do for the society what the environmental movement seeks to do for nature: to safe-guard and enhance our future" (p. 3). "The responsive community is the best form of human organization yet devised for respecting human dignity and safe-guarding human decency and the way of life most open to needed self-revision through shared deliberation" (p. 265). The establishment of moral coherence within a community is the moral foundation of the common good (Etzioni, 1991). The notion of common good has a dynamic element: "the community and individuals working toward a telos: a common purpose or goal" (p. 132). Etzioni's vision can be reflected in what we call here the designing community. Community that has the purpose of creating its future system as a moral commu-nity, serving the shared purpose, the common good. MacIntyer's (1984) concep-tion of a moral community reflects Aristotle's vision of civic virtue, according to which people aim to build their moral community, serving the collective "telos," the common good, by engaging in coherent shared activities. The search for the common good cannot be an individual enterprise. For such enterprise, the com-munity provides the only legitimate context. In social systems design, it is the designing community—all of the stakeholders—that provides the only legiti-mate context for social systems design.

The sociologist Ferdinand Tonnies (Lemming, 1994) distinguishes two ideal types of social system. "Geselshaft social organizations are characterized by relationships that are impersonal, atomistic, and mechanistic relationships; valued as means to an end" (p. 56). In a Gemeinshaft, "relationships among members are valued as ends in themselves, and the actions of individuals proceed from and express underlying communal identification" (p. 56). A social engi-neering expert design group (first-generation and some second-generation design approaches) is a geselshaft. A community of user-designers is a gemeinshaft: they design their own system and continue to live in the system they design. A community (Peck, 1993) is a state of "being together with both individual au-thenticity and interpersonal harmony so that people become able to function with a collective energy even greater than the sum of their collective energies" (p. 472). Such a community has the capacity to evolve in wisdom, in effective-ness, and in maturity. Group consciousness becomes a way of life. Members of the community speak their minds honestly and openly. Issues are aired fairly; members feel that they are heard. They feel that everyone has equal power and responsibility.

In presenting an array of strategies of search conferences, Weisbord (1992) proposed a set of core values that are pertinent to designing communities. He suggests that in design situations (1) the knowledge of ordinary people is an extraordinary source of information; (2) people can create their own future; (3) people want to have opportunities to engage their heads and hearts and partici-pate in the creative process; (4) everyone is equal in the design situation; (5) given the opportunity, people learn to cooperate; (6) the design process empowers people to feel more knowledgeable and in control of their future; and (7) diversity is highly valued and appreciated.

Making use of the collective intelligence of designers is one of the key characteristics of the third-generation design approach. Pinchot and Pinchot (1993) suggest that the collective use of intelligence is a key imperative in organizations. In discussing community in the workplace, the authors suggest that the challenge of intelligent organizations is to establish strong communities so that everyone can contribute. Such a community serves as the vessel of vision and values, and guides the work of the group. It "combines freedom of choice and responsibility for the whole—everyone's relationships are full of choice and also collaborative, educational, vision sharing and value driven" (p. 216). People take responsibility for the qualities of the system as a whole. Responsibility develops naturally in contexts that "contain lots of freedom, empowerment, and practice in collaborative management" (p. 222). Aspects that promote such a community include individual and collective freedom, responsibility shared through collaboration, the view of everyone as of equal value, diversity of contributions, propagation of—and loyalty to—shared values and goals, consensual self-management of the whole, strong integrating focus, recognition of the value of each individual and the power of collective intelligence, everyone's responsibility for adding value, long timeframe thinking, and freedom to innovate, take risk, and self-direct. Social systems that are strong communities "have the potential for involving everyone in shaping the organization's movement into the future and coming to terms as a group of what is truly needed by the society" (p. 232).

6.2.1.2.   Equity and Diversity

In an egalitarian spirit, intelligent communities treat everyone as being of equal value, as a high-potential member of the community (Pinchot and Pinchot, 1993). "Everyone can bring his or her intelligence and talent to bear on the organization's challenges, large and small, local or global" (p. 233). At the same time, unique individuality is nurtured in intelligent communities. The combined honoring and encouraging of these two qualities in designing communities enable designers to see each design issue from many different viewpoints. It enables designers to learn from each other and generate creativity. Equity creates the climate that allows diversity to flourish and to generate the richness and robustness needed in the design of social systems. One brittle way to achieve a sense of community is by building on similarities. But narrowly defined communities are ineffective for two reasons: "First they lack the diversity of talent and ways of thinking that are necessary to solve many cross-disciplinary and cross-cultural problems; and second, their narrow definition of how members should think, act, and feel limits the personal growth and thus the capabilities of all their members" (p. 234).

The core ideas highlighted in the preceding chapter and in the previous section of this chapter unmistakably establish as a requirement the nurturing of both diversity and equity in the design community. The power of creative conversation in design is determined a great deal by the variety of viewpoints and the diversity of participants. The greater the diversity the greater the likelihood of finding breakthrough design solutions. The authors of "breakthrough thinking" in systems design, Nadler and Hibino (1990), set forth as one of seven principles of design the principle of diversity. Outstanding designers are diverse people who seek many different sources of information in their efforts to find design solutions. In a designing community participants are welcomed for their diverse viewpoints as well as for their similarities in that they bring into the design conversation a rich array of perspectives. Churchman (1971) has called for the sweeping in of the greatest possible variety of viewpoints and positions from the greatest possible variety of fields and disciplines. Tsivacou (1990) suggests that many of the existing design methods, "when driven to rational consent, tend to suppress differences in the name of effectiveness" (p. 547). Such methods reduce the power of design when applied in the context of complex social systems in which conflicting views exist, especially in situations of deep change. In such contexts design methodology has to satisfy two demands: "(a) the establishment of a climate of dialogue which allows not only for free expression of desire but the expression of differences, as a pool of invention; and (b) the enactment of criteria for the legitimation of choices in case of strong disagreement" (p. 547). One of the core principles of evolution is requisite diversity, which is an essential condition of the robustness of all lifeforms. Monocultures create barren spaces in nature as well as in the ecology of society.

6.2.1.3. Stewardship Transcends Leadership

The search for authentic design solutions questions our current notions of leadership, which is associated with taking initiative, controlling, and knowing what is best for others (Block, 1993). The act of leading cultural and organizational change by determining the desired future, defining the path to get there, is alien to the user-designers approach as well as to the community of designers notion articulated in this chapter. The often-noted search for strong leadership means that we place responsibility for our systems in the hands of others, which then reduces our opportunity to determine our own future. "The attraction of the idea of leadership is that it includes a vision of the future, some transforming quality that we yearn for" (p. 12). That is what much of the current organizational literature is promoting by calling for "buying into" the vision of the leader. No wonder leaders believe their "key task is to recreate themselves down through the organization" as they wonder: "How do I instill in others the same vision and behaviors that have worked for me?" (p. 13).

The alternative to leadership is stewardship. Stewardship asks each one of us to be responsible and accountable for the outcomes of the system we design. Stewardship is shared accountability, which is fueled by a shared commitment to service. Partnership and empowerment are integrated and distributed among all participants of the design effort and among all stakeholders of the future system. "There is pride in leadership, it evokes images of direction," says Block (1993, p. 41), and "there is humility in stewardship, it evokes images of service. Service is central to stewardship." Stewardship "promises the means of achieving fundamental change in the way we govern"; it is "to hold something in trust for another," most importantly for future generations. "We choose service over self-interest most powerfully when we build the capacity of the next generation to govern themselves" (p. xx). When we serve, we build capability in others by supporting their ownership and empowerment, and their right to participate at every level of the system. The kind of stewardship described here is the only viable governance of designing communities: stewardship transcends leadership.

6.2.1.4. Communities and Their Health

The community, says Nelson (1989), as a social system is a shared space, a shared identity and character. In a community we not only share values but also share ways by which to discard old values and take on new ones. (These functions are critical to a designing community.) As a teleologic (designing) social system, a community is in search of a purpose that gives direction to it. In a community the individual has "the power to participate in common choice while still retaining the sense of being unique and valuable" (p. 361). A designing community is not an organization of formally related identical individuals. It is a social structure of unique individuals who play specific roles by which they contribute to the overall design task.

The shared vision of designers provides the energy that is required to induce transcendental change, leading to the rebirth of a social system. Vision is essential to the health of a community. However, a designing community faces two dangers. The first is that a quest for vision does not guarantee a safe journey. The design journey is always laden with risk. The second is that of facing the consequences of introducing disequilibrium by the pursuit of a new vision, against the conventional perception that equilibrium and balance are the desired healthy conditions.

> Good health in communities does not consist of natural states of health. Good health is comprised of unnatural states which require constant energy and risk. This means that health is not something that can be achieved once and for all, requiring no further investment in human energy and intention. (p. 366)

Only healthy designing communities can make judgments about how to best use their collective intelligence and resources as means to maintain their health.

6.2.2. Design Teams: The Core Units of a Designing Community

From all that we have learned about what design is and how it works, it has become (at least implicitly) evident that the core unit of design action is the design team. Here I want to address this issue explicitly. There are several reasons why it is proposed that the communication function among members of a designing community are carried out in teams of designers. Reasons include (1) the nature of social systems design, (2) the nature of the communication mode of design, and (3) the nature of teams.

6.2.2.1. The Nature of Social Systems Design Invites Teamwork

Systems design carried out in the third-generation mode is not directed by an expert but it emerges from the intensive, creative and dynamic interaction of members of design teams. Rowland's (1995) musical metaphors—the orchestra versus the jazz ensemble—well represent the contrasting design modes of the user-designer and the expert. The orchestra is the metaphor for expert-driven design. The conductor makes all the decisions, which the large number of players follow. In contrast, in the jazz ensemble the small group decides what to play and how to play it. They improvise around a basic plan, react to each other as they play, and challenge each other with new ideas. They explore opportunities and design new patterns. The play implies a high degree of interaction and cooperation. These are also the kinds of behavior that creative and interactive design implies. If we engage a user community in design, then we better learn to play design in the mode that the jazz ensemble uses.

6.2.2.2. The Nature of Design Communication Invites Teamwork

In Chapter 5 we defined "design conversation" as the basic and most effective mode of design communication. Design conversation weaves into a pattern two types of dialogue: the generative and the strategic. Before the design group engages in the substantive (strategic) tasks of design, it must create a flow of shared meaning, a commonly held set of values and a collective worldview developed in a social milieu of friendship and stewardship. This process transforms the people who come into the experience of design into a community of stakeholders. Once the groups attained a state of shared consciousness, then and only then do they turn to addressing design tasks in the strategic dialogue mode.

The design conversation reviewed here cannot be brought to life in a large group but only in small teams.

6.2.2.3.   Teams and Teamnets

In a designing community the core units are the design teams. These teams are arranged in a system of teams, often called teamnets.

6.2.2.3a.   Teams and Their Characteristics.   Katzenbach and Smith (1993) suggest that the potential impact of single teams and networks of teams on the performance of organizations becomes increasingly clear. Teams should be the basic units of performance in organizations. In situations that require "real-time combination of multiple skills, experience and judgments, a team inevitably gets better results than a collection of individuals operating within confined job roles" (p. 15). Furthermore, teams are more effective and more flexible than larger organizational groupings. A team melds together the skills, experiences, insights, and creative potential of its members and can engender a high level of commitment to a shared goal. The authors define a team as "a small number of people with complementary skills who are committed to a common purpose, performance, and approach for which they hold themselves mutually accountable" (p. 45). The authors find that (1) the number of team members should be less than ten. (2) Complementary skills should include technical and functional expertise, decision-making skills, and interpersonal skills. (3) The best teams invest much time and effort exploring, shaping, and agreeing on individually and collectively held purposes. (4) Teams are committed to a common approach that defines how they work together to accomplish their purposes, agreeing on who will do what, what skills need to be developed, and how to make and modify decisions. (5) A group becomes a team when it can hold itself collectively accountable. At its core, team accountability is about the sincere promise we make to ourselves and others, promises that underpin two critical aspects of teams: commitment and trust.

6.2.2.3b.   Teamnets.   Lipnack and Stamps (1993) suggest that teamnets bring together two powerful organizational ideas: "*Teams,* where small groups of people work with focus, motivation, and skill to achieve shared goals; and *Networks,* where disparate groups of people and groups 'link' to work together based on a common purpose" (p. 7). Teamnets are networks of teams. Applied to small groups it means "networked" teams. Teamnet, when applied to large groups, implies more teamlike networks. In the context of a designing community the application of the two kinds of arrangement depends on the size of the designing community.

Reflections

Reflecting on the discussion above in the context of the various dimensions of the design experience explored in Chapter 5, the significance of design teams and teamwork in design becomes quite obvious. In a designing community we shall have a number of design teams—which create a network of teams—possibly coordinated by a core design team. At the beginning of design, all teams engage in the generative dialogue mode of design conversation that will go on across all teams. Then they engage—in a strategic dialogue mode—in specific design tasks that might be shared tasks as well as tasks assigned to specific teams. It will take time to find the best approach by which to engage the design teams of the designing community. We shall further discuss the work of design teams and examples of their arrangements when we explore the creation of the designing system.

6.2.3. A Definition of an Authentic Community

Our exploration of the idea of a designing community now concludes with the self-definition of a working designing community. The definition is prefaced by a quote from Peck (1993), who says that a community is

> a way of being together with both individual authenticity and interpersonal harmony so that people become able to function with a collective energy even greater than the sum of their individual energies. (p. 272)

The following statement was developed to guide the life and work of the International Systems Institute (ISI). The Institute is a not-for-profit research and educational organization, dedicated to the improvement of the human condition. Its purpose is to provide learning programs and resources for the development of authentic designing communities.

Our challenge is to learn to become an authentic community of scholarly practitioners and practicing scholars, to apply what we learn in all aspects of our lives, and help others to learn to develop their own authentic communities.

An authentic community is a group of individuals who have developed a deep and significant commitment to each other and to a shared vision and purpose. Members of the community (1) feel that they belong together and believe that they can make a difference in the world by pursuing their shared vision and purpose; (2) communicate with each other openly, honestly, and creatively; (3) organize themselves with total absence of hierarchy and bureaucracy as equal partners in service and in mutual assistance—they govern themselves by shared stewardship; (4) apply maximum flexibility in their shared work, taking full advantage of their unique and collective potential, knowledge, skills, creativity, and intuition; (5) take responsibility for the continuing development of their individual and collective capabilities; (6) nurture and practice genuine and authentic participation in achieving their common purpose and in creating the common future of their community; and (7) become bonded, knowing that they can rely on each other, trust, honor, and support each other, share values, aspirations, and hopes, and live by a collectively defined code of ethics.

The statement above was formulated some time ago. It reflects some, but not all, of the ideas about a designing community developed in this section. The task of making use of the ideas introduced in this section becomes your task as suggested in the activity below.

Activity #44

(1) Review the text of this section and select core ideas that best define community in general and a designing community in particular. Organize the core ideas in clusters. (2) Use the core ideas you selected and develop a statement of guidance for the creation of an authentic designing community. This community can be your family, a group you work with, or a system of your choice.

Reflections

My hope for a better human future is grounded in the expectation that it is possible to create the type of genuine and authentic communities explored in this work. This hope has emerged as a shared hope as I worked with and learned from many people and groups throughout my life. I have—we have—also learned that this hope can be transformed into an intention. But intention in itself does not create the future, it does not create genuine communities; only design does. The ability to design communities is to be developed by learning the whats and hows of systems design, by the acquisition of a design culture. In Chapter 2 we worked with the crucial role of design culture in the overall human experience. In the next section, we revisit the concept of design culture and explore how it can and should be developed in designing communities.

6.3. Developing a Design Culture

In the preceding section it was suggested that in an age of increasing systemic complexities, rapid societal changes, and constantly emerging new realities that affect all aspects of our lives, the responsibility for the design of our social systems cannot be left in the hands of the "experts." The right and responsibility of design lies with those who serve the system, who are served by it, and who are affected by it. We called these groups collectively the stakeholders of the system. The stakeholders can exercise this right and assume responsibility only if they learn what design is, how design works, and how design is carried out. In an age of constant change, the ability to design is a commodity of the highest value. Therefore, what is required today is the development of individual and collective capability to acquire a design culture that becomes a shared pattern of behavior in our communities and eventually throughout the entire society. In the idea of design culture we now have a viable and authentic definition of the meaning of social empowerment. It is this power that can bring to life true participative democracy and enable the society to create its own future.

The notion of design culture was explored in Section 2.5, Chapter 2. It was suggested that to respond to the emerged requirements of our age, the two traditional cultures—science and the humanities—have to be complemented by what Cross (1984) and Warfield (1987) called the "third culture," the culture of design. This section begins with a brief review of systems design and design culture, followed by a description of why we need a design culture and how it can be developed.

6.3.1. Design Culture and Its Implications

Systems design in the context of human activity systems is a future-creating disciplined inquiry. The stakeholders of a system engage in design in order to create a system based on their vision of what that system "should be." The stakeholder community—characterized as the designing community—embarks on the future-creating journey by enculturating itself in design thinking, knowing, behaving, and working. Design culture is a learned pattern of behavior, shared by members of a designing community. Design culture enables the collective creation of novel phenomenon. It integrates (1) design's own distinct ways of thinking and knowing; (2) design concepts and principles that constitute design inquiry; (3) methods and means by which creativity is applied in such actions as envisioning, imaging, inventing, assessing, and creating design solutions; (4) the use of the "language" of design, and modeling, as design's own specific form of expression and representation; and (5) conversation and consensus building as the special social communication behavior of a designing community. Finally, design culture is manifested in action that aims at the creation of a system that realizes the aspirations and expectations of the designing community.

6.3.1.1. The Concept of Design Culture

The understanding that design culture is an essential component of the wholeness of human experience has emerged only recently. Popper (1974) made a distinction of three worlds. The first world is the world of things, the second is the world of subjective experience, and the third is the world of our creations of all kinds, which have their own autonomous laws. The disciplined inquiry by which the three worlds are explored are science for world I, the humanities for world II, and design for world III. Worlds I and II (Christakis 1987) are the domains of descriptive science, while design is prescriptive. Design "starts with a need conceived in world II, perhaps from an observations of world I, and conceptualizes some kind of innovation that might mitigate or relieve the need" (p. 16). Design culture is now recognized as the third culture, which complements the cultures of science and the humanities. As already reasoned in Chapter 2, a lack of any one of the three cultures leads to a grave loss of substance and value in the quality of human experience. But today, design culture is not yet part of the general human experience. While our schools provide literacy and competence in the cultures of the sciences and the humanities, they provide design learning only to a few professions. Today we do not provide general education in design.

6.3.1.2.   The Rationale for Building a Design Culture

A rationale for enculturation in design has emerged only recently from a recognition of the need for attaining both design literacy and design competence. These two are the two branches of design culture. They are the two sides of the design coin. We have by now recognized that in an age of an explosive growth of all kinds of design, we are at the mercy of those who design for us. We individually, and collectively as a society, are uninformed design illiterates. When bad designs are thrown at us the best we can do is complain. (And we have certainly no capability to participate competently in the design of systems in which we live.) Faced with this debilitating predicament, design literacy comes to our rescue. It will enable us to understand what design is, what design does, how it does it, and what the impacts of designs are on our quality of life. Design literacy can create informed users of products and systems in the creation of which technical expertise is required. But design literacy is only one side of the coin of design culture. The other side is design competence. When it comes to social systems—the systems we inhabit—it is we the people in the systems who are the experts. Only we, the stakeholders of our systems, have the right—and carry the responsibility—to design our systems. But we can exercise this right and fulfill our responsibility only if we become competent in systems design.

6.3.2.   Building a Design Culture: Conceptualization

During times of relative stability—characteristics of previous eras—piecemeal adjustments were able to bring our systems in line with the slow rate of change in the societal environment. But in a time of accelerating and dynamic changes and transformations—characteristics of our current era, when a new stage has unfolded in societal evolution—piecemeal adjustments or improvements of the old designs will create more problems than are solved. The emerging "new realities" of the massive societal changes of the postindustrial information/knowledge age require continuous design activity at all levels of the society. They require the redesign of existing systems or new designs of our systems so that they will become in sync with the new societal realities as well as with our own aspirations (Banathy, 1992c).

We can relegate design decisions, as we do today, to others who "represent" us and make or legislate decisions for us. Or we can empower ourselves by (1) acquiring design literacy and using it to make informed judgments and choices of designs, (2) developing design competence, and (3) assuming responsibility for designing the systems in which we live. The building of a design culture enables us to create a participative democracy about which we talk so much today, but which—in a true sense—is not yet part of the human experience.

6.3.2.2. The Systems Complex of Design Enculturation

Building a design culture in a society is a process of enculturation through learning. This process can be accomplished by the design, development, and implementation of a systems complex of design learning. This complex would operate at various societal levels. It should have the potential to provide learning resources and programs by which to offer design enculturation. It would be a gross underconceptualization of the need for a design culture if we would limit its development to specific designing communities. It is not only selected designing communities that need to be empowered by design learning; we must also find a way to incorporate design culture into the culture of each and every community and into the overall society.

We should think of the task of design enculturation as one that can be accomplished by the design and development of a systems complex of design education. The systems complex is comprised of learning systems that operate at various societal levels at which it builds and nurtures design cultures.

A systems complex of design enculturation might operate at four levels. Level (A) is the level of genesis of the design education system. This is a level where the entire complex and its component systems are conceptualized and their design is originated. At Level (B), systems provide for the professional development of practitioners of various social service and human development systems. At Level (C), we should offer learning resources for general design education. And at Level (D) we provide design education to the overall society as lifelong learning. These levels and the interaction of systems operating at these levels are portrayed in Table 6.1 and are now discussed here. This proposal for a systems complex for design enculturation is highly speculative. It intends to trigger contemplation, conversation, and exploration.

6.3.2.3. Tasks to Attend to at the Four Levels

There are four tasks to attend to to accomplish a comprehensive development of a design culture. These tasks are proposed next.

TABLE 6.1
The Systems Complex of Design Enculturation

Level (A)	Builds learning resources and programs for design enculturation to be used by academic (higher educational) and R & D institutions and systems at Level (B).
Level (B)	Uses learning resources/programs built at Level (A) and builds learning resources/programs for systems at levels (C) and (D).
Level (C)	Uses learning resources/programs built at Level (B) for the design enculturation of children and youth.
Level (D)	Uses learning resources/programs built at Level (B) for the design enculturation of the general public.

6.3.2.3a. Level (A): The Genesis Level.   The genesis level is where the design of the whole systems complex originates. Systems operating at this level are (1) schools and R & D centers of social sciences and systems science; (2) schools of social and health services and schools of education and human development in higher education institutions; and (3) social and health services and education/human development R & D Centers. These systems are called upon to design and develop their own design culture and design and develop systems that build design cultures at Level (B). They would also provide process models for designing/developing enculturation systems at Levels (C) and (D).

6.3.2.3b. Level (B): The Enculturation of Practicing Professionals.   The development of resources and programs for design enculturation of human and social service professionals becomes the responsibility of systems at Level A. The various schools and R & D agencies, operating at Level (A), would design and offer learning resources and programs that focus on the development of professional competence in comprehensive systems design. This is a new territory that has to be conceptualized by creating the appropriate knowledge base in design and design learning and testing and developing design enculturation systems. In turn, those who receive this training will engage in the design and offering of systems of design enculturation for children and youth in formal/informal settings at Level (C) and in formal/informal settings of the society at large at Level (D).

6.3.2.3c. Level (C): Systems of Enculturation for General Education. Professionals at Level (B) design and offer design enculturation learning opportunities and resources for children and youth in formal/informal educational settings. As the outcome of these long-range enculturation programs we shall develop informed design literates as well as prepare user-designers of future designing communities. Level (C) programs will become part of new systems of learning and human development.

6.3.2.3d. Level D: Design Enculturation of the Society at Large. This enculturation will be attained at the long-range as the result of systems operating at Level (C). But the society needs to attain design competence now. We can no longer afford a society, the members of which are design illiterate and incompetent in participating in the design of its own systems. Thus, we must design/develop learning resources and programs through a variety of informal and formal arrangements by which people in our communities can acquire competence in social systems design.

6.3.3. Building a Design Culture: Realization

The design and development of the systems complex of design enculturation invites (1) the consideration of how the various systems of the complex can be designed and developed, (2) the development of "user languages" (languages that are readily understood by the users) at the various levels of the complex, and (3) the contemplation of strategies of implementation and institutionalization. In closing, the modest beginning of an R & D program is reported, aimed at the creation of design cultures.

6.3.3.1. The Building of the Systems Complex

Realization of the systems complex becomes an issue of systems design. The design of the enculturation complex unfolds as we (1) provide a definition and specification of the system complex; (2) define the learning agenda at the various levels of the complex; (3) design alternative representations of the learning systems of the complex; (4) test these and select the most desirable and feasible design alternative; (5) provide a systems description of the selected design; (6) formulate a plan for its development and implementation; and (7) develop and install the system.

This text is an example of a kind of learning resource that could be used at Level (B) for the design enculturation of people serving in social and health systems and educational and human development professionals. As individuals and groups work with the text and apply what has been learned in developing the many activities, they will be well on the road to developing design literacy and acquiring competence in the design of social service systems.

6.3.3.2. The Creation of "User Languages"

The issue of creating user languages is a major hurdle that has to be overcome before we design and develop learning resources. The term "user" means the users of the enculturation learning systems, those who want to become competent in design and who eventually will become designers in designing their own systems. Therefore the issue becomes: What language should we use in the learning resources and programs designed for the various levels of the systems complex? One thing is sure; the language must be user friendly.

The issue of user language is not treated lightly in the research literature. The various scientific and professional domains develop their own languages to enhance communication among scholars and practitioners of those domains. Reflecting on the lack of user languages and its consequences for the larger society, Warfield (1976) remarked: "It is ironic that what began as a means of communication should become a means whereby communication is frustrated" (p. 59). The use of the language as a means of communication of a specific scientific or professional field is restricted to that field. It often becomes difficult to grasp the language of a specific field for people who work in other fields. And specialized language becomes rather incomprehensible for laymen. Warfield calls for the development of language that enables cross-field communications as well as for the development of language that enables the use of findings of science and the professions by the laymen. For de Zeeuw (1993),

> user languages imply paying attention to actual users, that is people "in situ," people
> in the here and now, each with his or her own history, and his or her own peculiarities.
> Such peculiarities have to be transformed into resources for action. User languages
> will emphasize differences between actors, in terms of intended use. (p. 13)

He concludes, "to be able to create actors–and 'educated persons'—one needs to construct user languages" (p. 16).

Reflecting on the construction and introduction of user languages for the systems complex of design enculturation, it is assumed that people operating at Level (A) are capable of working with the language of the systems and design sciences. Their task is to transform systems and design science language for users at Level (B), to be used by the various professionals who operate in the fields of social and health services and education and human development. Now, those at Level (B) have the more difficult task of constructing the language of design enculturation for children and youth and constructing design language for adults.

6.3.4.   A Strategy for Realization

The design of the systems complex requires a major inquiry effort. It will have to be attended by a network of cooperating R & D agencies and academic institutions that devote their people and resources to the design effort. The network approach ensures a distribution of the design effort among various groups of organizations. A concentration of this effort in just one (or in a few) agencies would significantly limit the results of the effort and its impact on the creation of a designing society. A highly distributed effort, on the other hand, would allow for broad coverage and built-in multipliers; above all, it would meet the test of uniqueness in attending to a variety of (unique) conditions endemic to various geopolitical areas and the sociocultural characteristics of various user communities.

A modest effort has been already initiated toward the definition and elaboration of the idea and practice of building design cultures by the International Systems Institute (ISI). ISI, a member of the International Federation of Systems Research, constitutes a network of systems and design researchers, educators, and practitioners who serve as institute fellows. Representing several dozen higher education and R & D institutions from over a dozen countries, the "ISI community" has held over 20 research conversations. Meeting biannually in Austria, periodically in other countries, and annually at the Asilomar Conference Center in California, ISI fellows explore a range of systems and design inquiry issues, with the overall purpose of creating models, methods, resources, and programs for systems and design learning, and the development of design cultures.

Reflections

The design and development of the systems complex of design enculturation are immense tasks with far-reaching consequences. We are talking about nothing less than a great societal venture by which to empower individuals, families, communities, and the great variety of social and societal systems to give direction to their own evolution by purposeful design and to shape their own futures. The accomplishment of comprehensive enculturation would lead us into a new era of a creating society.

Activity #45

(1) Review the text and identify a set of core concepts of design culture and the building of design cultures. (2) Draft a brief explanation of what design culture is and why we need it. Discuss these two issues with someone who is a novice to systems design. (3) Select a section from this book that directly deals with how to design a social system and transform a couple of pages of the selected text either to the language of children or the language of laymen. Enter your findings in your notebook.

6.4. The Systems Complex of Design

The designing system is the entity that designs the system we want to bring about. In the "good old days," when the first generation of design approach reigned, the notion of a designing system in social/organizational settings was not even known. The top people in the system hired the design expert who came in and did his social engineering and produced the "blueprint" (for the system). If it was accepted by those who hired the expert, they directed their staff to implement it. A wall separated the expert from people in the system. The blueprint was thrown over the wall for others to repair or build the system.

The second-generation design approach wasn't much different. The expert/consultant came in, planned out the inquiry, used selected people as sources of design information and design verification, and presented the design to the decision makers. There are other kind of second generation approaches. For example, the board of directors goes for a three-day retreat, develops a vision for the organization, sets purposes, and establishes a budget. Attempts are made to "sell the plan to people" who are then directed (coerced) to put it into place. A variation of this is to be found in many organizations. The CEO develops a vision, becomes really enthused about it, holds a series of meetings, goes around and shares his vision, and with best of intentions asks people to "buy into it." The relationship between the first- and second-generation design approaches is continuous. The second generation is a benign modification of the first. And the "sell" or "buy-in" alternatives are modifications of the second-generation approach.

However, by now many of us know that social systems design just does not work this way. It does not work the way it used to work. The days of social engineering, the days of top-down driven designs, should be over. As you have worked through this text you most likely gained insights as to why it does not work the old way, or why it should not work anymore as it used to work. We shall see clearly in the rest of the sections in this chapter the radical difference, the discontinuity, the "liberating" difference between the user-designer/stakeholder design mode and the expert designer approach.

Three interacting and integrated ideas created the demand for this radically different approach to social systems design. These three were explored in the first part of this chapter. These are the ideas of the user-designer, the designing community, and the design culture as the shared culture of the designing community. The interaction and integration of these three create the requirement for the establishment of a designing system. This system is the organizational arrangement of the designing community of user-designers, who develop their own design culture by their very involvement in the design inquiry.

It has taken us almost a half century to understand and appreciate that the design of social systems, the design of human activity systems, is the sole right and responsibility of the stakeholders: those who serve the system, who are served by it, and who are affected by it. But these stakeholders can exercise this right and responsibility only if they empower themselves individually and collectively by learning how to design. They are challenged to develop design literacy and design competence and forge a design culture as the culture of their designing community. A designing community is created by sharing responsibility for design through collaboration, by viewing everyone as being of equal value in the creation of the design, by having loyalty to shared values and purposes, by maintaining consensual self-management of the enterprise, by drawing on the individual and collective intelligence of the group, and by involving everyone in a genuine and authentic way in designing their system and shaping their collective future.

In the previous chapters, we developed an understanding of what systems design is and how it works. In this chapter we aspire to understand how a community of user-designers can prepare for design and how they can get ready to engage in the design or redesign of a system of their interest. In this section, I map out the whole systems complex of the systems design. On the map I mark those systems of the complex that have already been explored and define those systems of the complex that are involved with the design and operation of the designing system (see Fig. 6.1).

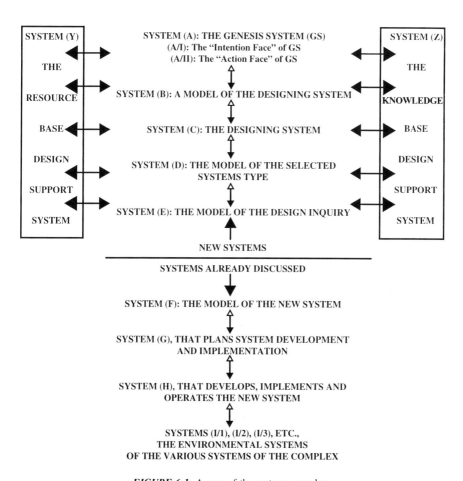

FIGURE 6.1. A map of the systems complex.

6.4.1. Defining the Systems of the Systems Complex of Design

In identifying and defining the systems complex of design, a distinction is made between several kinds of systems:

1. Human activity systems that engage in the design and implementation of the system to be designed.
2. Abstract or conceptual systems, e.g., systems models that are conceptual representations of the systems that are designed, and are created by the various real-world designing systems.
3. Design support systems include (1) a knowledge-base system, which supports the disciplined inquiry of design and (2) a resource-base system, composed of people, materials, financial, and facilities support.
4. Environmental systems can be real or abstract, systemic or general environments.

6.4.1.1.   System (A): The Genesis System

System (A) is a two-faced system. System (A/I) is the system of "intention" and system (A/II) is the system of initiating design "action."

6.4.1.1a.   System (A/I): "Intention Face" of the Genesis System.   System (A/I) is a human activity system whose members want to initiate design. Their shared perceptions, ideas, and aspirations motivate and generate interest and willingness to engage in design. This system becomes the nucleus around which the designing community will be built. To initiate design, members of this system should understand what design is and how it works. They should understand what systems are involved in design and how they are related.

6.4.1.1b.   System (A/II): "Action Face" of the Genesis System.   The "intention" system is transformed into the "action" system by engaging in design learning with the use of learning resources and programs available in the knowledge-base system, marked on the map as System ("Z") (see Fig. 6.1). Here is a legitimate role for experts who are knowledgeable about and competent in design. They are to develop design learning resources and programs and can be of help in guiding design learning.

Designers in System (A/II) are faced with three tasks: (1) developing design literacy and design competence; (2) creating a map of the design journey, such as the map in Fig. 6.1; and (3) engaging in the design of the model of the designing system: System (B). By accomplishing these tasks, designers not only devise the model of the designing system, but even more importantly, they are well on their way to developing their own design culture.

6.4.1.2. System (B): A Model of the Designing System (DS)

This is an abstract system, a conceptual representation and description of the model of the designing system. In creating the model, designers arrange design activities in the framework of the design architecture introduced in Chapter 2 and portray the design tasks of designing the designing system. Tasks include: (1) creating an image of the designing system (DS); (2) formulating its core definition/purpose, and stating its systems specifications; (3) defining the system of functions of the DS; (4) designing the system that has the organizational capacity and human capability to carry out the functions, meet the systems requirements, and accomplish the purposes of the DS; (5) testing the various alternatives that emerge from tasks (1) to (4); (6) selecting the most promising alternative; and (7) describing the model of the DS. To describe the DS it is appropriate to use the systems-environment, functions/structure, and process/behavioral models. In the course of the inquiry, designers (1) constantly make use of the knowledge-base system (Z) and (2) apply the range of resources available in the resource-base system (Y). As designers complete these design tasks, they continue design learning and develop and refine their design culture.

6.4.1.3. System (C): The Designing System

The designing system (C), is a human-activity system. It is an implementation of the system model created as System (B). This action is a sterling example of self-creation and self-organization and the continuing evolution of a human activity system directed by purposeful design. The designing community, which created the model, now brings the model to life by its own existence. It will now engage in the primary task of designing the new or target system that the designing community aspires to create. But the design of that target system depends on the availability of two models. Model (D) is involved in the selection of the type of system designers wish to create, e.g., purposeful, heuristic, purpose-seeking. Based on the characterization of the selected system type, designers create the model of the design inquiry, System (E), which is a representation of the design strategy and the methods the designers will apply as most appropriate to use in the design of the "target system," System (F).

In the course of attending to these tasks, designers constantly connect with, and make use of, the knowledge base in System (Z) and the human and material resources in System ("Y"). The outcomes of systems (D) and (E) are required as conceptual inputs to the design of System (F). A description of systems (D) and (E) is the subject of the last two sections of this chapter. Again, the crucial outcome of the design of these systems is the continuing development of design competence and the ongoing creation of the design culture of the designing community.

6.4.1.4.   System (D): The Selected Systems Type

In Chapter 2, we reviewed the various systems types from which designers may select the type they wish to create, such as purposeful, heuristic, or purpose-seeking. The selection of a certain type is a task of the designing system: System (C). It is a prerequisite and is an essential input to the formulation of the design inquiry employed in the design of the target system. Section 6.6 addresses this task.

6.4.1.5.   System (E): The Model of the Design Inquiry

The requirement for a specific design inquiry approach to the design of a social system emerged from an appreciation of and respect for the criteria of uniqueness and authenticity developed in an earlier section of this chapter. Section 6.7 describes the approach to designing the inquiry system.

6.4.1.6.   System (F): The Model of the Target System

The design of this model is the primary task of people in the designing system. Carrying out this task has been the topic of all the previous chapters.
The line drawn on the middle of Fig. 6.1 marks the division between what is new material (above the line) and material that we have already worked with (below the line). Below the line we have several systems that mark the continuation of the overall tasks assigned to the systems complex of design.

6.4.1.7.   System (G): Plan for System Development and Implementation

Once the model of the new system is available, the designing community is responsible for the preparation of a plan for the development, for developmental testing and implementation of the model, and, based on the plan, for bringing the design to life, that is, turning it into a real-world operating system. The plan provides (1) a detailed description of these activities that also reflects the unique context and the nature of the system to be operationalized; (2) a plan that introduces in great detail the human, information, material, financial, and facilities resources required for the development and implementation of the system; and (3) plans for their acquisition.

6.4.1.8.   System (H): That Develops, Implements, and Operates the New System

This is the last and crucial task of the designing community. Based on the plan of System (G), the designing community develops, tests, and implements the new system. In fact, this crucial task is accomplished as the designing community transforms itself into becoming the new system. Members of the designing community now live in, work in, serve in, and operate the system they themselves designed. The success of the new system is guaranteed by the fact that it was created by those who serve the system and who are served and affected by it.

6.4.1.9. System (Z): The Knowledge-Base Design Support System

The design of social systems, as disciplined inquiry, requires an extensive knowledge base that must be developed and made available from the onset and that should be constantly enriched as the design unfolds and calls for new knowledge. System (Z) should be responsive to the information/knowledge requirements of the many aspects of the context and content of the various object systems and the many dimensions of the intellectual technology of systems design. Without exception, all systems of the system complex will have to have access to System (Z).

6.4.1.10. System (Y): The Resource-Base Design Support System

As the design activities of the systems complex unfold, they will require increasing support in terms of people, materials, facilities, and financial means. The lack of availability of any one of these resources constrains design, while, on the other hand, their unrestricted availability enhances design. For example, how much of the ideal system can we attain depends a great deal on the availability of resources.

6.4.1.11. Systems (I/1), (I/2), (I/3), etc: Environmental Systems

Each and every system, abstract or real-world, portrayed in Fig. 6.1 and described in the text above has its own environment. Each has a systemic environment, the environment with which the particular system of the complex constantly interacts, and the general (larger) environment that embeds the particular systemic environment. Some of the environments overlap and may constitute the context of more than one system. Then, the whole complex has its own, even larger environment that is more than the sum of the various environments.

Activity #46

(1) Identify and describe sets of core ideas that emerge for you in working with this section. (2) Using Fig. 6.1 and the text that describes the various systems of the systems complex, describe the interactions that the arrows represent. (3) Speculate about the characteristics and the functions of the various systemic and general environments. Describe your findings in your workbook.

Reflections

As you review the first four sections of this chapter, most likely you will appreciate the great differences between the first two generations' approach to design (the expert and the consultant) and the user-designer/designing community approach to social systems design. In a consultant/expert-driven design the systems that are placed above the line in Fig. 6.1 do not exist. Whatever front-end system work is done in the expert-driven design modes is done by the consultant/expert.

As we think through the implications of what the emergence of design culture might mean as a vital component of the culture of our families, our organizations, our communities, and the larger society, a new horizon opens up for us. We can capture the image of the liberation of individual and collective human potential and the recognition of the right and acceptance of the responsibility for designing our systems and our communities and taking responsibility for the shaping of our individual and collective futures.

The next three sections address some of the organizational and methodological aspects that inform designing communities in conducting their preliminary design activities.

6.5. Designing the Designing System

The systems complex of design is rather a novel notion and has not been elaborated in the design literature. Furthermore, the issue of designing the designing system is a largely unattended topic. Nadler (1981) provided a set of propositions that guide thinking about it. Ackoff (1981), Banathy (1991), and Christakis (1995) have briefly described approaches for designing the design system. The reason for such a dearth of knowledge is quite obvious. The first- and second-generation design approaches are expert driven. Experts have their own approach to design and have no reason and no intention to empower their clients to do their own design. But once we have taken the position, as we have, that it is unethical to design a social system for someone else and the design of a social system is the right and responsibility of people who serve the system and are served and affected by it, then designing the designing system becomes an unavoidable and absolute requirement.

Figure 6.1 displayed the systems complex of design. In this section, the first three systems in the figure are addressed.

6.5.1. System (A): The Genesis System

We called System (A) the "genesis" system. It is the "seed" of the whole design complex; it is the origin of the "intent" to explore the design of a new system or redesign a system. This system has two component systems: System (A/I) and System (A/II).

6.5.1.1. System (A/I): The "Intention Face" of Genesis

This system is a human-activity system. People in this system, for whatever reason, initiate the exploration of creating a new social system. This intention emerges from a vision of a system we wish to create and a realization that "We cannot address a problem from the same consciousness that created it: We have to think anew," as Einstein said. Thus a prerequisite to the transformation of the vision into its realization is the development of new consciousness, a new way of thinking. This new way of thinking is systems and design thinking. Review Section 4.5 of Chapter 4 on the initiation of design. As you do, you might wonder: Are we doing the same thing over again here? The answer is: "No" and "Yes." Let us reflect on this.

6.5.1.1a. First the "No." Look at Fig. 6.1. The line in the middle of the figure separates the systems we have discussed and worked with up to this point from those we have not. It means that, in fact, we started the exploration of the design of social systems at the middle of the systems complex of design. But why? In discussing the "logic" of this book at the onset, it was suggested that we begin with an understanding of what social systems design is and how it works by consulting the existing knowledge base in the design literature. However, in the existing knowledge base there is practically no guidance about how to get ready for design and how to create a designing system. The existing literature generally is a reflection of the second-generation design approach. The few sources that reflect the third-generation approach do not elaborate on the issues of "getting ready" for design or creating the designing system. The rationale of addressing these issues is grounded in the assertion that social systems should be designed by their own stakeholders. But the designing system is a social system that is to be owned by the stakeholder designing community. Now questions arise: What is the system that can engage in the design of the intended new social system? Who should design this system? How should it be designed? We already answered these two questions. But there is a third question: What is it that designers should know, understand, and be able to do in order to be engaged in designing the designing system? These questions are answered in this section.

6.5.1.1b. Now the "Yes." Yes, the designing community of the genesis system has to contemplate the same question that was raised in Chapter 4: Should we initiate design? This contemplation takes some time. They want to know, and have to know, what they are getting into. The exploration of this question is rarely an up-front yes or no answer. Here, at the genesis stage, as well as later

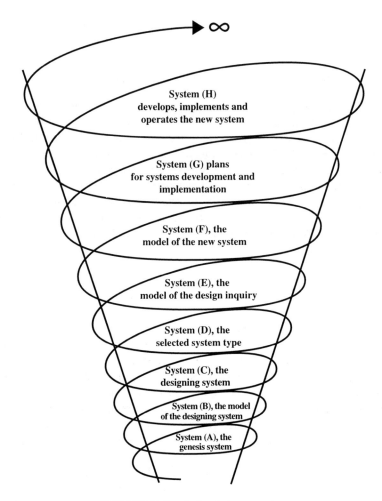

FIGURE 6.2. Developing design expertise.

on, uncertainty and ambiguity reign. Expectations are often laced with a fear of change. The design genesis community already at this stage has to get some solid orientation about the issues of what design is and how it works, so that they can become increasingly confident that their only reasonable choice is to engage in design.

Now, think about the issue of "how we become experts" in designing our systems as a "spiraling and continuously unfolding" learning, doing, and experiencing process. This process, in fact, never ends. The designing community is becoming ever more knowledgeable about and competent in designing as they are involved in design learning and applying learning as they engage in design and as they learn from the design experience.

The "spiral of developing expertise," depicted in Fig. 6.2, reflects Fig. 6.1. There are eight systems in the complex and they are now considered as learning opportunity systems in the spiral figure. We find the genesis system at the bottom of the spiral. As the spirals move upward, they become larger. They indicate cumulatively more depth and breath in design knowledge, design competence, design experience, and confidence in design ability. Thus, the spirals integrate into a "wholeness of an experience of emerging design expertise."

Having explored how the genesis system fits into the larger scheme of social systems design, and assuming that a decision was made to engage in design, we move into (A/II) of System (A).

6.5.1.2. System (A/II): The "Action Face" of the Genesis System

The action face of the genesis systems involves the initiating of design learning and the mapping of the design journey.

6.5.1.2a. Initiating Design Learning. This activity involves an intensive orientation about social systems design, the development of design literacy, and an engagement in the kind of design learning that is relevant to working with the tasks of the genesis system. These tasks require design knowledge and competence that enables the designing community to (1) work in teams in a design-conversation mode to explore intention/resolution issues; (2) understand the process of social systems design so that they can lay out the map of the design journey; and (3) engage in the design, modeling, and implementation of the designing system. All these learning activities are placed in the functional context of working with the four tasks of System (A/II). They are not presented in an instructional mode but in the form of doing while learning and learning while doing.

The question now emerges: What resources/programs might be accessed that enable such design learning? In the short range, we can rely on the assistance and guidance of design professionals who introduce and guide the use of design-learning resources. Here and only here do design professionals have a legitimate role to play in the course of a design program. They are accustomed to playing the expert role of designing the system, or at least driving the design activities. They have to learn now to play a new role of providing guidance to the design learning of the designing community. In the long run, we expect that the designing community will be able to work with user-friendly design-learning resources and programs, which we discussed earlier in the development of a design culture. Design professionals might play the role of introducing these learning resources. The more we advance in developing a design culture, the less we depend on outside help. Over the long run, the task of design professionals becomes the development of design-learning resources and design-learning systems.

As Fig. 6.1 shows, parallel with activities in System (A/II), work should be initiated in knowledge-base design support in System (Z). Arrangements, resources, and programs that are required to support design learning in System (A) should be made available, developed, and deposited in System (Z) from the very onset of the initiation of the design program. The same kind of development has to commence in resource-base design support, System (Y). This system provides facilities, material, and human resources required during the activities of System (A).

6.5.1.2b. Mapping the Design Journey. This mapping is an integrated learning and doing experience. It provides for a second learning and doing experience that enables the designing community to learn more about design as they draw the map of design activities from A to Z, as shown in Fig. 6.2. During this design activity, the designing community develops an understanding of what design is and how it works at a level of knowledge and competence that enables them to develop not only a map of major design activities but also to have a grasp of how these activities are carried out. The manner in which this task is accomplished is in the form of guided workshops that involve intensive conversation and documentation of findings. In the course of this activity they initiate learning about the creation of the designing system.

6.5.2. Creating the Designing System (B)

The notion that there is a designing system that is to be designed is rarely understood and seldom discussed. In contemplating an approach to creating the designing system, we can draw upon what we have learned about designing social systems. Still there is a significant difference between the design of the target system and the design of the designing system that designs the target system. The designing system is an "enabling system" that provides organizational arrangements and a design approach to the design of the target system. The design tasks of designing this enabling system include creating a vision, formulating core ideas and core values, describing the image of the design system, and, based on the image, engaging in the creation and modeling of the designing system. The text below provides some examples of working with these tasks.

6.5.2.1. Creating a Vision

An example of a vision: The designing system is a community of designers who developed their own design culture, which empowers them individually and collectively as a community, to design their own lives and the systems in which they live, to shape their own futures, and to design just systems of learning and human development that empower future generations to design their own lives and shape their own futures.

6.5.2.2. Core Ideas about Designing the Design System

A set of core ideas are introduced here as an example of ideas that the designing community would generate, ideas that will guide their thinking about designing the designing system. First, consider that the designing system is a human activity system. It is a system of interacting and interdependent components, operating in and interacting with its systemic environment, from which it gets its resources and information required for design. The interaction with the environment is coevolutionary and cocreative. Furthermore, the designing system has a purpose: it works through the input, transformation, and output processes; it is guided by negative and positive feedback. The designing system is created by the disciplined inquiry of systems design, which produces the system model of the designing system. The model then has to be implemented and operationalized as the designing system, System (C). Designers should seek to design a robust system that allows for flexibility and continuous improvement and redesign.

Another core idea is recognition of the uniqueness of the designing system. Nadler (1981) reinforces our earlier proposition that every design situation, every design environment, every designing community is unique. Therefore, a specific designing system reflects this uniqueness. A design system operating successfully and effectively in one specific environment cannot be transferred to another effort or another organization. Understanding this proposition overcomes one of the most dangerous assumptions in systems design: the notion of transferability. Organizations pay a high price for trying to adopt the simplistic idea of direct transfer.

Multiple perspectives, such as the technical, the organizational, the cultural, and the personal should be considered from the very onset of designing the design system. The competent use of both generative and strategic dialogue, which are two operating modes of design communication, is also a critical concern. Creativity is the main approach to seeking design solutions. The notion of sweeping in the widest range of perspectives and information and knowledge will help to avoid the underconceptualization of the designing system. The search for a range of alternatives at each point of making design decisions serves the same purpose. We have to keep in mind the idea of user-friendly language, language that is compatible with the everyday language used by the designing community. The dynamics of design call for the application of a multidirectional, free flow of unrestricted feed-forward and recursive interaction, and constant and recursive feedback.

6.5.2.3. Core Values That Guide Design Decisions

Core values in systems design are of two kinds. One set of values reflects the aspirations and expectations that members of the designing community have individually and collectively about the system they wish to create. The second set is the values they hold about the design of the designing system. Here this second set is highlighted as an example of a generic set of values that the designing community might consider in the course of their design conversation as they collectively decide on values that will guide decisions about the characteristics of the designing system they wish to build. The example reflects perspectives developed in this work.

The designing community that shares the perspectives of the third-generation design approach most likely would value the genuine and authentic participation of the community and the activation and use of their collective intelligence, the development of their full potential through continuous individual, collective, and organizational learning, and the building of a shared design culture. They would value the liberation of individual and collective potential, the nurturing of freedom of choice, and the taking of responsibility for the design. They would hold that if these values are realized in the designing system it would not only enhance the sustainability of the designing system, but it would add value to it. Furthermore, the designing community would place high value on the search for and pursuing the ideal and the ethics of the process and product of the design. In their ethical stance they would be especially sensitive to the effect of design on future generations. They would aspire to attain an aesthetic quality of their design and nurture creativity and personal and collective uniqueness. They would be sensitive to and would honor diversity and individual and group differences. They believe that if all these values are activated in personal and collective behavior and manifested in the designing system, they would attain the best that could be attained in the design inquiry. More importantly, they would be on the road to building a genuine, responsive, committed, competent, and responsible designing community.

6.5.2.4. The Systemic Image of the Designing System: An Example

The image provides a broad-stroked picture, a "macroview" of the designing system. It is devised as a set of essential markers that stand for the qualities we seek to realize in the designing system, arranged in systemic relationships. The image is rooted in the vision and manifests the core ideas and core values of the designing community. The markers are clustered according to the nature of the qualities we seek to realize in the designing system. These may be qualities of the design process, systemic/organizational qualities, individual and collective human qualities, qualities of the community and its culture building, and aesthetic and ethical qualities. The markers of these qualities should be internally consistent and mutually reinforcing. What is crucial to understand is that these qualities are not and should not be considered in isolation but in their interaction and integration. Integration enables the emergence of the overall quality of the designing system and the designing community. Figure 6.3 depicts an example of an image of qualities we might seek to realize in a designing system. The image is speculative and serves the purpose of contemplation and conversation. It is created by the designing community as the first systemic representation of what they believe they "should become" as a community. The image becomes the basis for engaging in the design of the designing system as well as the design of the target system. Keep in mind that in a user-designer mode the designing community eventually becomes the target system; they design their own future system.

The image markers arranged in the circle of Fig. 6.3 imply qualities that designers could seek to bring about as they design the designing system. The markers, as well as their arrangement into sets of markers, are speculative. Two important considerations should be noted. The markers should be internally consistent. They imply a design philosophy of what one might consider to be important qualities in the deign of designing systems. The second consideration is the interacting, mutually reinforcing nature of the various markers. It is from

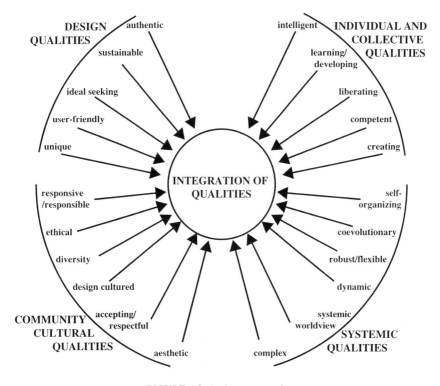

FIGURE 6.3. An image example.

such interaction that the essential overall quality of the system emerges. Designers should test for both internal consistency and emergence. In so doing, they might find some "missing" or redundant markers and might find some "counteracting" relationships. This testing is important to the future successful unfolding of design inquiry.

6.5.2.5. Transforming the Image of the Designing System into the Model of the Designing System

Transformation is accomplished by carrying out five tasks: (1) formulate the purposes of the designing system (DS), (2) design the system of functions, (3) design the organization that carries out those functions, (4) model the DS and its systemic environment, and (5) prepare a plan for the implementation of the DS.

6.5.2.5a. An Example of Possible Purposes of the DS.   The DS should

1. Be self-organizing, be systemic in nature and behavior, be robust and flexible, employ multiple perspectives, and be capable of coevolving with its environment.
2. Enable the designing community to carry out the design of the authentic and sustainable design of an ideal model of the desired target system.
3. Devise a design inquiry that responds to the unique conditions and requirements of the design situation of the target system and select the systems type for it.
4. Provide programs and opportunities for individual and collective development of design competence and organizational learning.
5. Liberate the human potential of designers and ensure the use of individual and collective intelligence.
6. Nurture creativity, uniqueness, and diversity.
7. Crèate arrangements for the coordinated use of required knowledge and resources.
8. Ensure high quality of working conditions for members of the designing community.
9. Encourage and nurture cooperative and helping relationships among the designers and facilitate their personal and professional development.

6.5.2.5b. Designing a System of Functions.   Given the purposes formulated by the designing community, designers now face the question: What are functions that have to be carried out to attain the purposes of the designing system? This inquiry leads to considering alternative functions, evaluating them, and selecting those that best respond to the stated purposes. A system of key functions emerges from this inquiry. Each key function has its own system of functions, and each function has its own system of component functions. This inquiry produces a systems complex of functions. The system of functions becomes the first systems model of the designing system.

6.5.2.5c. Designing the Enabling Systems of the DS. The systems complex of functions tells us what activities have to be carried out to achieve stated purposes. Now the functions become the basis for designing the system that has the capability to manage the design effort and for designing the organization that has the systemic capacity and human capability to carry out the design.

6.5.2.5d. Modeling the Designing System and Its Environment. The outcome of the design of a social system is a prescriptive representation of the future system. In the present case, the question is: What is that system that can carry out the design of the desired future system? The designing system is presented now in three complementary systems models: the systems-environment, the functions/structure, and the process/behavioral models. These three models were briefly described in Chapter 3. In working with the systems-environment model, designers will also model the systemic environment that has the capacity to support the designing system.

6.5.2.5e. Formulating the Implementation Plan. This task takes us beyond design. It brings the design to life. The plan addresses the development of the design system and the arrangement of resources and programs required for its implementation and operation.

6.5.2.6. The Dynamics of Design

The five design functions described above are not carried out in a linear way. They are arranged in a systemic architecture that provides the recursive interplay and the integration of the five tasks that create the dynamics of systems design. Figure 6.4 presents this architecture. (The generic form of this architecture was introduced in Chapter 3.)

The arrows in Fig. 6.4 indicate how the architecture can be used to portray the dynamics of design. The process initiates from the exploration space and spirals through the four tasks in the design solution space. Each space in the architecture is connected with the knowledge/resources space. The spirals go through the testing experimentation space. Finally, the models are displayed in the modeling space. The two-headed arrows indicate the dynamics of recursiveness and mutually influencing feedback and feed-forward.

6.5.3. Designing Systems (C) in the Systems Literature

One can find only a few descriptions of designing systems in the literature. Ackoff (1981) portrays a description of a circular organization for design partici-

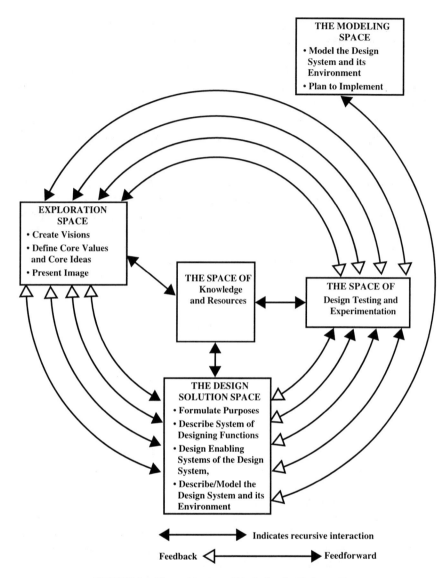

FIGURE 6.4. The architecture of designing the design system.

pation in corporations, called interactive design boards. The boards are constitut-
ed at each level of the organizational hierarchy and each board has representation
from the next higher and lower levels. This arrangement enables a free flow
of design information and an integrated design activity up and down through
the designing system. This system was adapted to the design of a research

and development organization (Banathy 1987c). In another example, Banathy and Jenks (1990) and Banathy (1991a) describe a comprehensive organizational arrangement for the design of systems of learning and human development.

It is suggested that the design of designing systems, described in this section, presents a far more comprehensive approach to the creation of designing systems than previously described in the design literature and it implies a far more substantive designing system than the examples mentioned here.

Reflections

The first five sections in this chapter cumulatively provide an approach to preparing for and engaging in social systems design. But as I reflect on the deeper meaning of the process unfolding through these sections, I begin to see implications that go far beyond being just an approach to social systems design. The consideration of stakeholders as user-designers, the idea of a genuine designing community, the development of a design culture, the building of a systems complex of design, and the creation of a designing system; all these cumulatively establish organizational capacity and individual and collective capability to carry out social system design. But there is much more to it. If we engage in the system-building activities of designing a designing system, these activities empower us individually, and collectively as a community and a society, to design our own lives and the systems and the communities in which we live, and thus shape our own future. And this is precisely what the true meaning of democracy is: "power to the people." This is what true empowerment is, which today is only a slogan without any substance. Today, we empower our political representatives, who make decisions for us. But tomorrow, if we develop a design culture, then, democracy: meaning power to the people, will become a reality in the form of a truly participative, creating, designing society.

Activity #47

(1) Describe the core ideas of building a designing community. (2) Study Fig. 6.3. Consider the markers of the image and ask: Which marker relates to what other markers? How do the markers reinforce each other? What are some "missing" or "redundant" markers. What are markers that might counter and reduce the effectiveness of others? Record your answers in your workbook.

6.6   Coordinating Design Methods with System Types

The basic premise of this section is that when engaging in the design of social systems, the design methods chosen should be coordinated with the type of system we wish to design and the characteristics of the design situation. This theme is developed here in three interdependent components that jointly constitute the argument for the need to select design methods that are appropriate to both the system type and the characteristics of the design situation. First, we consider the relationship of methodologies and system types. Next, a framework is constructed that enables us to make distinctions among the various types and explore their characteristics. Third, the implications of this inquiry is examined. It is important to note that the selection of methods is only one aspect of the considerations one has to contemplate in designing the design inquiry. (The text of this section is a partial adaptation and further elaboration of an article (Banathy [1988c].)

6.6.1. Methodology and System Types

In developing the theme of this section, the first task is to show the relationship between design methodology and system types. This relationship is explored as follows: define methodology, identify key factors in selecting methods, and introduce dimensions by which to make distinctions among various types of systems.

6.6.1.1. Methodology

A methodology in disciplined inquiry is defined as a set of coherent, related, and internally consistent methods applicable to pursuing disciplined inquiry. In conclusion-oriented disciplined inquiry of a particular discipline, methodology is clearly defined and it is to be adhered to rigorously regardless of the problem or area of investigation in which it is applied. In the conclusion-oriented domain of scientific disciplines, methodology is the hallmark of a discipline. In decision-oriented disciplined inquiry, on the other hand, one selects methods and methodological tools or approaches that best fit the specifics of the inquiry. Moreover, the methodology judged to be appropriate may consist of any one or a combination of mathematical, computational, heuristic, experimental, and exploratory methods (Klir, 1981).

6.6.1.2. Selecting Methodology

The goal of the inquirer is, therefore, to select or define methods and tools that (1) have internal consistency, (2) are consistent with, and reflect, their theoretical base, and (3) best fit the nature and the context of the inquiry. This position was further defined and refined by Flood and Jackson (1991) and Jackson (1992, 1995). Their principle of complementarity holds that the various methodologies are to be grounded in theoretical positions. (The relationships between philosophy, theory, methodology, and use of methodology were discussed in Section 4.2, Chapter 4.) In their discussions on systems metaphors (that represent different system types) and in setting forth the idea of a "system of systems methodologies," the authors show a range of systems methodologies. Complementarism offers guidance and tools that could be used to put methodologies to work on the issues and contexts for which they are most suitable.

Design inquiry is a specific type of systems inquiry in the context of which there is a variety of approaches and methodologies available that need to be matched with the various aspects of design inquiry to ensure "goodness of fit." These aspects include the design issue at hand, the context and content of design, the available resource and knowledge base, the designing community, and the type of system designers wish to create. In this section the exploration is limited to an investigation of "goodness of fit" between design methods and system types.

6.6.1.3. System Types

In order to advance the exploration, a set of continua that differentiates various types of human activity systems is introduced here. The continua enable us to develop a typology of human activity systems. Designers should be sensitive to the variations in types and select design approaches, methods, and tools that are most appropriate to the type of system they wish to design. In developing this section, I have used the work of several systems scholars, including Ackoff (1981), Banathy (1988c), Flood and Jackson (1991), Jantsch (1980), Jackson and Keys (1984), Jackson (1992), Sutherland (1973), and Vickers (1983).

Five continua that provide dimensions on which to differentiate system types are proposed. Each continuum is constructed from opposites: mechanistic versus systemic, unitary versus pluralist, restricted or simple versus complex, closed versus open, and dominating or autocratic versus liberating and empower-

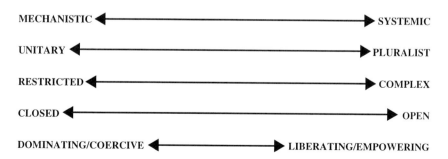

FIGURE 6.5. Five continua of system types.

ing. These continua are shown in Fig. 6.5. The ten descriptors in Fig. 6.5 are defined as follows:

- *Mechanistic* implies a system in which the parts are of primary significance, are stable, and operate in a fixed relationship.
- *Systemic* indicates dynamic relations among the components of the system, where the interactions, and whatever emerges as a result of those interactions, are the significant properties of the system; the whole organizes the parts.
- *Unitary* refers to a system where there is a clearly designated or prescribed singleness of purpose or goal.
- *Pluralist* is a system in which there is a diversity or possibly even conflict as to purposes or goals.
- *Restricted* denotes a system with few, clearly defined variables and permanence of state status.
- *Complex* indicates a system with a large number of system variables, interactions, and components, and multiple levels of decision making.
- *Closed* refers to a system with well-defined and guarded boundaries and limited and highly regulated interactions with the environment.
- *Open* does not mean complete openness and no boundaries, but a great deal of interaction and exchange between the system and its environment, coupled with flexible or even fuzzy boundary conditions. There is mutuality of influence, even coevolution between the system and its systemic environment.
- *Dominating* means an autocratic system, with little regard to the desires and purposes of people in the system. People are there only to serve the purposes of the system.
- *Liberating and empowering* means a system in which people are invited to make unique contributions, to participate in decision making and use their individual and collective creativity and intelligence.

The dimensions described above will now be used to construct a map that will help us travel the territory of system types, describe the various types, and, with the use of the map, explore the general characteristics of the various types.

6.6.2. A Map for the Characterization of System Types

Using the ten descriptors that compose the continua as the vertical dimension, a set of five system types can be contemplated and named on five lines of the horizontal dimension. This two-dimensional framework enables us to create a map that displays five major system types, distributed on the five continua. The vertical order of the five continua is of no particular significance. The sequence

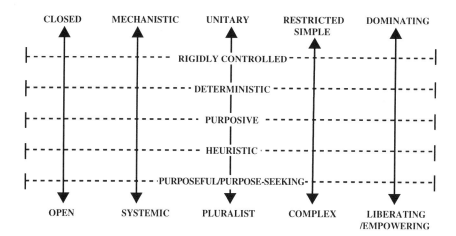

FIGURE 6.6. A map of system types.

of designating the system types, however, is guided by the position the system types take on the various continua. Figure 6.6 presents the map.

The map depicted in Fig. 6.6 travels the territory of system types and describes their characteristics by

1. Providing a general definition of system types.
2. Characterizing the degree of freedom available to people to participate in decision making.
3. Describing the nature of the structure and the systemic relationships of the type. The description is coupled with (1) metaphors that might be used to reflect on the nature of the system type, (2) organizational examples of the particular type, and (3) the mode of overseeing.
4. Speculating about methodologies that might be appropriate in working with the particular system type.

The source of metaphors is Morgan's (1986) *Images of Organizations.*

6.6.2.1.   Rigidly Controlled Systems

1. In general, these systems are rather closed and have only limited and well-guarded interactions with their environments. They are restricted and have few components. They are unitary in purpose with clearly defined goals, and behave mechanistically. People in the system are dominated by the hierarchy and they are coerced to comply.

2. People in these systems have practically no operational freedom. Objectives and ways and means of operation are clearly prescribed and rules and

standard operating procedures regulate behavior. Decisions are made at the top. There is little or no room for self-direction and no room for creativity here.

3. These systems have rigid structure, and well-defined stable relationships are established among the various system components, with little dynamics and only minor changes over time. (1) Metaphors for this type are: organizations as machines (emphasizing efficiency) and organizations as instruments of domination. (2) Examples: man–machine systems, assembly-line work groups, autocratic organizations, and some primitive socioeconomic systems. (3) The mode of "overseeing" is "foremanship" and "supervision."

4. Methods that might be appropriate include deterministic methods that rely on quantification and mathematical representations or simulation models, traditional operational research methods, certain systems engineering methods, linear systems analysis, etc. The design approach to this type is systems that are designed by the engineer or expert.

6.6.2.2. Deterministic Systems

1. Moving one notch away on the map from the mechanistic/unitary/restricted/closed/dominating end of the continuum, the next type is defined as deterministic. Systems of this type are more open to their environment, but still closely guard their boundaries. They have more variables and the relational arrangement of their components is more complex than that of the rigidly controlled type. They still operate in a mechanistic mode with clearly defined purposes and goals. Domination still prevails.

2. Goals are clearly set, and operational objectives are prescribed. Decisions are made at the top. Ways and means of operating are still regulated, but there might be some discretion in using methods/tools. Thus, people in this type of system have some limited operational freedom. There is not much room for systemswide application of one's creativity and intelligence.

3. The structure of deterministic systems is more complex than that of the rigidly controlled systems, and relationships among the components are also increased. Some minor relational and structural changes can be expected to happen through time. This type of system is still a steady-state system. (1) Metaphors are: (still) organizations as machines; the Darwinian view of population ecology (survival, fight for resources and competition); and the dominating metaphor. (2) Examples: machine bureaucracies, professional bureaucracies, centralized (national and religious) educational systems, government agencies, military organizations, and small business. (3) Mode of "overseeing" includes "supervision" and "administration."

4. Applicable methodologies might include: operations research and systems analysis methods, living systems process analysis, information theory–

based engineering, negative feedback–driven methods, first-order cybernetics, systems dynamics, management by objectives, etc. The design approach to this type is the first-generation approach, designed by the expert mode.

6.6.2.3. Purposive Systems

1. With the purposive system type we reach the midrange of the mechanistic versus systemic and the open versus closed continua. On the other hand, on the restricted versus complex continuum, purposive systems may be placed close to the complex end. On the unitary versus pluralist continuum, purposive systems are still unitary as to purpose and goals. On the dominating versus liberating continuum, we place this type in a middle position.

2. Purposes are set for the system and strategic goals are prescribed; however, operational objectives and methods and means of operation can be self-selected. There are occasional rewards for some inventiveness. But a systemswide use of one's creativity and intelligence is usually limited to making suggestions.

3. State changes are gradual, influenced primarily by environmental changes that indicate the need for adaptation. Structural changes happen gradually over time and are coupled with changes in systemic relationships. Increasing complexity in this system type indicates multilevel hierarchy and multiple embeddedness. Purposive systems usually embed deterministic and rigidly controlled systems. (1) Appropriate metaphors include: organizations as organisms (managing organizational health and environmental relations), organizations as brain (information processing and learning). (2) Examples: Corporations and (post) industrial production systems, techno systems, public service agencies, and public education. (3) Mode of "overseeing" is "management."

4. Methods that are relevant for this type may include those based on second-order cybernetics (e.g., viable systems approach) systems analysis, social engineering, consultant-driven systems design, and living systems process analysis. Having some freedom for selecting operational objectives, consensus-building methods also come to play in purposive systems.

6.6.2.4. Heuristic Systems

1. The overall purpose of heuristic systems is still defined, but heuristic systems tend toward being pluralist in that they can formulate their own goals and objectives. They are complex and systemic in their functional and structural arrangements and are open to, and coevolving with, their environments. They tend to be liberating and empower people in the system, inviting their genuine participation and making use of their collective intelligence.

2. Overall policy is set for heuristic systems, and within the policy frame-

work, goals, objectives, ways and means, and methods of operation are self-selected. Relying on participation, the creativity and intelligence of people in the system are encouraged and engaged.

3. Significant relational changes and some structural changes might occur through time. State changes are evolutionary and are directed by design. Emergence, ambiguity, and uncertainty surround state changes. (1) Relevant metaphors include: organizations as brains; holographic organization in which the organizational "DNA" is distributed in all parts and in all members of the system; organizations as cultures, with collectively shared values, perceptions, and meaning. (2) Examples include: corporations developing new ventures, innovating and renewing education systems, high-tech organizations, R & D agencies, nontraditional health care systems, and environmental protection/renewal systems. (3) Mode of "overseeing" is "leadership."

4. Methods are grounded in second order cybernetics, soft systems and critical systems theory, and systems view–based organizational theories. Applicable methodologies include participative social systems design, soft systems methodologies, critical systems heuristics, total systems intervention, double-loop organizational learning, interactive management, and consensus-building methods.

6.6.2.5. Purposeful/Purpose-Seeking Systems

1. These are complex, ideal-seeking systems guided by images of the future that the stakeholders of the system shape. These systems are open to the environment as well as shaping their environment, they are coevolving with it. They are pluralist and thus able to seek and explore new purposes. They are systemic in their arrangements and behavior. They exhibit such behavior as self-transcendence and self-transformation. They have a tendency for cooperation and even integration with other systems. They often reorganize at higher levels of complexity. They nurture the liberation of people's potential and their empowerment by enabling them to attain design competence. They exemplify "creating democracy."

2. Policies, purposes, goals are formulated based on images of the future that people in the system shape collectively. There is constant search for ways and means to pursue the ideal. The authentic and competent contribution, the creativity, and the intelligence of all members of the system are constantly invited, sought, and nurtured.

3. Significant structural changes may occur through time. State changes may be independent of prior states. Discontinuity and reorganization at higher levels of complexity are expected to happen and are directed by purposeful design. Ambiguity and uncertainty are used creatively. (1) Metaphors: organization as culture, holographic organization, the learning/self-creating organization.

(2) Examples: organizations seeking and designing new institutional roles in the private and public sector, systems seeking societal renewal through the design of integrative community systems, self-creating systems of learning and human development, wellness systems, high-tech enterprises, artistic/creating enterprises, peace development systems, and alternative security systems. (3) Mode of "over-seeing" is shared leadership and "stewardship."

4. Methods in general are rooted in the emerging theory of dynamic systems, critical systems theory, soft systems theory, unbounded systems thinking, and the theory of social evolution. Methodologies that are most useful include soft systems methods, participative ideal systems design, systems heuristics, total systems intervention, interorganizational linkage methods, double-loop organizational learning, interactive management, and various consensus-building methods.

From the way the five types are presented here, it may appear that a preference is assigned to heuristic and purposeful systems. However, presentation of the five types is descriptive, not prescriptive. Still, there is a general desire that we secure more freedom for people to develop and use their full potential and their creativity and intelligence. Obviously, there will be components embedded in complex systems that have more restrictions in the way they are to operate, such as an accounting or a shipping component. But there is no reason why the embedding systems would not fully attend to the desires and interest of members in these components or why their initiative, creativity, and intelligence should not be fully engaged.

6.6.3. Implications

Issues discussed here include (1) the implications of embeddedness of system types and (2) additional characterizations of heuristic and purposeful/purpose-seeking system types.

6.6.3.1. Embeddedness of Systems Types

In complex systems we find various layers of embeddedness of systems that represent several of the types described above. It is important to recognize the potential of such embeddedness and the need to approach the various embedded and embedding systems with the use of methods that are appropriate to the particular system type. For example, an industry-based corporation may have embedded in it a large number of assembly line (rigidly controlled type) systems that are embedded in several production management (deterministic type) systems that are embedded in corporate management (purposive type) systems.

Another arrangement of embeddedness can be observed in a large research and development organization. Specific projects in such an organization operate as deterministic systems. Project goals and objectives are set by contractual arrangements with freedom limited to selecting means/methods. At the next level, projects are clustered in programs that operate as purposive systems. Programs may cluster in divisions that may be the heuristics type. The R & D organization that embeds these program areas behaves as either a heuristic or purposeful/purpose-seeking system. It is in constant search of new domains of research, new niches in the environment. In such an organization, designing a project system that has clearly defined programmatic, financial, legal, and time boundaries requires a design approach and methods that are very different from those that might be used in case the task were to design or redesign the entire R & D organization.

6.6.3.2. Further Characterizations and Their Implications

In a multiyear research project (Banathy *et al.*, 1979), we studied the design research literature extensively and explored a large number of design approaches/models. In the literature we did not find any comprehensive statement that addressed the issue of systems types. Furthermore, while we identified several sets of design models that are appropriate to rigidly controlled, deterministic, and purposive system, only a few appeared to be relevant to heuristic and even fewer to the purposeful/purpose-seeking system type. Based on the findings, or rather the lack of findings, I have become increasingly aware of the need to address the issue of "matching" system type with design methods. Inasmuch as our interest in this work is the design of social systems, which are mainly the heuristic and purposeful/purpose-seeking types, it seems to be appropriate to further explore and understand these types. An excellent source of understanding these types is Huber's (1986) article. The text below draws upon some of his ideas. The exploration that follows is tentative and speculative.

Heuristic and purpose-seeking systems have to cope with the explosive increase of relevant knowledge, environmental complexity, and turbulence. Furthermore, the changes faced by these system types increase with respect to their intensity, variety, and dimensionality. Consequently, designers who work with these system types need to (1) establish a space in the design inquiry where knowledge can be continuously infused and organized to serve the inquiry; (2) build a mental (conceptual) model of the environment that is relevant to the system to be designed; and (3) in the case of purpose-seeking systems, which coevolve with the environment, include in their design, arrangements and methods by which to shape or influence the environment. In view of the multidirectional and dynamic interactive nature of these systems types: (1) decision making becomes increasingly more complex and frequent, (2) the need for innovation and change accelerates, and (3) information distribution becomes more diffused.

The characteristics described above have some implications for design:

1. Persons contributing to design decisions will have to be involved in larger numbers than heretofore; thus, the making of design decisions will be more diffuse and the boundaries of the design decision system will become fuzzier and more permeable.
2. The heterogeneity of the design group will increase, as it will have to include people with various types of expertise, and will have to have an all-inclusive representation from the various constituencies.
3. The task of processing information in the course of contemplating design solutions will require the availability of effective decision methods with techniques of communication technologies built into them (Christakis, 1995).
4. The all-inclusive participation of stakeholders invites the availability of various communication and consensus-building methods and techniques (Warfield, 1990).

By their nature heuristic and purpose-seeking systems are adaptive to and are coevolving with their environment. The constantly changing environments of heuristic and purpose-seeking systems require frequent reconsideration of goals, perceptions, functions, and components and their structure. This situation calls for frequent organizational experimentation and redesign (Nystrom *et al.*, 1976). This state of "guided fluidity" invites continuous organizational learning (Jackson and Keys, 1984). The consequence of continuous organizational experimentation and learning suggests that the system would become ever more competent in the use of a variety of design features and "enacted" environments (Weick, 1979). Another consequence would be that the "organization would remain flexible, and thus would be less resistant to adopting unfamiliar features or engaging unfamiliar environments" (Huber, 1986, p. 13). Design will take on the characteristics of learning and—on the other hand—organizational learning becomes the genesis of continuous organizational design (Argyris and Schon, 1978).

Reflections

Reflecting upon the issues raised above, some additional design inquiry tasks can be set forth: (1) Formulate perspectives and criteria for the selection of methods that are sensitive to the systems type and characteristics; (2) develop an inclusive inventory of design methods and display them in the knowledge base; (3) experiment with various relevant methods and test for their internal consistency and applicability; (4) pay attention to the embeddedness and coordination of various system types; (5) define and model the systemic environment and inter-action with it.

Activity #48

(1) Describe core ideas you identified in this section. (2) Select a system of your interest, and in view of the description of system types, define the type that best matches your selected type and describe your reason for selecting it. (3) Review the design inquiry task in the "Reflections" above and integrate them into a comprehensive guidance for selecting design methods.

6.7. Designing the Design Inquiry

Various types of social systems were defined and characterized in the pre-ceding section. It is one of the tasks of the stakeholders who operate the design-ing system (DS) to select a certain system type before they engage in the design of the future desired (target) system. Their second task is to design the design inquiry process that they would apply in the design of the target system. A differentiation is in order. When designing a social system we ask what that system should be; in designing the design inquiry we ask how we should design that system. In designing the design inquiry, designers are to review design approaches, models, methodologies, methods, and tools. Then they select or develop those which are most appropriate to: (1) the systemic context of the design situation, (2) the desired future system, (3) the selected systems type, (4) the design competence and characteristics of the designing community, (5) the perceived complexity of the design task, (6) the time constraints if any, and (7) the knowledge and resources base available to support the design effort. The assumption here is that the design is carried out by the stakeholder community of the future system.

The process of designing the design inquiry requires a framework, an archi-tecture, within which designers create the design inquiry. In this section, I discuss design inquiry as a system of inquiry components, followed by introduc-tion of the inquiry architecture within which these components are considered, selected, and displayed as a model of the design inquiry program.

6.7.1. The System of Design Inquiry

Terms such as approaches, models, methodologies, methods, and tools were used to describe design inquiry in the previous paragraph. At times we don't have a clearly agreed upon interpretation of these terms. Therefore, I define these terms here.

6.7.1.1. Approach

Approach is the most general term. A design approach is a system of internally consistent principles of design inquiry, such as the ten principles of breakthrough thinking (Nadler and Hibino, 1990). Examples: the ideal systems approach, social engineering, soft systems approach to design, a critical systems approach, breakthrough thinking approach, critical heuristics of social systems design, etc.

6.7.1.2. Model of Design Inquiry

A model of design inquiry is an epistomological or process model, a descriptive representation of a system of internally consistent, theory-based design methodologies, with their component methods and tools. Examples: the design models of Ackoff, Banathy, Checkland, Christakis, Flood and Jackson, Jones, Nadler, Nadler and Hibino, and Warfield.

6.7.1.3. Design Methodology

A design methodology is an internally consistent system of methods selected to be used in a design situation. It could be subsumed by a design model. It incorporates several design methods and tools. Examples: transcendence (transcending the existing system), creating the image of the future system, transforming the image into a design of the future system. Other examples: creating a rich picture of the problem situation, exploring relevant systems, designing a conceptual model, selecting a methodology for means/methods analysis, etc.

6.7.1.4. Design Methods

A design method addresses a particular design task and is a constituent of a design methodology. Example: Creating an image of the future system might involve methods for envisioning the future system, establishing the boundaries of the inquiry, considering major options, formulating core ideas and values, and designing the core image of the system.

6.7.1.5. Tools

Tools are specific techniques applied to address the most detailed aspects of design inquiry. Nadler and Hibino (1990) describe over 20 tools. Warfield and Cardenas (1994) presented tools such as: ideawriting, nominal group technique, delphi, option fields analysis, option profile, tradeoff analysis, etc.

6.7.2. Architecture for Designing the Design Inquiry System

The architecture for designing the design inquiry system is adapted from a generic inquiry architecture (Banathy, 1993b). The architecture used here is a specific case of the generic, offered as a framework for designing the design inquiry program. The outcome or the product of this inquiry is a description or model of the design program, denoted in Fig. 6.7.

As pictured by the architecture (Fig. 6.7), the inquiry proceeds from the contextual space in which designers set the stage for the design of the design inquiry program. From this space the process moves into the design solution space, in which the various design approaches, strategies, methods, and tools are considered and selected. These are tested and validated in the design program

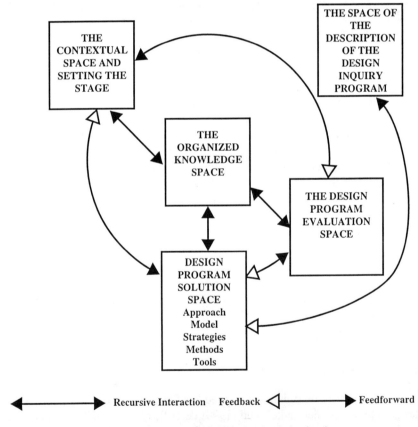

FIGURE 6.7. Architecture of designing the design inquiry.

evaluation space. The validated design inquiry program is then described in the space marked as the model of the design program. The organized knowledge space, in the middle of the figure, is where all the information and knowledge relevant to design of the design inquiry program is deposited. Figure 6.7 shows arrows that connect the various spaces. There are multiple arrows that indicate a spiraling process guided by the type of selections that are to be made in the design solution space. The two directional arrow heads indicate the recursiveness of the feedback/feed-forward process.

6.7.2.1. Description of Design Spaces and Activities

The five design spaces, introduced in Fig. 6.7, and the design activities that connect these spaces are described next.

6.7.2.1a. The Contextual Space. The contextual space displays design information that was generated in the course of activities accomplished in the first four systems of the systems complex of design described in Section 6.4. This includes the characterization of the relevant systemic environment of the design inquiry, a portrayal of the desired future system, the selected systems type, the design competence and the sociocultural and socioeconomic characteristics of the designing community, the perceived complexity of the design task, the time constraints of the design of the future system, and knowledge and resources available to support the design effort. Much of this information can be derived from findings of the previous design work. Still, further exploration is often needed.

The second major task to accomplish while working in the contextual space is the formulation of organizing perspectives that will guide our thinking in designing the design inquiry program. These perspectives make explicit the values and qualities that the designing community wants to realize through the design activity. A source of these could be the values/qualities elaborated in the image of the designing system presented in Fig. 6.3, Section 6.5 of this chapter.

6.7.2.1b. The Organized Knowledge Space. This space is inhabited by the following types of knowledge: (1) findings of all the inquiry accomplished heretofore in the systems complex of design; (2) knowledge about design in general and specifically about design approaches, models, methodologies, methods and tools that are applicable to social systems design; (3) knowledge about systems and design thinking; (4) knowledge about various types of social systems and their characteristics; (5) knowledge of the target system and its environment; (6) knowledge about ways that target system can be characterized and modeled as a system; (7) information about organizational capacity and human capability available to the design effort; (8) information about resources available to support the inquiry effort. In this space, knowledge and information will be continuously acquired and displayed as design work calls for them in the design solution and evaluation spaces. Thus, this space of the architecture is open and evolving.

6.7.2.1c.   The Design Solution Space.   This space is the territory where the designing community engages in formulating solution alternatives that are tested in the design-testing space. This inquiry goes on in the five domains shown in the figure. Designers consider available options in terms of design approaches, models, methodologies, methods, and tools. Alternatives developed in these domains are never considered in isolation but always in their interrelationship and interaction. What is implied here is that if the designers select an ideal design approach as their model of design, the methodologies, methods, and tools should be internally compatible and consistent with each other. As the designers proceed toward the selection and development of a design inquiry program, they may find they need additional information or knowledge; thus they call for these from the design knowledge space.

6.7.2.1d.   The Design Program Evaluation Space.   Here we define the criteria by which to evaluate the various alternatives that emerge from the design inquiry solution space. In addition to the values and qualities that are stated in the contextual space, designers should develop both external and internal criteria for the testing of design alternatives.

External criteria might include probing into the general validity of the approach or model examined, e.g., articulated theoretical base, evidence of testing and successful use in social systems such as the one designers intend to design, or internal consistency with other design choices made.

Internal criteria assess a "goodness of fit" with the overall design effort and with the kind of system designers aspire to build. The following criteria were used in the design of a design inquiry program developed for the institutional design of an R & D laboratory (Banathy, 1987c):

1. General assessment of use, such as making a judgment if the alternative identified is realistic and feasible as well as effective to use in the institutional context of the laboratory.
2. The inquiry power of the alternative. Does it allow going beyond the existing boundaries of the system and its systemic environment? Does it offer flexibility of use? Does it allow continuous review and modification? Is it sensitive to evolution and "emergence"?
3. Is there a systemic match, e.g., is the alternative considered appropriate to use in the context of a system such as the laboratory? Can the design

inquiry alternative fit in with, and transfer into, the general inquiry approach of the laboratory, or does it require a change?

4. Does the alternative considered nurture unrestricted exploration and the use of creativity? Is it sensitive and supportive toward new initiatives?

The formulation of internal criteria should be ongoing during the design of the inquiry program. Its guidance should be derived from the knowledge base as well as from the contextual and design solution spaces.

6.7.2.1e.   The Space of Displaying the Design Inquiry Program.   This space is the "resolution space" of the inquiry, in which designers display the description of the most promising design inquiry program they devised and which they intend to use in designing the desired future system.

6.7.3.   The Dynamics of the Inquiry

The design of the design inquiry program is not accomplished in a step-by-step, linear fashion. It is carried out through the dynamics in recurring spirals and recursive, mutually influencing interactions among activities and spaces. It happens as we explore and re-explore the various spaces and integrate information, knowledge, experience, and evaluation relevant to the emerging inquiry alternatives of the design inquiry program. More specifically, as we proceed with the design inquiry, and as solution alternatives emerge in the design solution space, we continuously revisit the contextual space to gain more insights about the issues our design inquiry should address. The same is true with the knowledge space. As we proceed with the design of the design inquiry program, we draw increasingly, in a more focused way, on the sources of design knowledge. We often find that we need new knowledge to inform and enlighten our selection of solution alternatives. We also repeatedly move into the design testing space to evaluate (based on the external and internal criteria we identified) and test the emerging inquiry alternatives. This testing shapes the model of the inquiry program; it validates or questions it. This process is the main source of attaining confidence that there is a "goodness of fit" between the inquiry program and the criteria we established. The knowledge space also offers sources of design testing. In the knowledge space, we deposit information about design testing and evaluation means and methods. The design inquiry that intersects all these spaces eventually converges as it moves into the modeling space, where it displays the product of the inquiry as a representation or a model of the design inquiry program.

Another aspect of the dynamics of the inquiry is manifested in the dynamics of divergence-convergence, operating in the design solution space. It relates to the search, creation, and selection of inquiry alternatives. Initially, we seek to consider a number of alternatives; thus we operate in a divergent mode. An assessment of the relevance of alternatives and their evaluation against stated criteria eventually lead us to converge and select the most promising, most appropriate design inquiry program.

Activity #49

(1) Describe the core ideas of this section. (2) You are not expected to develop a comprehensive design inquiry program. Still, to create your own understanding and construct your own meaning of designing a design inquiry program, and apply what you have learned in a functional context of your choice, it is suggested that you speculate about the task of developing a design inquiry program for a system of your interest and outline briefly how you would go about the design of such a program. Enter your findings in your workbook.

Reflections

The design of a design inquiry program, portrayed in this section, is an integral function of the comprehensive design approach that is carried out by a stakeholder community. It is an integral part of the process during the course of which a designing community is getting ready for the design of the desired future system. As we have noted, this process of "getting ready" has received only scarce attention in the design literature. In the first- and second-generation design modes, it was the design expert or consultant who, when invited, was already ready to move in with his or her toolbox.

The very moment a position is taken that it is the prerogative, the right, and the responsibility of the stakeholder community to design its own system, we must embrace the idea that the designing community has to learn to engage in and carry out the design of a design inquiry program that best fits the design situation.

The idea of engaging a stakeholder community in the process of getting ready for design, however, goes far beyond the issue of readiness for design and the creation of a designing community. I suggest that if a community—any community—attends to the tasks described in this chapter, then that community in fact develops its own design culture and transforms representative democracy into a creating/designing democracy. By so doing, the community empowers itself to create and shape its own future. Getting ready for design and developing a design culture is individual and collective empowerment at its most robust. Such empowerment gives meaning and substance to guiding our future.

7

Evaluation and Value Adding

This chapter is devoted to the exploration of several issues that advance social systems design and advance the society. The first two sections address preventive evaluation. The question is: What should we avoid in systems design. Here, we attend to the issue of how misconceptions and underconceptualization might devalue the design effort, and then we become aware of a set of pitfalls. In Section 7.3 we explore various approaches to design evaluation.

In the next four sections, we see how design adds value. First, we explore how the designing community seeks evidence of honoring the values they wish to realize in designing the design system and in designing the desired future system. Then we ask the big question: What values can design add to our lives as individuals, to our organizations, communities, and the society? More specifically, in the context of societal evolution, we ask: What is the role of social systems design in societal evolution? In Section 7.7, we arrive at the high point of our inquiry. We claim that the acquisition of a design culture and the development of design competence will empower us individually and collectively to design our lives and enable us to shape our future. That is what empowerment is about. Such empowerment will transform what we now call representative democracy into "a creating democracy," into a designing society.

7.1. Misconceptions and Underconceptualization

There are two major types of failures in social systems design. Components of the first type are misconceptions and underconceptualization. The second type, called pitfalls, involves inappropriate or incompetent use of design approaches and methods. The two types constitute "early warning" and serve as "preventive evaluation." This section highlights the most salient misconceptions and underconceptualizations in social systems design. Pitfalls are discussed in Section 7.2.

7.1.1. Misconceptions

Misconceptions about design have two major sources. General misconceptions fail to distinguish between conclusion- and decision-oriented disciplined inquiry. Specific misconceptions fail to recognize the differences between the various modes of decision-oriented inquiry or to confuse them with each other.

7.1.1.1. The General Source of Misconceptions

As Simon (1969) pointed out, in the field of systems design the most general misconception is a failure to differentiate between the approaches and methods of science and those of the professions. Natural and behavioral sciences describe what things are and how they work. They make theories based on predictions. They are organized in the compartmentalized domains of the scientific disciplines. Their business is knowledge production, and their salient intellectual processes are analysis and reduction. They are concerned with "what is." On the other hand, the professions construct and reconstruct systems, creating them or shaping them according to stated purposes. They are focused on "what should be." They are in the business of design. Design inquiry is decision oriented, science is conclusion oriented. In design we use knowledge developed in the various scientific disciplines. The salient intellectual process of design is synthesis. The tasks encountered by the practitioners of the design professions invite ways of thinking and methods of acting that are markedly different from those in use in the various scientific disciplines. A failure of differentiation between these two modes of thinking and their methods of practice often leads to an uncritical adoption of the methods of scientific disciplines in design inquiry. The traditional scientific "mind-set," when adopted in design, leads to an overemphasis on the analysis of the problem situation. It locks us into and keeps us within the boundaries of the existing system. It often causes a "paralysis of analysis," as we drive deeper and deeper into finding out what is wrong. Designers often remark that "getting rid of what is not wanted (what is wrong) does not give us what is desired." "We can become so fascinated with understanding the prison we are in as to distract ourselves from studying the way out" (Frantz, 1995).

Furthermore, the reductionist tendency of the same scientific mind-set guides us to find solutions by "fixing" parts, not recognizing that the "optimal performance of a part does not prevent the bankruptcy of the whole." So, in addressing a problem situation, we travel on a piecemeal, incremental, and disjointed path toward hoped-for solutions, which, however, seldom come about. In the social systems domain, we have witnessed the gross failure of this approach in recent efforts of the educational improvement movement as well as in approaches to reform the health care system.

The general misconception described above can be avoided in two ways. First, design learning, offered to the stakeholder community, should bring into the awareness of stakeholders the danger of using an analytical/reductionist orientation in design. Next, this awareness can be further nurtured by portraying the differences between the three cultures—the cultures of science, humanities, and design.

We must, however, recognize the role of science in design scholarship and practice. First, in design we make use of the knowledge produced by science. Then, while being aware of the danger of misuse of the scientific method in systems design, we should always point out that the methods of science have a very important place in systems design. This place is the use of the methods of science in doing research on design and the production of knowledge about design. Such knowledge production is not only the prerogative of design researchers but also of design practitioners. As stakeholders engage in design, they reflect on and test the effectiveness of the use of various design methods and tools. They will recognize what works and what does not. They will record their findings and, thus, enrich their own design knowledge and will make contributions to the general knowledge-base about design.

7.1.1.2. Specific Sources of Misconception

The most salient specific source of misconception is confusing systems design with other modes of decision-oriented disciplined inquiry, or failing to make distinctions between design and those other modes. This type of confusion and failure occurs quite frequently. When I discuss systems design with social systems professionals and bring up solution-focused systems design, the typical remarks are "It is exactly what we are doing," or "There is nothing new for us in this design stuff," or "You are only using different terms for the same thing." One reason for these remarks is that we fail to clearly spell out contrasting features between systems design and other modes of disciplined inquiry by citing specific examples. The other is that the "hearer" often responds with a "no difference" statement in defense of his or her well-known method, or thinks that we attack his or her prestige as a professional problem solver.

Manifestations of the misconceptions described here include confusing or failing to differentiate between: (1) improvement and design, (2) invention and design, (3) planning and design, (4) piecemeal fixing and comprehensive design, (5) adjustment and design, (6) restructuring and design, and (7) designing *for* the future, rather than designing *the* future. We have already discussed these differences in Chapter 2 and continuously throughout the book when there was a call for it. Some of the consequences of these misconceptions are: trying to fix the existing system even when design is called for, pursuing traditional and incremental systems planning, treating social systems as closed systems, working out from the existing system, speculating about the future rather than designing it, rearranging parts within the existing systems and/or reassigning responsibilities within it, (re)designing parts rather than the whole system.

We can avoid the type of misconceptions mentioned here by purposefully setting forth examples that demonstrate differences and make distinctions between systems design and other modes of decision-oriented inquiry modes. In fact, demonstration of what systems design is not begins to develop an understanding of what systems design is. We know well from learning theory that learning occurs by making new distinctions. But by engaging in a clarification of what design is—and what design is not—by making distinctions, we have to be careful not to leave the impression that other modes of decision-oriented disciplined inquiry are inferior to systems design. All those mentioned above are appropriate under circumstances that call for their use. But they are not systems design.

7.1.2. Underconceptualization

Warfield (1990) suggests that the large-scale systems failures of recent years indicate clearly that the system of beliefs that supports thoughts about systems design is grossly underdeveloped and underconceptualized. He further suggests that "correcting the underconceptualized base of systems thinking is like curing alcoholism. Unless the admission is made that the alcoholic needs treatment, treatment is unlikely even to be started, much less to be effective" (p. 6). Warfield finds that the base of this cure should be approached from two major directions. He calls for and sets forth in his work approaches for the development of a science of generic design, which should be grounded in a high-quality knowledge base and should meet appropriate scientific criteria. Parallel with this development he says that we should focus on design education in all professional studies as a subject of its own.

In the remaining part of this section I introduce a set of specific aspects of social systems design where underconceptualization is most rampant, and therefore, grossly undermines the effectiveness of finding design solutions. I label the set the "seven cardinal sins" of underconceptualization.

7.1.2.1. Underconceptualized Boundaries

The underconceptualization (UC) of setting narrow boundaries is probably the most damaging specific shortcoming in social systems design. There are multiple sources and instances of this type of UC.

1. The most frequent is a failure to transcend the existing system or the existing setting of the design situation. This practice locks designers within the boundaries of the existing system where they spend much effort in analyzing what's wrong. This "staying within" may lead to some improvement of the existing system or some restructuring, but it will never lead to a new design.
2. Even if designers transcend the existing system, they often set the

boundaries of the design inquiry too narrowly. This leads to limitations in exploring design options and design alternatives.

3. The same self-imposed limitation prevails in case designers set narrow boundaries for the system to be designed. (Note: we make a distinction between setting the boundaries of the inquiry versus setting the boundaries of the new system.)

4. A lack of broad definition of the space of the systemic environment will result in lack of compatibility between the future systems and will result in the possibility of not having an adequate resource base for supporting the future system.

5. The narrow definition of time boundaries, the "hurry up to show results" syndrome, is only too frequent in design situations. This UC is one of the greatest sources of having "cost-regret" in design. Rather than narrowing or restricting, designers should keep boundaries open and push them out as far as possible. They should "paint the largest possible picture on the largest possible canvas."

7.1.2.2.   Underconceptualizing by "Shifting Down"

In the course of design it often happens that designers shift from the whole system level down to the level of the subsystems and focus on designing around lower-level objective(s). This is a very typical general example of underconceptualization. It is a tempting proposition, particularly if the characteristics of a subsystem are well known. This shifting down and focusing on a subsystem is called "suboptimization." Reflecting on this practice, we often say that "the optimal performance of a part does not prevent the bankruptcy of the whole."

7.1.2.3.   Underconceptualization of Perceptions, Beliefs, and Values

An extremely high price is paid if designers limit attention to issues related to perceptions, worldviews, beliefs, and values. These issues are often dismissed as being "personal issues or agendas" leading to unnecessary discussion that prolongs the process. Nothing can be further from the truth. A dismissal of such personal issues/agendas will cause a paralysis later on in the design process. Proceeding with design without engaging in design conversation that hears everyone's personal issues and—through the conversation process—aims to create a common frame of thinking, shared worldviews, and values as basis for making decisions is dangerously counterproductive. It will lead, in the course of design, to situations when designers are incapable of forging collective decisions. They will find themselves entangled in constant arguments, the reason being that they failed to deal with underlying assumptions and the underlying issues of their design decisions.

Even more importantly, designers cannot transform an existing system into a new and different future system without transforming themselves. They must transcend their old ways of thinking, the old paradigm that is reflective of the existing design, and they must leave the old mind-set behind. They must conceptualize a new paradigm, develop a shared way of thinking, and acquire new perspectives and a new worldview. Einstein's statement comes to mind again: "We cannot address a problem from the same consciousness that created it. We have to learn to think anew."

7.1.2.4.   The Ideal System Issue

Another critical issue of underconceptualizing is striving for less than the ideal, or compromising the ideal. There is a great temptation to rationalize or compromise by saying: "Be a realist." "There are too many constraints." "We have to show results quickly." "You are chasing dreams." "It would never work." "We have no time for it." "It will cost too much." It is the ideal systems approach to design that gives power to design inquiry. At times of dynamic, revolutionary, and continuous societal changes and transformations, anything less than the design of an ideal system and a continuous pursuit of the ideal leaves us behind. Anything less is a waste of time. The ideal system could be revolutionary, but the journey toward it can be evolutionary. Nothing less than the ideal is worth the effort.

7.1.2.5.   Underconceptualizing the Use of Knowledge and Information

This type of underconceptualization (UC) is manifested in the UC of the knowledge base that is needed and the UC of the role of information in design.

7.1.2.5a.   The UC of the Knowledge Base.   There are several sources of this UC: (1) failing to consider that systems design needs a rich knowledge base and multiple knowledge sources, (2) relying on findings of the analysis of what is wrong with the existing system, (3) failing to sweep in all relevant knowledge and all relevant positions, and (4) failing to acquire additional knowledge in an ongoing basis as the unfolding design inquiry requires it. Designers can avoid this UC by organizing the knowledge base as "living" and developing, as an open information/knowledge system that will access, analyze, develop, and display information and knowledge as an ongoing knowledge-generating process in a variety of areas that are needed to inform and enlighten design decisions.

7.1.2.5b.   The Underconceptualized Notion of Information.   Banathy, A. (1995) suggests that we often run into difficulty in the use of a theoretical and methodological design framework that rests on an underconceptualized notion of information. The author suggests that "design is an informational container in which life organizes itself" (p. 1). It is the job of designers to build such containers. These containers, says the author, are woven from three types of information strands. One strand accounts for information-in-action, called referential information, or "referential as experienced." The second strand conveys some description of what the system is or is to become, taking the form of information-as-knowledge, called nonreferential, or—"as observed," information (Kampis, 1991). The third strand, the state-referential information (Banathy, 1995), imposes an apriory—preplanned—condition (a state determination) on either or both of the previous strands as a means of control.

The problem is that state-referential cannot fully account for the processes of the two other types. And nonreferential cannot account fully for referential information processes. "The informational strands of the container in which life organizes itself have specific 'containment' relationships to each other." Furthermore, "if we try to hold social systems in state-referential containers, the creative dynamics will spill on the floor, and create a real mess" (p. 2). This happens, for example, if we fail to transcend the existing system and stick with the state determination of the old system. The underconceptualization of information processes, described here, is one that is most often observable in current design programs.

7.1.2.6.   Underconceptualizing System Representation or Modeling

Even the most successful, the best "goodness of fit" design effort comes up short due to this UC. An underconceptualization of the representation or modeling of the new system results in a less than complete, a less than comprehensive, and a less than multimodal description/representation/modeling of the new system and its systemic (supporting) environment. How can system development and implementation proceed with a less than complete and comprehensive "blueprint" of the new system? Designers can avoid this UC by describing/representing the new system by the use of a three-dimensional systems model, by the use of a system-environment model, a functions/structure model, and a process/behavioral model (Banathy, 1992a).

7.1.2.7.   Underconceptualizing Stakeholders Involvement

Even when people adhere to the notion of design by the stakeholders of the future system, there is a strong tendency to limit their involvement. A less than full involvement can be easily rationalized. I find in the literature the phrase: "Yes, ideally that's the way to do it *but* . . . " People often say it is very difficult to arrange for a comprehensive involvement and it is costly in time and effort. We can use surrogates to stand for a class of stakeholders. Yes, it is time-consuming, costly, and complex. But the question is: What price shall we pay for a less than comprehensive involvement? In developing perspectives for the establishment of the designing system, shown in Figure 6.3 of Chapter 6, I introduced an image of key markers of a designing system. Several of these are pertinent to stakeholders' involvement. A review of the markers will help us assess the price we shall pay for a less than comprehensive involvement. The price we pay is a design outcome that will be less authentic, less sustainable, less intelligent, less ethical, less diverse, less creative, less competent, less user-friendly, and less liberating. Is there a general prescription of how to go about stakeholder involvement? How to decide on the degree and depth of involvement? How to create a communication system for the stakeholders that integrates their contribution to the design? We can answer these questions only in a very general sense. Much more work must be done in this area. For example, the image of the designing system introduced in Section 6.5 of Chapter 6 can guide us in our thinking about this issue. We should remind ourselves that each and every design situation is unique. Therefore, each and every design situation will invite an approach to stakeholder involvement that is unique to the particular situation.

Reflections

In the previous chapters of this book we have worked with design ideas in an affirmative sense. Two sections of this chapter are devoted to warnings about what to look out for in social systems design to avoid self-inflicted problems. In this section, we first reviewed some misconceptions that relate to confusing systems design with other much better known modes of disciplined inquiry. I suggested that by learning to make these distinctions we not only avoid misconceptions but we also learn about design itself. In the main part of the section I introduced the "seven cardinal sins" of underconceptualization. I am sure there are some others. The point I was trying to make is to alert you to the hidden dangers of misconceptions and underconceptualizations that you should always look out for.

Activity #50

(1) Review the text that sets forth a set of misconceptions that relate to confusing design with other modes of decision-oriented inquiry. Discuss with someone who is not familiar with social systems design the difference between design and some of the other modes of inquiry. Make a report on your discussion in your workbook. (2) Review the seven examples of underconceptualization and, based on your experience and the insights generated while working with this text, describe how you would go about avoiding these underconceptualizations. (3) Speculate about other types of misconceptions and UCs. Enter your findings in your workbook.

7.2.   Pitfalls in Systems Design

In the first section of this chapter it was suggested that experience with social systems design has revealed that the quality and effectiveness of systems design is much impaired and "devalued" by three kinds of sources. One source is misconceptions about social systems design or a lack of differentiation of design from other modes of decision-oriented disciplined inquiry. Another source is underconceptualization, which places constraints or limits on the conceptualization of design solutions. The third source is what we will call "pitfalls." A pitfall, as defined by the Webster (1979), is "a hidden or not easily recognized danger or difficulty." In the domain of disciplined inquiry, "a pitfall is a conceptual error into which, because of its specious plausibility, people frequently and easily fall" (Majone and Quade, 1980). A pitfall is a hidden mistake or a lack of clarity and understanding that may undermine or even destroy the inquiry. It could be a fallacy, which is a mistake in reasoning. But pitfalls should not be confused with blunders. In the first part of this section I introduce statements from the design literature that seem to relate to pitfalls in systems design. The second part presents additional notions of pitfalls that are related to the various tasks of social systems design.

7.2.1.   Selected Statements from the Design Literature

The statements introduced here bring to the attention of designers errors that might happen in the course of design. They point to potential failures to be aware of and suggest ways to avoid them. The term "pitfall" seems to be a suitable umbrella for these.

Ackoff (1981) warns us that we should watch out for four major potential errors in the course of design. In introducing Ackoff's potential errors, suggestions are made for how to avoid them. One of these errors is the use of information in decision making that is inappropriate or irrelevant. This can be avoided by establishing a criteria of relevance in screening input information. Another is the use of faulty decision-making processes. As discussed earlier, we should ensure that the decision rules of design are well established and are relevant to the purposes and uniqueness of the design situation. Often, decisions made in the course of design are not taken into account in later decisions. This pitfall is very likely to occur when designers follow a linear process and fail to establish feedback loops that connect present decisions with those made earlier. Finally, there might be a lack of sensitivity to environmental changes. This can be avoided by anticipating such changes and by continuously scanning and connecting with the systemic environment.

Nadler (1981) points to several sources of pitfalls:

1. Making solution decisions at a point where the views of the designers are

still too divergent and thus a compromise is adopted that nobody likes. The results are unachieved purposes, defensiveness, even hostility, and ineffective procedures.

2. Transferring or adopting solutions from one situation to another without considering the uniqueness of the particular situation.

3. Failing to develop supporting links with valued elements of the real-world context.

4. Failing to relate the purposes and objectives of the system to be designed with relevant systems in the systemic environment. The design system should always be considered as a component of its embedding system:c environment.

5. Transferring a designing system that operates successfully in one setting into another design effort. Each and every design situation is unique and each and every designing community should design its own design arrangements.

6. Failure to consider the various elements and dimensions of the design effort as being interdependent, where changes in one element or dimension can bring about change in others.

7. As mirror reflections of potential failures, Nadler and Hibino (1990) suggest that a typical failure is a tendency of not tolerating failures. Such an attitude discourages experimentation with novel solutions and also discourages risk taking.

Hammer and Champy (1993) catalogued a list of the most common errors that lead organizations to fail at systems design:

1. Often people try to fix a process instead of changing it. The authors suggest that in design it is "useful to recall the old saying that hanging a sign on a cow that says 'I am a horse' doesn't make it a horse." "Fixing" is a great temptation. In defending fixing, people argue that the infrastructure is already there to support the fixing process.

2. Neglecting people's values, beliefs, and expectations is a pitfall that destroys the authenticity of design.

3. Settling for minor results is taking the easy path, the path of marginal improvement, which sacrifices the attainment of viable design solutions. It throws us back where we have been, reinforcing the mentality of disjointed incrementalism.

4. Quitting too early or scaling back the design effort at the first sign of facing some difficulties. Failing to persevere and failing to devote adequate time to design have very high opportunity costs.

5. Placing constraints on the scope of design effort leads to a scaling down of the value of the design solution.

6. Assigning someone to lead the design effort who doesn't thoroughly understand systems design. Whoever leads the effort should not only

truly understand what design is about but must have deep seated commitment to the effort.

7. Skimping on the resources devoted to the design effort. The saying that you can't get something for nothing holds here. We can't gain the benefits of design without investing in it the kind of resources that are needed to support it.

7.2.2. Pitfalls We Might Stumble into in the Course of Design

The pitfalls introduced next are arranged in the following sets: designation of designers, preparation for design, the design process from envisioning to image creation, and the transformation of the image into its detailed representation as the model of the future system. The exploration of these sets of pitfalls is based on the portrayal of design as developed in this work.

7.2.2.1. Pitfalls in Designating the Designers

We have devoted much attention to the issue of who should be the designers of a social system and how one can establish a designing community that involves all the stakeholders of the system to be designed or redesigned. With this understanding in mind, pitfalls that we should be aware of in designating the designers include (1) letting someone else design our system for us, which we called "the throwing of the design over the wall" syndrome; (2) top-down design, when people in authority attempt to dictate design by decree; (3) attempts to convince stakeholders to buy into a vision or a solution the creation of which they have not been involved in. All these pitfalls, or any combinations of these, counter and even undermine the realization of the values a designing community might aspire to bring about through design. We discussed a number of these values in Chapter 6. Most prominently, these values might include the authenticity of the design and its maintainability, the use of the collective intelligence of stakeholders, and the attainment of a system that can learn as an organization. These values and others will become the victims of the the pitfall of "letting others do it" for us. On the other hand, these pitfalls can be avoided by the use of the user-designer, a participative designing approach, and by the establishment of a designing community that includes all those who serve the system, are served by it, and are affected by it.

7.2.2.2. Pitfalls in Organizing and Preparing for Design

It is the initiation phase of design, namely, the processes of organization and preparation for design, that might make all the difference in the successful accomplishment of social system design. Therefore, it is of special importance that we become aware of potential pitfalls that might undermine the success of the design inquiry during this preparation, or getting ready, phase:

1. A pitfall that too often goes unnoticed is the employment of the existing bureaucratic structure and the use of its standardized arrangements, formalized hierarchical relationships, and operational processes. This approach will tend to keep the inquiry within the boundaries of the existing system and will impose on the process the mechanics and dominance of the existing hierarchy. This pitfall can be avoided by establishing an "adhocracy"-type designing community that is different and separate from the existing bureaucracy and then empowering this community to carry out the design effort.
2. A pitfall that in fact ensures the failure of design is neglecting or short-circuiting design learning, and a failure to develop design competence in the designing community. Critical preconditions of successful systems design are the introduction of the type of intensive design learning that is relevant to the functional context of the design effort, and the development of individual and collective design competence.
3. The next pitfall is failing to design and establish the designing system. This pitfall undermines the design effort. The designing system is a social system in its own right. As such it should be designed and empowered as fully committed to a focused, disciplined inquiry.
4. Another pitfall is the adaptation of a general design model that prevents the establishment of a design inquiry that responds to the uniqueness of the systemic context and the unique conditions of the design situation.
5. A similar pitfall is selecting design methods without matching them to the type of system designers wish to design.
6. Often we fail to define the primary system level—out of several possible levels—at which the future system is to be designed. This leads to much confusion in the course of the design inquiry.
7. A pitfall of far-reaching consequence is the failure to marshall the necessary financial, material, and facilities resources that will support the design effort or acquire the knowledge base that is needed to inform the design program.

7.2.2.3. Pitfalls That Might Undermine the Front End of Design

At the front end of social systems design we engage in design tasks that take us from the formulation of a vision of the future system to the creation of its fist systemic image. As we already noted, designers often commence with design by focusing on the analysis of the problem situation. This is a most dangerous pitfall. Problem analysis focuses attention on finding out what is wrong with the existing state of affairs or what is wrong with the existing system. Even if those wrongs are righted, it leaves us with a corrected system that is based on and that manifests the old design, which we should leave behind. Analysis locks us into the old design. It is this old design that we should transcended to create a new design. We often say that getting rid of what is not wanted does not give us what is desired. Having a problem focus derails design that should focus on exploring and finding design solutions. Another disabling pitfall is rushing through the front-end design tasks to get to what many think is the "real" task of formulating the purposes and goals of the new system. This rushing through, and paying only lip service to the front-end tasks, results in failing to establish firm foundations for building the new design. It results in a failure to give in-depth and detailed attention to creating visions of the future, establish the boundaries of the design space, explore major design options, create a common and shared frame of reference, and forge a shared world view. Failure in exploring and articulating collective and shared core values and core ideas means building on sand rather than on firm foundations. These values and ideas are the very bases for creating solution alternatives; these are the bases for making design decisions throughout the design inquiry. The vision and the core values and ideas are the very bases of designing the first image of the future system. The time designers might gain by rushing through the front-end task of design will be lost many times over as we continue the inquiry. Lacking a collectively chosen shared value and image-base for making design decisions, designers will engage in endless arguments that will delay and even derail the design effort. They will pay a very high price for their impatient rushing through the front part of the design inquiry. The term for this is "cost-regret." Such a rushing through is what cost-regret is all about.

7.2.2.4.   Pitfalls That Endanger the Successful Completion of Design

During the second phase of the design journey, designers are called upon to transform the image of the system into the detailed description or model of the future system. This journey is endangered by another series of pitfalls. (1) In transforming the image into the core definition, formulating the purposes and goals of the future system, designers often fall into the trap of presenting very narrow formulations. They often fail to formulate purposes that serve the larger society and other relevant systems in the systemic environment. They often fail to address quality-of-life goals that will benefit stakeholders or address aesthetic purposes. (2) In the course of defining systems specifications, designing the system of functions that respond to purposes and specification, and designing the organization that carries out the functions, the design inquiry is endangered by another set of pitfalls. These include the creation and exploration of limited sets of alternatives, the consideration of inadequate numbers of design iterations, failure in formulating and reformulating criteria for the evaluation of alternatives, inadequate collection of evidence of "goodness of fit," and, consequently, premature convergence on a design solution. Such inadequate design solution is presented as the model of the new system. (3) Here we face two more pitfalls. One is the failure to formulate a detailed in-depth narrative description, a multi-dimensional model of the system we designed. The other is a failure to design a model of the systemic environment or the embedding system of the new system, which is capable and is ready to support the new system.

7.2.2.5. Tinkering with the Design Community

In the course of the entire design inquiry process a crucial pitfall is a lack of calling upon the individual and collective creating power and potential of designers and failing to make full use of their individual and collective intelligence. Furthermore, most design solutions are seriously devalued by a failure to build into the design, and transfer into the new system, the acquired organizational capacity and learned design capability that is needed to engage in the continuous design/redesign of the system and that enables continuously ongoing organizational learning. What is implied here is that upon the completion of the design inquiry, which in fact is never completed, the designing community becomes the operating community of the new system. It becomes the community that develops, implements, and institutionalizes the design and that becomes stakeholders and stewards of the system it has designed.

Reflections

We can now reflect briefly on the concept of devaluing or the devolution of design. Misconceptions about design, underconceptualization, and a host of pitfalls are the enemies of social systems design. In reflecting on this, we can rightly say that "we met the enemy and it is us." These enemies of design are our own creation. We allow these enemies to exist by a lack of genuine understanding of what design is and how it works. We allow these enemies to exist by not taking the time to fully internalize a systems and design view, by not fully developing our own design thinking and core ideas about design. We allow these enemies to exist by not taking the time to become fully aware of the existence of all the danger and warning signs that represent the causes and sources of devaluing design, and by not posting these signs for ourselves and our designing communities so that they are constantly in view.

In the next two sections we build on the insights we gained in the first two sections of this chapter as we move into an exploration of how to avoid a devaluation of design by a thoughtful and well-designed evaluation, and how to build into our inquiry a system by which we collect evidence that our design solution will realize the values we seek to bring to life in the design of our system.

Activity #51

Two tasks are proposed here. First, review this section and prepare a list of pitfalls introduced in the text. Reflect upon these pitfalls and think about ways, and make a list of actions to be taken, to avoid them in designing of your system of interest. It should be mentioned that the pitfalls that I noted in this section are only examples. They are not all that exist. I wanted to sensitize you to the idea of recognizing and dealing with pitfalls. Thus, the second task suggests that you on your own identify pitfalls that you can uncover, pitfalls that have not been mentioned in the text of this section. Enter your findings in your workbook.

7.3. Approaches to Design Evaluation

Evaluating design requires us first to understand the processes by which design efforts may be undermined, devalued, e.g., misconceptions, underconceptualizations, and pitfalls that endanger social systems design. Second is to create a well-conceptualized and -developed system of design evaluation. This will be treated in this section. The other approach is to bring into focus ways by which we can collect evidence that tells us whether the design we created will bring to life the values we wish to realize. This issue is addressed in the next section.

In the first part of this section, we will explore design evaluation from several viewpoints found in the design literature. The second part introduces an image of a comprehensive design evaluation system.

7.3.1. *Exploration of the Design Literature*

First, some views on design evaluation are explored, including views that reflect ways of thinking about design evolution. This exploration also considers perspectives on the evaluation of emerging solution alternatives (let us label this "microevaluation"). Then looking at the wholeness of an emerged solution, the question is asked: Does the design solution stand the test of "systemness?" (let us call this "macroevaluation").

7.3.1.1. Macroviews on Design Evaluation

In a search for views on design evaluation we find a wide range of statements, varying as to their relevance to social systems design. Jones (1980), an early pioneer of design scholarship, considers evaluation as the means by which convergence on a preferred design solution can be achieved. He discusses evaluation aims and methods by which to attain those aims. A lead aim is to recognize whether we have attained an acceptable design. Methods used include formulating an evaluation objective(s) that should be able to assess the attainment of an acceptable design and identifying fail-safe directions for the attainment of that objective(s). Jones develops the concept of establishing a zone between what is acceptable and what is not and specifying criteria that indicate whether the design solution is on the fail-safe side of the zone of acceptability. A companion aim of design evaluation is to further refine the "acceptability" of design solutions with the use of the evaluation criteria. Methods include the identification of objectives that the solution should satisfy, the measurement or estimate of the degree to which the design solution in question satisfies the objectives, and the selection of the solution that best fits the criteria.

Another early pioneer of design scholarship, Sage (1977) describes various ways of evaluating design solutions of large-scale systems. His "worth assessment" approach seems to be the most relevant to social systems design. This approach arranges multiple objectives and their assessment criteria in an organized form. It also establishes worth connections between criteria and performance consequences and contemplates interdependence among various assessment criteria. More specifically, the worth assessment procedure (1) formulates overall objectives and attributes, (2) constructs a hierarchy of the defined performance criteria, (3) selects appropriate performance measures, (4) develops connections between the measure(s) and the worth indicated by the measure(s), and (5) establishes relative importance within the hierarchy of criteria. Sage suggests that in assessing the worth of a consequence, we can stipulate only conditional worth, since we can only assume that the outcome in fact will occur. Worth assessment always reflects the preferences of the designers.

In Cross's compendium (1984), several authors discuss design evaluation and reveal to us a way of thinking about design evaluation rather than proposing methods for it. The thinking considered here brings us directly into the domain of social systems design. Luckman (1984) suggests that design progresses from very general considerations of design solutions to solutions that are increasingly more specific. Evaluation and selection accomplished at earlier points in time are indicators of the direction in which to go at the succeeding levels. Any selection decision made at an earlier level provides input for the next. Thus "it is better not to single out one feasible solution, but rather keep several ideas open to allow more thorough exploration of the next level. By doing so, the designer is leaving open the chance of using feedback to the earlier levels" (p. 85). This approach enables designers to build up the total solution in a dynamic, interacting, integrated, and internally consistent way from earlier (sub)solutions. Levin (1984), very much like Luckman, is concerned that each selection decision we make limits the field within which later selections can be made. It means that earlier decisions impose limits on the designer's discretion of making later decisions. This observation leads us to suggest the need for a continuous move back and forth among decision points, an ongoing "revisiting" of earlier decisions made, leaving open the possibility of reconsideration. Rittel and Webber (1984) suggest that design is an argumentative process, in the course of which the solution emerges gradually as a process of incessant judgment, subject to critical argument. Designers terminate the process not for reasons that are inherent in logic, but because in the course of design they arrive at a point where they say: "That's good enough," or "This is the best I can do within the limitation of the project," or "I like this solution," etc. In design there are no true-or-false answers. Assessments of "proposed solutions are expressed as 'good' or 'bad' or 'more likely,' or as 'better or worse' or 'satisfying'" (p. 144). The authors also say that while in the scientific community people are not blamed for postulating hypotheses that are refuted, in systems design no such immunity is tolerated. In design the aim is not to find the truth, "but to improve some characteristics of the world in which people live" (p. 144). Designers are liable and accountable for the consequences of their actions. Darke (1984) suggests that systems designers do not start with an explicit list of factors or predetermined performance limits. Rather, they have to find ways to reduce the variety of potential solutions to as yet imperfectly understood design situations "to a small class of solutions that is cognitively manageable" (p. 186). Designers focus on a solution set that they value. Such a selection "rests on their subjective judgment" (p. 186).

Nadler (1981) suggests that the aim of evaluation is to provide information about performance and assure accountability. Evaluation occurs in relationship to every purposeful activity. Most evaluation relates to the values and objectives we wish to attain. Some approaches revolve around experiments or pilot projects. Although rational thinking in making judgments "seems to govern most approaches, affective considerations are increasingly voiced." This observation is particularly relevant to social systems design. An evaluation strategy, says Nadler, should determine the purpose of solution finding and it should define the intended users of the outcome of evaluation. Designers should identify relevant values and goals and their priorities and develop methods for obtaining measurement of design solutions. It is preferred that users of the evaluation be able to operate the measurement system. The measurement system should be implemented and the outcomes analyzed and assessed for the significance of findings. Designers should make use of the results in making design choices. They should also evaluate the evaluation system itself against its own objectives and make operating adjustments or redesign it as needed.

Checkland and Scholes (1990) propose a set of five "Es" for the overall evaluation of the solution. The solution should be judged (1) for its Efficacy, meaning: Does it work? (2) for its Efficiency: Are minimum resources used? (3) for its Effectiveness: Does it attain the goals and expectations? (4) for its Ethicality: Is it the moral thing to do? and (5) for its Elegance: Is it aesthetically pleasing? The authors suggest that it is the task of the designers to establish performance indicators as criteria for measuring the five Es. But these measures of performance should never be defined in a vacuum; but should be considered as part of a description of the system. Rowland (1994) suggests that

> designing and evaluating are two complementary parts of one process: comprehensive systems design. They share similar concepts; strategies are instruments; requirements are criteria; purposes are constructs, and situation and image are context. Viewing these concepts as similar offers a bridge between knowledge of design and evaluation and reveals that the two combined, both conceptually and temporally, create a powerful perspective from which to think and act. (pp. 19–20)

7.3.1.2. Microviews: Evaluating Alternatives

Design decisions made in selecting solution alternatives are a critical aspect of design evaluation. There is a host of decisions to be made in the course of a design program, and each decision point involves the formulation of alternatives and the making of choices among those alternatives. I labeled these choices as the microevaluation of design inquiry. I introduce here the views of three design scholars on this subject.

Ackoff (1981) suggests that design choices are always based on a comparative evaluation of sets of solution alternatives. The amount of effort that goes into such comparison "should depend on (1) the potential cost of selecting less than the best of the set, (2) how apparent the relative effectiveness of the alternative is, and (3) the cost of carrying out a sufficiently careful evaluation" (p. 195). Ackoff suggests that no matter how superior an alternative may appear to be, it should be chosen only if strong evidence supports it. A well-designed experiment, introduced in a context in which the solution alternative is intended to be used, might be the best way of testing it. Evaluation of alternatives requires time and resources. But the cost of making less than the best choice may be higher.

According to Nadler (1981), making choices among alternatives addresses the generation of alternatives and the selection of one. These two tasks are interrelated since alternatives are organized in usable groupings for selection. Grouping should contain alternatives that are equivalent in detail so that they can be fairly compared. Making choices among alternatives involves three primary tasks: generation, organization, and selection. There are also secondary tasks. These are attached to the appropriate primary task. One secondary task is related to information and knowledge acquisition, as explained next. In generating alternatives, in case our current knowledge or information is inadequate, we should acquire additional information or knowledge. The same goes for the organization of alternatives. If our current knowledge is inadequate, we should gather whatever information or knowledge we need to accomplish the task of organization. The third task is the actual selection of alternatives. This primary task has two kinds of secondary tasks. First we ask the question: Is it possible for us to select the best alternative based on our current knowledge? If not, we should obtain the information/knowledge required so that we can move toward the selection. If we have the information/knowledge required, or if we attained it, then we can move toward selection. In making the selection we ask: Would it be useful to formulate specific selection criteria? If so, then we are to generate alternative selection criteria, organize the alternatives, select the most appropriate one, and use it in selecting the best specific solution alternative.

Warfield (1990), in his seminal work, *A Science of General Design,* presents the tradeoff analysis method (TAM) as an approach to selecting alternatives. TAM is applied in the course of an ongoing design inquiry at decision points when a selection is to be made from a set of competing alternatives. How we get to that point is described in detail by Warfield. TAM is applied when a set of competing alternatives is available and designers are ready to make a choice. The first TAM task is to generate ideas for formulating an evaluation criteria that is appropriate to make a judgment of the alternatives. This idea generation can be accomplished by one of several consensus-building methods, such as the nominal group technique or idea writing. These methods were described in Section 4.7 of Chapter 4.

The process of TAM unfolds as a test for dominance is applied by organizing the alternatives and the evaluation criteria in a matrix. Alternatives are introduced in the rows and criteria items in the columns. Each alternative is scored on each of the criteria by a number, the highest rating being "1." The ranking is arrived at by the use of the prioritizing group technique. The scores assigned to alternatives enable designers to compare them and gradually arrive at two or a small set of alternatives having the highest scores. Given the reduced set, TAM calls for difference ranking between any two alternatives on each criteria. Given a judgment of the relative significance of the criteria items, ranking is assigned to the two alternatives we compare. The ranked alternatives are scaled on each item and added together to arrive at the final score of each of the alternatives. The TAM process, reviewed here briefly, appears to be extremely complex. Indeed it is. But in the generic design program developed by Warfield, called interpretive structural modeling, its application is aided by a software program. TAM is introduced here to show the advancement that has been made in design evaluation in general and specifically in the evaluation of design alternatives.

7.3.1.3. Evaluating the "Systemness" of Design Solutions

When we look at the whole design process and the outcome produced by the social systems design inquiry, we ask: Does the process meet the criteria of "systemness?" Here I introduce an approach that addresses this question. According to Churchman (1971), an entity has to meet nine conditions to be considered a social system. The understanding and use of these conditions in examining a social system inquiry enable us to compare alternative designs of a system and judge their systemness. Checkland (1981) adopted Churchman's scheme in evaluating the systemness of conceptual models produced by the use of soft systems methodology. Jenks and Amsler (1993) applied Churchman's systems conditions in assessing the adequacy of social systems designs. Churchman's nine conditions are briefly described as follows.

1. Is the system teleological? Does it exist to serve a purpose?
2. Does it have a measure of performance? Are expected performances identified and are relevant measurements available and are they carried out?
3. Are the clients, the stakeholders of the system, identified people, whose interests and values are to be served by the system?
4. Does the system have teleological components which coproduce the expected performance of the system? Do those components have measures of performance that are related to the performance of the system?
5. Is the system's environment clearly defined? Is the relationship, the mutual interaction patterns between the system and its environment, defined?
6. Does the system have identified designers who serve the interest and values of the stakeholders? How are these interests and values known to the designers? Who is involved in validating the design?
7. Does the system have a decision maker? According to Churchman, the client stakeholders, the designers, and the decision makers can be the same (this is also the position taken in this work).
8. Do the designers intend to change the system so as to maximize its value to the client/stakeholders? Do they maintain fidelity between the preferred/ideal design and the operationalized design?
9. Is there a guarantee that the designers intentions are realizable? According to Checkland (1981), this guarantee can be attained by the continuity of design. (This is also the position taken in this work since we say that organizational learning and design inquiry never end.)

The questions proposed by Churchman can be further elaborated in terms of formulating a variety of relevant inquiry tasks. These tasks could be operationalized by the development of diagnostic instruments that probe into how well the system meets the nine conditions. The operationalized tasks then can be integrated into the overall process of design inquiry. The designers will focus on these nine conditions when they compare the emerged competing overall solution alternatives during the convergence phase of the inquiry.

7.3.2. *An Image of a Comprehensive Design Evaluation System*

An image of a system is its first broad-scope representation, defined by an internally consistent set of markers. An example of key markers of a comprehensive design evaluation system is proposed here as follows: (1) design evaluation inquiry ought to be interactive and integrated with design inquiry; (2) it should be dynamic; (3) it should be inclusive in the sense that it should go beyond evaluating the adequacy of design choices and decisions made and should probe into the adequacy of the methods and tools of the design inquiry. The three markers are described next.

7.3.2.1. Interactive and Integrated Evaluation

The assumption that underlies integration is based on the idea that every time designers face a choice, they clarify how they will assess their decision. Every design choice should be complemented by a process for evaluating its adequacy. This complementary relationship is recursive. The evaluation criteria at any decision point have to be appropriate to the specific design solution under consideration. We ask: Is a criterion appropriate to measure the adequacy of the solution item? Conversely, we ask: Is the solution item (that we are to evaluate) defined with sufficient clarity and detail so that it can be readily evaluated? These recursive questions have the added potential to improve and refine both the design inquiry and the outcome of the design effort.

7.3.2.2. The Dynamics of Evaluation

The assumption that underlies the second marker is that integrated design evaluation inquiry is dynamic and as such it is multidirectional and recursive. At each decision task, feedback loops reach back and inquire into the compatibility and internal consistency of the presently made decision with decisions previously made. This reaching back is also recursive in the sense that the present decision might inform and influence decisions made earlier. In a previous section it was suggested that in the course of design we should keep somewhat open decisions so that those decisions will not overly constrain decisions to be made later. In addition to the dynamics of feedback loops, feed-forward loops reach ahead. This reaching ahead enables designers to make judgments in the present that are influenced by anticipated future decisions. Another property of design dynamics is that at each decision task event designers reach into the organized knowledge base for usable knowledge that might provide them with new insight and enlightment.

7.3.2.3. Inclusiveness

The third marker posits inclusiveness. Inclusiveness means that while evaluation probes the adequacy of design solutions and the adequacy of the evaluation approach and criteria used, it should also provide for the evaluation of the design methods and tools used in the course of the inquiry. As we have seen in Chapter 6, in designing the design system, the designing community addresses the design of design inquiry itself. But the choices of methods and tools made are not carved in stone. The real test of those choices comes when we apply them in the course of the design inquiry. The evaluation of design methods used enables designers to continuously refine and improve the design inquiry. This also implies hands-on design learning.

Reflections

As we review this section in its entirety, we cannot escape recognizing a good deal of diversity, tentativeness, and the evolutionary and unfolding nature of design evaluation. This tentative nature of the current state of affairs in design evaluation becomes quite evident as we review the discussion on approaches to the evaluation of alternatives. This state becomes understandable once we realize that a call for a well-defined, designed, and explained design evaluation approach to social systems design inquiry has emerged only recently as we have moved into the third-generation design mode. In this mode stakeholders take responsibility for the design of their system and they search for design approaches that will help them to validate their designs.

Activity #52

Review this section and address the following tasks. (1) Identify the core ideas implied and organize them in macro-, micro-, and system-level categories. (2) Reflect on these and construct a statement that represents your own thinking about design evaluation. (3) Select a decision task event in the functional context of your system of interst and describe and evaluation approach to it.

7.4.   Values and Qualities to Realize in our Design

Designers of a particular system of interest always aim at realizing certain values in their creation, whether the nature of the systems they design is technical, sociotechnical, socioeconomic, or sociocultural. For example, in designing a physical product, design values might include fidelity to the intended purpose, reliability, safety, ease of use, durability, economy, affordability, aesthetics, etc.

In social systems, the purpose for which the system is designed implies two kinds of values: (1) values that are generic to any social systems and (2) values that are specific to the system to be designed. This section characterizes values that are generic to any social system. At the end of the section I provide a brief example of values that are system specific.

In the domain of social systems, there are values that seem to be generally desirable properties that cut across all types of human systems. In Section 6.5 of Chapter 6 we discussed values and qualities designers might aspire to realize in their own designing systems as well as in the system they design.

Above I used "values" and "qualities" interchangeably. This needs clarification. Designers collectively will articulate values that they hold and wish to realize in their design. Once these values are manifested in the design, then the values become the definable qualities of the system. So in the text that follows I will use the term "quality." This section was conceived under the umbrella of the evaluation and assessment of design. The key questions we address here are: What qualities do we wish to realize in our design and does our design realize those qualities?

7.4.1.   Qualities to Be Manifested

Qualities are organized in four categories: (1) qualities that the design inquiry itself manifests, (2) individual and collective human qualities of people in the system, (3) qualities manifested in the collective of the designing community, and (4) qualities of systemness that designers wish to realize in the system they design.

7.4.1.1.   Qualities of the Design Inquiry

In design inquiry various qualities are sought. One is that the design is authentic, meaning that it is carried out by the genuine participation of the stakeholders. The quality of sustainability is related to authenticity and it also seeks the quality of competence in the stakeholders as a condition of sustainability. The quality of being user-friendly means the use of technical language that is meaningful to the designers. The design approach should manifest the quality of being disciplined, which is a technical quality meaning that up-to-date design technology and multiple perspectives are used. Ideal seeking/ pursuing is a special quality of the design of social systems. The design should also manifest multidimensionality, attention to the uniqueness of the design situation, and the uniqueness of the designing community. Designers also seek to manifest aesthetic qualities in their design.

7.4.1.2.   Individual and Collective Human Qualities

The individual and collective human qualities we seek to realize might include the purposeful involvement of the individual and collective intelligence of the stakeholders, the development of their human potential through continuous individual and collective learning, the attainment of competence in the design of their own lives and their systems, their liberation and emancipation from dominance of any form, and the full activation and engagement of their creative potential and ability.

7.4.1.3. Qualities of the Designing Community

The designing community seeks to attain high ethical qualities, knowing well that only ethical people can design ethical systems. In their ethical stance, they especially seek the quality of being sensitive to the effect of the design on future generations. As a community, they wish to become responsive to the aspirations of stakeholders of the system and responsible for the design they create. The designing community will seek to manifest the quality of diversity in their own community as they recognize the intrinsic strength of diversity. They seek among themselves the qualities of unconditional acceptance and respect for each other. Their dominant aspiration is to become a learning system and to develop their own design culture as a collective quality of their community. The designing community will regard having a shared systemic worldview, a quality of the highest order.

7.4.1.4. The Quality of Being Systemic

Designers seek to realize systemic qualities in their design. These qualities are grounded in an understanding of systems concepts, principles, and models, and their manifestation in the behavior of social systems and in their internal and external relationships. These qualities can be realized in both the process of design and the design product.

7.4.2. *Defining Desirable Qualities*

The qualities reviewed above under the four headings, are now briefly defined and explored.

7.4.2.1. Qualities of the Design Inquiry

When we design social systems, inquiry qualities are those that are most in view of the designers. Except for the qualities of purposefulness, viability, and design competence, most of the other qualities defined here are seldom made explicit.

7.4.2.1a. Attainment of Purpose and Viability. The two overall qualities we seek in design inquiry are that the design manifests the attainment of the purpose we seek to bring about and its viability, in the sense that the system can be brought into existence to operate successfully in its systemic environment.

7.4.2.1b. Authentic and Sustainable. The system is authentic if, and only if, it is designed by all who serve the system, are served by it, and have a stake in it. Design is authentic when people in the system can incorporate their individual and collective values and ideas in the design and can exercise their autonomy and responsibility for their participation in and contribution to the life of the system. Genuine participation and functional competence acquired in the course of the design will ensure the sustainability of the system, because people who participate in the design will take part more effectively and with a greater commitment when the time comes to implement the design and operate the system. People in the system "own" the design.

7.4.2.1c. Ideal Seeking. Design is always normative. Designing human activity systems in a constantly changing world requires that we set forth an image of the system we design that is the best we can create, that reflects our highest aspirations and expectations. Even if we cannot attain it, the ideal will always attract and inspire us as we continuously pursue it.

7.4.2.1d. Multiperspectives and Multidimensional. The design inquiry should manifest multiple perspectives, such as the technical, the cultural, the organizational and the personal perspectives of the designing community. Furthermore, the design should reflect all of the domains of existential experience, including the sociocultural, ethical (self-realization, social, and ecological ethics), socioeconomic, wellness (physical, mental, spiritual), learning and human/social development, aesthetic, scientific and technological, and political/governance.

7.4.2.1e. User-friendly. To enable the full participation of stakeholders, the technical language of design used in the course of design should be meaningful to, and compatible with, the everyday language use of the designing community.

7.4.2.1f. Uniqueness. Each and every design situation, design environment, and designing community is unique. An essential quality of the design is that it always takes into account and honors uniqueness.

7.4.2.1g. Aesthetics. This quality comes into play in two senses. Through their participation the designing community can bring into the design individually and collectively their aesthetic values and ideas so that the design, once implemented, will be aesthetically pleasing. If this quality is explicitly sought and realized in the course of design, then the designers' involvement in the design inquiry will become in itself an aesthetic experience.

7.4.2.2.   Individual and Collective Human Qualities

In the first- and second-generation design modes the issue of defining and realizing certain human qualities as part of the design effort was not even considered. However, this state of affairs changed radically in the third-generation mode, when we designated the stakeholders as being the designers of their system. This mode of design calls for the activation and use of all the individual and collective capabilities possessed by the designing community. Design becomes not only system creation but also an individual and collective learning and human development experience. Some of these key human qualities are defined next.

7.4.2.2a.   Individual and Collective Intelligence.   These are design qualities of the highest value. It is in the nature of authentic design to call upon the intelligence and talents of each and every member of the designing community. The activation of collective intelligence means the application of individual intelligence in a purposeful and coordinated way. Systems design calls upon a wide scope of different capabilities, and the designing community is to create conditions in which these capabilities are offered freely, are exercised in a cooperative and collective way, and are directed toward common ends.

7.4.2.2b.   The Development of Individual and Collective Potential.   Nurturing such development becomes a specific design task. Design in the third-generation design mode provides resources and unique opportunities to members of the designing community for continuing individual and collective learning and human development. In the design scholarship community we often remark that such learning and development might be more important than the product of design itself.

7.4.2.2c.   The Attainment of Competence in Design.   The attainment of design competence is a special kind of learning that is an important by-product of social systems design. Few people would question that in an age of constant change, the ability to design is one of the highest individual and collective human qualities. Involvement in design offers people the opportunity to learn to design their own lives and the systems in which they live. Design competence empowers them to shape their own futures and, collectively, the futures of their families, their systems, and their communities. Ultimately, that is what true participative democracy is about.

7.4.2.2d.   Liberation and Emancipation from Dominance.   These are key markers of social systems design in an authentic, participative mode. In earlier design modes, design was directed from the top of the organization or attempts where made to "sell" the vision to the boss. People in the system were expected to comply and implement someone else's design. However, in a true participative design mode, there are no chiefs, no designated authorities. Dominance stifles creative involvement. Design flourishes only if there is equity among the participants. Everyone has the same right and responsibility in making contributions. This may be the most difficult idea to accept by those in authority. Liberation and emancipation from dominance becomes among the highest qualities to be realized in social systems design.

7.4.2.2e. The Activation and Continuing Development of Creativity. Creativity is central to design. It brings forth something that does not yet exist. We cannot understand design unless we understand creativity. The quality of creativity can be realized once we understand what it is and how it works. Once we are aware of the internal and external conditions that are to be met to activate and sustain creativity, and once we know the barriers that prevent it as well as the ways to overcome those barriers, then—and only then—are we in the position to use the individual and collective creative potential of the designing community.

7.4.2.3. Qualities of the Designing Community

We seek to understand the qualities that the designing community should posses to become a viable and effective designing system. Clearly it should embrace and posses the qualities described in the other three realms. But beyond those, there are special qualities of the designing community, which are described next.

7.4.2.3a. The Community's Quality as an Ethical System. Only an ethical system can design an ethical system. Individual members, as well as the collective of the designing system, should manifest the quality of high moral and ethical standards. It might be useful for the community to establish a code of ethics for itself as a guide. A particular ethical consideration is concern about the impact of design on future generations. Thus, at every design decision point we should ask: How will the system we are considering affect future generations? Substantive evidence of such concern is the involvement of children and youth in the design inquiry. Symbolic evidence is an empty chair in design sessions reminding us of the unborn. We have used this symbolic metaphor in many of our conversations over the last decade. It is a powerful reminder, and it triggers a new way of thinking.

7.4.2.3b. The Community Is Responsive and Responsible. These qualities are part of being ethical, but deserve special emphasis. Being responsive means that designers respond to the aspirations and desires not only of the stakeholders but of the larger community and future generations. The quality of responsibility means that the designing community collectively takes responsibility for the design it created.

7.4.2.3c.   The Quality of Diversity.   The presence of diversity is an essential condition of evolution of life. The loss of diversity in life, such as in monocultures, creates fragile conditions. Only diversity can generate the broad multifaceted design intelligence required today in designing complex systems that represent cultural and racial diversity and operate in ever-changing and diverse environments. Exclusion of diversity from design or from a designing community and seeking only the involvement of designers with similar viewpoints builds brittle designs that will not weather the reality of our dynamic and diverse world.

7.4.2.3d.   The Quality of Unconditional Acceptance and Respect.   This quality is a core requirement of maintaining viable designing communities. It is a requirement that is more than a condition of civility and humanness. In the designing community only such acceptance of and respect for each other ensure that team members will be willing to offer solution ideas and make creative contributions spontaneously and without any fear of being rebuffed or ridiculed.

7.4.2.3e.   The Cultural Quality of Designing.   An essential quality is having a shared design culture of the designing community. This is a higher-order quality in that it includes several of the qualities mentioned in this set and in the set of desired human qualities. It is from the interaction of these qualities that design culture emerges and becomes an enriching "lived" individual and collective cultural characteristic of the designing community.

7.4.2.3f.   The Quality of Becoming a Learning System.   The designing community is to become a learning system. This quality ensures that arrangements, operations, and structures are in place by which to build organizational capacity and human capability to engage in continuous organizational learning. This quality will ensure continuous coevolution and creative interaction with the environment, and readiness to engage in the ever ongoing renewal and, if indicated, the redesign of our system.

7.4.2.3g.   The Quality of Being Guided by an Explicit Worldview.   A worldview attributes meaning to what we observe and experience. It shapes our behavior, our actions. This quality is another higher-order quality that emerges from an interaction and integration of a systems view of the world, a systems view of the systems in which we live, and a design view of change and development. These views guide the emergence of designing consciousness. They will become explicit in shared core values and ideas and the collective worldview of the designing community. They constitute the philosophical basis of creating design solutions.

7.4.2.4. Qualities of Being Systemic

These qualities emerge in design as a realization of the systems and design views described above. The essence of this manifestation is the application of systems concepts, principles, and systems models, in both the process and product of design. The qualities, briefly named below, are by no means all inclusive but they are indicative.

7.4.2.4a. Desirable Qualities in the Nature of Social Systems. The qualities we seek here include wholeness—the indivisibility of the system; teleology—the seeking and fulfillment of purposes; openness to the environment, uniqueness in context and content; seeking complexity in purposes, functions, relations, components, and structure; robustness of built-in vitality and strength; and requisite variety, where the variety of the system matches the variety of the systemic environment.

7.4.2.4b. Desirable Qualities in the Behavior of the System. These behavioral qualities include the ability of the system to be self-referential—to know and understand itself—self-organizing, and self-directive; and to connect self-defined purposes with purpose-serving functions, components, their interactions, and structure. Another behavioral quality is conscious attention to both negative feedback, which reduces deviation from the existing norm by making adjustments in the system, and positive feedback, which increases deviation and calls for changing the whole system.

7.4.2.4c. Desirable Qualities in Dynamic Relationship to the Environment. These include the system's continuous awareness of and interaction with the environment; the seeking of coevolutionary and cocreative relationship with it; and expansionism, that is, relating to larger and larger spheres of the environment to become ever more relevant to the system's contexts and to make those ever more relevant to the system.

7.4.2.4d. Desirable Qualities of the Internal Dynamics. These include purposeful and guided emergence (rather than chance emergence) that comes about by the purposeful interaction and integration of functions and components that carry out those functions; recursiveness that guides interaction among functions and components; and purposeful attention to the use of feedback and feedforward loops.

7.4.3. An Example of System-Specific Qualities

A specific social system, such as the educational system, would have its own specific qualities that designers seek to bring about in their design. These specific qualities are in addition to the general qualities described above. For example, in designing systems of learning and human development, the values we want to realize might include the full development of the uniqueness and unique potential of the learner; the development of competence called for by the information/knowledge era; the attainment of a high quality of inner life—the ethical, moral, and spiritual; the development of social responsibility and cooperative competence; commitment to lifelong learning; and empowerment that enables learners to design their own life and participate in the design of systems in which they live. This example shows a set that is specific to a system type. Beyond these qualities, designers will seek qualities that are situation specific, values that are unique to the particular system of learning we design in a particular environment.

Activity #53

Three tasks might be helpful in constructing your own understanding and interpretation of transforming values (generic to social systems) into qualities manifested in the design. (1) Review the qualities described in this section and propose others that come to mind. (2) Speculate about qualities that you would seek to realize in the design of your system of interest. (3) Contemplate evidence that assures that the qualities you seek are realized. Enter your findings in your workbook.

Reflections

Design is value based. As we engage in the design of a system we do so with the anticipation that the system will manifest our aspirations and the expectation that it will reflect our values. In the course of design our values will be transformed into qualities that become the properties of the system we design. The overall quality of the system emerges from the interaction and integration of the qualities we seek to attain. That overall quality is more than the sum of the qualities. Therefore, we are to conceptualize the qualities as a system. First, we look for internal consistency and compatibility among the qualities that should be there to constitute the system we wish to create. Then we are to design them into a system by identifying relationships among the qualities that stipulate recursive interaction among qualities that through mutual influence reinforce the members of pairs and multiple pairs. Thus, qualities become strengthened and the overall quality emerges.

This section explored evaluation through valuation. As we state and explore qualities we wish to realize in our design, we must ask the question: Have we succeeded in realizing them? This question can be answered positively only if we find evidence that tells us that in fact a specific quality has been attained. This kind of evaluation, namely evidence seeking, should parallel the evaluation approach developed in Section 7.3.

7.5.   From Evolutionary Consciousness to Conscious Evolution

We are at a critical juncture of societal evolution where unprecedented human fulfillment as well as a loss of direction, despair, and destruction, are equally possible. However, we are not at the mercy of evolutionary forces but have the potential and the opportunity to give direction to societal evolution by design, provided we create an evolutionary vision for the future and develop the will and the competence to fulfill that vision in our own lives, in our families, in the systems in which we live, in our communities and societies, and in the global system of humanity.

In this section, I present a systems view of societal evolution and point out evolutionary gaps that are potential sources of our destruction. I will then suggest that it is within our power to steer societal evolution toward a hoped-for future, provided we (1) develop evolutionary consciousness, (2) attain the will to engage in conscious evolution, (3) develop evolutionary competence, (4) create an evolutionary image of our future, and (5) bring that image to life by design. The section was adopted in part from my earlier works (Banathy, 1987b, 1989, 1993a).

In the closing paragraph of *Order Out of Chaos* (1984), Prigogine and Stengers noted that societies are immensely complex systems, highly sensitive to fluctuations and involving a potentially enormous number of bifurcations. This, he says, leads to both threat and hope. "The threat lies in the realization that in our universe the security of stable, permanent rules are gone forever. We are living in a dangerous and uncertain world that inspires no blind confidence. Our hope arises from the knowledge that even small fluctuations may grow and change the overall structure. As a result, individual activity is not doomed to insignificance" (p. 313).

7.5.1.   A Systems View of Evolution

A historical perspective on societal evolution may help us to formulate a systems view of societal evolution. Adopting the ideas of Curtis (1982), I plotted several evolutionary stages. In Chapter 3 we briefly reviewed societal evolution as we traced the evolution of design throughout the various stages of human evolution. In this section and the one that follows we explore the guiding role of design in social evolution.

7.5.1.1.   Stages of Societal Evolution

For the purposes of our present exploration, we can mark five distinct stages of human evolution. Even though the stages build on each other, they are discontinuous. One cannot extrapolate the characteristics of a stage from the one preceding it. Here we review these stages and explore our role in the evolutionary process.

Stage one spanned possibly a million years, during which time human consciousness evolved, coupled with the greatest human creation: speech. Speech made it possible for us to expand the boundaries of human experience in time and space, as oral tradition embraced the past. Magico-religious myth became the all-embracing paradigm of understanding. Hunting-gathering tribes became the integrating context of collective human experience.

With stage two, about ten thousand years ago, we entered into the agricultural age. Self-reflective consciousness lead to a new creation: writing. Writing enabled the further extension of the boundaries of human experience. This stage was marked by the flourishing of city-states and the philosophy and logic that emerged in Greek culture.

Some five hundred years ago, the Renaissance became the genesis of stage three. With print as the new technology of communication, the boundaries of human experience were extended into national states. In this era of "enlightenment," Newtonian science, the mechanistic/deterministic worldview, and science-based technology emerged. This stage culminated in the Industrial Revolution, which brought forth a quantum jump, a discontinuous explosion in technical evolution. The designer's product became separated from man's template (Csanyi, 1989) as products could be produced in the millions. Technologies of the Industrial Revolution were often used unwisely, resulting in destructive influences on our habitat.

Stage four began around the end of the last century. Through "instant" global telecommunications our spatial and time boundaries exploded, embracing the whole globe. Beyond national consciousness, the potential of global consciousness emerged. Strong reaction emerged against the deterministic and reductionist paradigm of Newtonian science, epitomized in existentialism, relativism, and quantum theory. This stage had a time span of less than a hundred years. (See Figure 4.1.)

Stage five is our current evolutionary stage. It emerged around the middle of this century. Its genesis is marked by three events: the introduction of the greatest destructive force, the atomic bomb; the creation of the United Nations, as a new hope for humanity; and, most significantly, the emergence of cybernetic/systems science and its product, the computer. These events lead to what we now call the information/knowledge age. While the machine age exploded our physical power, cybernetic technology exploded our cognitive powers. Systems science has emerged as the new paradigm for knowledge production, organization, and utilization.

7.5.1.2. The Emergence of Evolutionary Gaps

Recognizing the great disproportion of the time span of the various evolutionary stages—over a million years, then ten thousand, five hundred, one hundred, and now fifty—we realize that the synergictic effect of the speed and intensity of the development of stage four and stage five has resulted in a perilous evolutionary imbalance. Stages four and five brought forth the potential of a global human community, but our collective consciousness is still locked within ethnocentric, racial, and national boundaries, thus creating an evolutionary consciousness gap. Furthermore, during the last several decades, the technological revolution, while giving us earlier unimagined and unprecedented power, has accelerated to the point where we have lost control over it. We have simply failed to match the advancement of our technological intelligence with an advancement in sociocultural intelligence, an advancement in human quality and wisdom. This situation has created the second evolutionary gap in the sociocultural sphere.

7.5.1.3. A Systemic Image of Societal Evolution

A systems view of societal evolution helps us to draw some general conclusions from the historical perspective depicted above. A systems view provides us with a lens through which we can view societal evolution and capture a comprehensive vision of evolution. From such a comprehensive vision we may come to understand evolution as (1) a phenomenon of the constantly expanding boundaries of the space and time dimensions of the human experience; (2) the emergence and conscious design of new systems of technologies of communication that enhance such expansion; (3) the continuous unfolding of new relationships among human systems, leading to their reorganization at higher levels of complexity; (4) the emergence of new paradigms of knowledge organization and utilization and new ways of beliefs and thinking; and, from the systemic integration of all the above, (5) the creation and emergence of new images of mankind at higher levels of collective consciousness. Such emergence is the ultimate hallmark of a particular evolutionary stage.

An evolutionary stage is complete when there is an integration of converged idea-structures and a state of internal coherence is attained (Csanyi, 1989). Follett (1965) called this the attainment of "self-created coherence." Evolution is directed by the innate tendency of the whole to create unity within its parts and synthesize their differences (Lorenz, 1977). Collective consciousness emerges as a result of such synthesis. At the current evolutionary stage, we have yet to create a unity of consciousness. Thus, today we are confronted with an evolutionary crisis, a crisis of consciousness. This crisis is the major source of the current human and global predicament. This is a crisis that we created, and we are responsible for acting upon it. The current evolutionary imbalance and consciousness gap frame a true window of vulnerability for mankind, producing the potential for the greatest threat: self-destruction. We can boldly face the challenge of this threat by attaining the will and capability of conscious and purposeful evolution.

7.5.2. Conditions of Purposeful Evolution

The human race has profoundly changed the parameters of the evolutionary process. Our unlimited capacity for learning and the explosive rate at which we produce knowledge, artifacts, and systems have had an extraordinary impact on evolution. The question that confronts us is: For what purpose are we going to use this unlimited capacity for learning and our collective creative power? We can use this capacity and power to create a better future and give a hopeful direction to our evolution. This, however, is dependent upon meeting four conditions: (1) the development of evolutionary consciousness; (2) the attainment of a will of conscious evolution, and, based on it; (3) the acquisition of evolutionary competence through evolutionary learning; and (4) the activation of evolutionary competence in creating an evolutionary vision as a guiding image of the future. In this section the first three conditions are discussed, while the fourth is developed in the section that follows.

7.5.2.1. The Development of Evolutionary Consciousness

Consciousness emancipated human beings from the confines of sensory reality and placed us in a world we ourselves created (Laszlo, 1972). When it evolved, consciousness took over the direction of our evolution. "The means became the end: the self-maintaining biological species was transformed into a culture sensitive to knowledge, beauty, faith, and morality" (p. 99).

Understanding relatedness and interdependence in the global context is global awareness. Having the intent, the will, the capacity, and the capability to relate to all and integrate with all else in the global system of humanity is the hallmark of global consciousness. Developing individual and collective global consciousness is the common task of individuals, the various societal systems, and the whole human community.

7.5.2.2. The Attainment of the Will of Conscious Evolution

In evolution the most advanced state of existence is human consciousness. It is best manifested in those who are most developed in terms of their relationship to others and in their ability to interact harmoniously with all else in their sphere of life. Its highest form is "evolutionary consciousness," which enables us to collaborate actively with the evolutionary process. Salk (1983) says that evolutionary consciousness can motivate action toward "conscious evolution," by which we can guide our future, provided we have a clear vision of what we wish to bring about. Conscious evolution enables us to use the creative power of our minds to guide our systems and our society toward the fulfillment of their potential. Laszlo (1987) highlights our evolutionary responsibility. He says: "The evolution of our societies, and therewith the future of our species, is now in our hands. Only by becoming conscious of evolution can we make evolution conscious" (p. 122). Conscious evolution is enabled by "self-reflective consciousness" and it is activated by "creating consciousness" (Banathy, 1993a).

7.5.2.2a. Self-Reflective Consciousness.   Self-reflective consciousness is a process by which individuals, groups, organizations, and societies contemplate and make presentations of their perceptions of the world—and their understanding of their place in the world—in their individual and collective minds. These representations are developed on the basis of values we hold and the ideas we have about how the world works, leading to the creation of a cognitive map of "what is." Cognitive maps are developed, confirmed, elaborated, tested, disconfirmed, changed, and redrawn. They are "alive." They affect our behavior and they are affected by it. This mutual affecting is recursive and it is constantly evolving.

7.5.2.2b. Creating Consciousness.   The genesis of creating consciousness is self-reflection that brings forth understanding insights and aspirations. It is a process by which individuals, groups, organizations, and societies envision "what should be." This creating thrust is based on the belief that while the future is influenced by the past and present it is not determined by what was or what is. It remains open to conscious and purposeful intervention that can be guided by an evolutionary image of the future. A representation of that image is the normative cognitive map of a desired future, which we can create individually and collectively.

7.5.3. Evolutionary Competence Through Evolutionary Learning

Conscious evolution and evolutionary vision are activated as we acquire evolutionary competence through evolutionary learning. Evolutionary competence enables us to give direction to our individual and collective evolution through purposeful design, provided we individually and collectively learn specific knowledge, ways of thinking, skills, and dispositions that jointly and interactively constitute the domain of evolutionary competence. The key point made here is that the hope for a better future for humanity lies in individual and societal learning of understanding ways of thinking, skills, and dispositions that are necessary to attain conscious evolution and acquire evolutionary competence.

7.5.3.1. Evolutionary Learning

The greatest source of change in social systems is the learning (Boulding, 1985), both the development of new knowledge and know-how that the human race never had before. It is this source that we have to activate to attain conscious evolution and acquire evolutionary competence through evolutionary learning. Evolutionary learning is first explored here in the context of current practices in education. It will be shown that we face a major evolutionary task in education itself, namely, the reconceptualization and redesign of education so that it can engender the acquisition of evolutionary competence through evolutionary learning.

7.5.3.1a.   A Major Hindrance to Evolutionary Learning.   A major hindrance to the development of evolutionary competence is inherent in our current practice of education, which focuses on "maintenance learning" (Botnik and Maltiza 1979). It involves the acquisition of fixed outlooks, methods, and rules of dealing with known events and recurring situations. We are promoting already established ways of working in systems that now exist. Maintenance learning is indispensable for the functioning of a society, but it is far from being enough in times of turbulence, rapid change, discontinuity, and massive transformations— characteristics of our current era.

Our present learning agenda should be complemented with another type of learning that is even more essential at the current evolutionary stage, namely, "evolutionary learning." Evolutionary learning enables us to cope with change and complexity, renew our perspectives, and redesign our systems, often reorganizing them at higher levels of complexity. Evolutionary learning empowers us to anticipate and face unexpected situations. It will help us to progress from unconscious adaptation to our environment to conscious innovation, coevolution, and cocreation with the environment and the development of the ability to direct and manage change.

7.5.3.1b.   Contrasting Maintenance and Evolutionary Learning.   Maintenance learning leads us to operate in a "negative feedback" mode of error detection and correction. This type of learning is adaptive. It reduces deviations from existing norms and is useful in maintaining the existing state. But we live in an era when we must learn another type, innovative learning, which operates primarily in a positive feedback mode. It amplifies deviation from existing practices as it moves us in a double-loop learning mode (Argyris, 1978). In this mode, we become open to examining and changing our purposes and perspectives, transcending our existing state, and redefining and re-creating our systems. We now speak of the most significant learning of our age: design learning.

While maintenance learning reinforces already-learned ways of responding, it leads to reluctance to change and makes us unable to guide change. Evolutionary learning enables us to face unexpected situations and, even more, engage in purposeful change. We develop the will and capability to shape change rather than just coping with it or becoming its victims. Evolutionary learning calls upon and nurtures our creative potentials as it enables us to envision the future and bring those images to life by design.

Competition is rewarded in our current education practices. Students compete against each other for grades. Conscious evolution places a high premium on cooperation, upon the shared envisioning of desired futures and collective action for bringing the vision to life. Evolutionary learning involves both cooperation as a mode of learning, which recently has become a practice in education, and the purposeful learning and development of cooperative group interaction skills (Banathy and Johnson, 1977).

In our current educational practices, the student is placed in subject matter boxes and is taught in a lockstepped and reductionist mode. In evolutionary learning, we seek to think and act systemically, to seek and understand integrated relationships, grasp the patterns that connect, and recognize the embeddedness and interdependence of emergence in systems. In evolutionary learning we transcend the subject matter boxes and integrate them in functional contexts that are real and important to the learner. Synthesis becomes the primary mode of inquiry.

The contrast developed here can be best summed up by a metaphor I heard from Simon Nichols (personal communication, 1979) of the Open University, who was guiding a seven-year multinational project of children designing the future. He said that in our conventional educational mode we are driving children into the future by looking into the rearview mirror. The windshield is blacked out for them and teachers are doing the driving. Isn't it time, he asks, to clear the windshield and enable students to do the driving? Evolutionary learning opens up for us an unlimited horizon and develops competence in driving toward the future with a purposeful destination in mind and purposeful action.

7.5.3.1c. Acquiring Evolutionary Competence. The acquisition of evolutionary competence enables individuals, families, groups, organizations, communities, and the society to create positive images of the future and steer their evolution by purposeful design. A program of evolutionary learning will include domains such as the following:

- The nurturing of such evolutionary values as cooperation, trust, benevolence, altruism, love, and harmony, and the development of a universal set of values that generate evolutionary consciousness and an ever-maturing vision of the future.

- The fostering of evolutionary ethics that include self-realization ethics and social and ecological ethics.
- The attainment of cooperative group interaction skills by which we can increase our capacities for entering into ever-widening human relationships and for managing conflicts in a nonviolent manner.
- The acquisition of competence in systems thinking and practice, by which to understand complexity, grasp connectedness and interdependence, and perceive the notions of embeddedness and wholeness. The development of a systems view of the world and the attainment of the capability to relate functionally to the ever-enlarging social systems in which we are embedded.
- The development of competence that enables the creation of desirable images of the future and the learning of the skills to generate and evaluate design alternatives by which to bring those images to life.

7.5.4. *Creating Conditions for Evolutionary Learning*

At least seven conditions are proposed, the meeting of which enhance evolutionary learning, namely, (1) creating a climate of nurturing, (2) providing multiple learning types, (3) providing learning in functional contexts that are relevant to the learner, (4) creating broad-based learning resources, (5) engendering self-created meaning, (6) engaging in evolutionary imaging and designing, and (7) applying what has been learned to real-world contexts.

7.5.4.1. Creating a Climate of Nurturing

Evolutionary learning can flourish only in a climate in which nurturing and caring relationships are created and support and trust flows both ways between those who learn and those who foster learning. As Elise Boulding (1981) noted, our current educational practices expect compliance, which often engenders insecurity, resistance, and even fear. Nurturing builds confidence, encourages exploration, and secures conditions for creativity and evolutionary learning.

7.5.4.2. Offering Multiple Learning Types

Learning types that are conducive to evolutionary learning include (1) socially supported individual learning in which the learner is guided by others; (2) self-directed learning in which the learner has access to learning resources; (3) team-learning arrangements in which learners cooperate and share experiences in joint mastery of learning tasks; and (4) learning in social contexts (e.g., family, peer-group, organizational contexts) that focus on the development of individual and collective evolutionary consciousness and guide conscious and purposeful evolution.

7.5.4.3. Providing Learner-Relevant Functional Contexts

To become meaningful to the learner, evolutionary learning should be provided in the context of social systems, such as the learning group, family, organized youth groups, organizations, the community, etc.—systems in which the learner is a participant and that offer actionable task environments for applications of what has been learned. Only in such contexts can we expect the development of knowledge, understanding, dispositions, and skills that enable the emergence of evolutionary values and competencies by which to guide one's own evolution and make contribution to the purposeful evolution of systems in which one lives and works.

7.5.4.4. Providing Broad-Based Learning Resources

Education is much more than schooling, and learning is much more than education (Banathy, 1981). The development of children and youth and continuing development through life meshes intricately with learning opportunities available in all facets of life. Beyond the boundaries of schooling and formal educational settings, learning opportunities and resources are offered in the home, in religious organizations, in youth and civic groups, in cultural and community agencies, through various media, in high-tech networks, in the world of work, and in many everyday situations. A powerful potential resides in the notion of creating an alliance of all the societal sectors that have the capability to support learning. The development of design culture and evolutionary learning can become the focus for creating such an alliance that can tap into a vast reservoir of resources for nurturing evolutionary learning and developing and applying evolutionary competence.

7.5.4.5. Exploring Self-Created Meaning

Whatever learning task is attended to by the learner, it can be "owned" by the learner only if the learner can self-reflect on it, make sense of it, and self-create meaning from whatever is offered in the course of the learning experience. The learner can internalize and integrate into his or her cognitive map what is being learned only if he or she can "construct" from it his or her own meaning and understanding. The process described here is an essential condition to a meaningful learning experience.

7.5.4.6. Creating an Evolutionary Image

Learning is not completed without its application in contexts and situations that are meaningful and important to the learner. Having met all the conditions (of learning) described above, the application of evolutionary learning is accomplished by challenging the learner to create an evolutionary vision of the future, a vision that is then elaborated in an evolutionary image. This creation is the product of the evolutionary consciousness and evolutionary learning.

7.5.4.7. Bringing the Image to Life

The last condition that enables the completion by application is to challenge the learner to create the system that transforms the image into reality, that brings the image to life.

Reflection

What is emerging from an understanding of (1) societal evolution, (2) evolutionary consciousness, (3) the nature of evolutionary learning, and (4) the role of evolutionary competence is that all these appear to be essential prerequisites for engaging in conscious evolution. Beyond this, however, evolutionary learning will also have a powerful impact on education itself. The fostering of evolutionary competence through evolutionary learning has the potential to change the purposes, the content, and the method of education. Education, enriched by evolutionary learning, will become a means to develop an evolutionary and design culture that can attain two far-reaching consequences. It can create a new societal way of life and it can guide societal evolution by purposeful design. This guiding role of design is the topic of the next section. An image of a new societal way of life is envisioned in the closing section of the chapter.

Activity #54

(1) Identify core ideas about societal evolution that make sense to you. (2) Explore what evolutionary consciousness, evolutionary learning and competence, and conscious evolution might mean to you personally and to the life of systems in which you participate. (3) Devise for yourself an agenda for evolutionary learning. Enter your findings on these three items in your workbook.

7.6. Guided Evolution

Conscious evolution provides a sense of direction for cultural and societal processes by illuminating those processes with guiding images. And the faster we go, as we do at our current evolutionary stage, the further we have to look for signs and images to guide our journey (Jantsch, 1975). The envisioning of such an image was defined by Kenneth Boulding (1978) as a unified view of evolution that connects all reality from cosmic/physical through biological/ecological/ sociobiological to psychological and social systems. It seeks to understand the evolutionary dynamics through which systems develop, and grasp the principles underlying the unfolding of evolution over space and time.

The core idea of evolutionary guidance says Jantsch (1981) is that evolution is not the result of one-sided adaptation and a desperate quest for survival, but is an expression of self-transcendence—the creative reaching out beyond the system's own boundaries. "We humans are the integral agents of evolution, we spearhead it on our planet and perhaps in our entire solar system. We are evolution and we are, to the extent of our power, responsible for it" (p. 4).

Evolutionary guidance implies arrangements and operations that are built into various human activity systems at all levels of the society by which these systems are empowered to give direction to their own evolution and move toward the realization of their evolutionary vision.

In this section, I (1) define evolutionary guidance, (2) propose a tentative image of a generic evolutionary guidance system (EGS), (3) introduce an example of an EGS, (4) characterize the role of design in evolution, and (5) describe the process by which we may create evolutionary guidance systems.

7.6.1. Evolutionary Guidance

Evolutionary guidance is a dynamic process of giving direction to the evolution of human systems and developing in these systems the organizational capacity and human capability to (1) nurture the physical, mental, emotional, and spiritual development and self-realization of individuals and their systems; (2) extend the boundaries of the possibilities for freedom and justice, economic and social well-being, and political participation; (3) increase cooperation and integration among societal systems and manage conflicts in a nonviolent manner; and (4) engage in the design of societal systems that can guide their own evolution by purposeful design. By attaining these purposes, we can re-create and empower our social systems as evolutionary guidance systems (EGS). This re-creation requires a fundamental reorganization or our inner map of reality away from fear, distrust, and hostility. It requires a change in the way we perceive ourselves and our relationships with others (Harman, 1984). In this way, we can create a shared image of the global human future and, at the same time, maintain and respect the diversity of our many cultures and societal systems.

An image of man, as Markley and Harman (1982) noted, is a gestalt perception of humankind, both the individual and the collective, in relation to self, others, society, and the cosmos. As a new evolutionary stage emerges, the use of old images creates more problems than it solves. On the other hand, when a new image leads sociocultural evolution, as it did when stage three emerged during the Renaissance period, it can exert what Polak (1973) called a "magnetic pull" toward the future. As a society moves toward the realization of that image, "the congruence increases between the image and the development of man and society" (Markley and Harman, 1982, p. 5). This will lead to internal consistency and harmony. Today we are still extrapolating from the old image of an industrial society. Thus, we are locked into conflictual struggles between races and nation-states. The old image is not working for us. It causes widespread frustration, alienation, and social upheavals. It may lead us to the brink of self-destruction. We desperately need a new image. This image shall spring forth from the emerging global consciousness and shared evolutionary vision. Its realization will be enabled by the acquisition of evolutionary competence.

7.6.2. A Tentative Definition of Evolutionary Guidance Systems

Evolution means "unrolling." It is a process by which successive forms and content unfold creatively. This process cannot be understood without considering the multidimensional reality of which it is a projection (Bohm, 1983). Such multidimensional unfolding has to be designed and implemented in all of our human systems, from the family to the global human community. What follows is a tentative presentation of a system of interactive dimensions that may enable such multidimensional unfolding and constitute evolutionary guidance systems. Such systems should have

- A social action dimension, ensuring social justice and an increase in cooperation, leading to the integration of our societal systems.
- An economic dimension with a focus on economic justice and integrated and indigenous development.
- A moral dimension that strengthens self-realization and social and ecological ethics.
- A "wellness" dimension that nurtures the physical, mental, emotional, and spiritual health and well-being of the individual and the society.
- A function of nurturing the full development of individuals and social groups and enabling them to develop a design culture and attain evolutionary competence.
- A scientific dimension, manifested in ethical science that serves human and social betterment.
- A technological dimension of placing technology under the guidance of sociocultural intelligence, placing it in the service of the nonviolent resolution of conflicts, and the improvement of the quality of life for all.
- An aesthetic dimension in the pursuit of beauty, cultural and spiritual

values, the various forms of arts, the treasurers of humanities, and the enrichment of our inner quality of life.
* A political dimension of self-determination, genuine participation in self-governance, continuous action for peace development, global cooperation and integration, and governance for the improvement of human conditions.

The purposeful design of these dimensions, as interactive aspects in all of our societal systems, will provide a powerful agenda for the self-directed evolution of our human systems.

7.6.3.   What If? An Example of an Evolutionary Guidance System

What would happen if the idea of evolutionary guidance systems became reality? What would happen if our human activity systems engaged in purposeful evolution on all nine dimensions proposed above? Let me speculate about the "what if" in the context of the most basic human activity system: the family. A few core ideas might serve as examples of creating an evolutionary vision for family development. We are envisioning an evolutionary guidance system that would enable the family to shape its future and develop along the lines of the nine dimensions described above. The example is not a prescription. But it might generate further thinking, exploration, and conversation about the idea of evolutionary guidance.

The family today is basically a socioeconomic unit. Its primary concern focuses on the economic necessities of existence, health, and safety, and often not much more than that. What kind of families would we have if, in addition to the above, the family purposefully developed an agenda for its evolution in all the dimensions of evolutionary guidance proposed here?

7.6.3.1.   The Social Action Dimension

In the social action dimension, we might envision the creation of a family agenda for the development of social consciousness, the realization of this consciousness in the family, the establishment of the idea and practice of social justice within the family, and its promotion in the systems in which members of the family live, the community, and the larger society.

7.6.3.2.   The Economic Dimension

In this dimension, beyond attention to economic necessities, core ideas would guide the establishment of economic justice within the family, as well as the promotion of same in systems in which members of the family participate in the community and in the larger society. The family would actively participate in activities that will advance the socioeconomic development of the neighborhood and the community.

7.6.3.3. The Moral and Ethical Dimensions

In any society the family is called upon to establish the foundations of ethical and moral consciousness and behavior in members of the family and in the family as a whole. It is this domain in which the formulation of core ideas and the establishment of an action agenda for evolutionary guidance is specifically called for. Such an agenda will include (1) the purposeful shaping of moral character and a moral worldview for making moral choices; (2) parents as role models; (3) the nurturing of self-realization ethics that guides members of the family and the family collectively to develop to their full potential; (4) the realization of social ethics that promotes respect of all other individuals and cultures, develops genuine concern for others, and engenders cooperation within the family and with others in the society; (5) ecological ethics by which the family learns to live responsively and in harmony with nature and by which individual and collective actions are taken to create natural beauty in the environment of the family, even if on a very modest scale.

7.6.3.4. Nurturing Wellness

The physical, mental, emotional, and spiritual dimensions have always been a primary domain of responsibility of the family. The evolutionary challenge here is focusing on an all-encompassing wellness, which is much more than countering or even preventing illness and weakness. The challenge is to develop an agenda with the family and its members for: (1) the development of physical fitness as individual and collective activity; (2) mental wellness, which actively promotes individual and collective self-reflection and creativity and the attainment of continuous cognitive development; (3) emotional wellness, the foundation of which is love, compassion, altruism, caring, and sharing (for the development of this dimension the family offers the most ideal social context); and (4) the evolution and constitution of the family as a spiritual unit, and the nurturing of faith and beliefs in all its members.

7.6.3.5. The Human Development Function

In the industrial society the educational and human development function was mostly assumed by schooling and, with it, the school became identified as its sole "legitimate" institutionalized form. Even today, at best we ask families to support and cooperate with the school. We do not assign families the role of becoming a primary territory for learning and human development or at least of becoming a full partner in education. Within the evolutionary perspective, developed here, the family assumes responsibility for nurturing the full development of its members and takes responsibility for its own collective learning and development, including nurturing the development of evolutionary competence (as described earlier) that will enable the family to direct its own evolution and shape its future. If we reconceptualize the educational function as described here, we shall talk about equal partnership between the family, the school, and many other societal institutions that can become learning territories, and may constitute the total ecological habitat of learning and human development.

7.6.3.6.   The Scientific Dimension

The scientific dimension can have a very special role in the family as an evolutionary unit of the society. Recognizing that the now-emerging society is knowledge-based and that science is the breeding ground of knowledge, the family can offer the first context in which to develop interest in—and excitement about—science, discovery, and creativity. It can offer opportunities to "do" science appropriate to the family scale, and nurture the development of systems thinking and systems practice in the daily life and activities of the family.

7.6.3.7.   The Technological Dimension

This dimension offers an interesting agenda for the family. Earlier, I identified the critical evolutionary gap between our highly advanced technological intelligence and sociocultural intelligence and wisdom. The early recognition of this gap as a topic for family discussion and content and context for collective family learning can set the basis for developing the kind of human wisdom that is needed to get the upper hand, to guide choices, and to make wiser decisions about the use of technology. The use of technological devices in the home and in the community offers a very practical, real-life context of learning to make such choices.

7.6.3.8.   The Aesthetic Dimension

The aesthetic dimension offers the learning and nurturing of (1) the pursuit of beauty—what it is and how it can be created and appreciated; (2) the engendering of cultural values; (3) the enjoyment and creation of arts, literature, poetry, and music, and above all, the enrichment of our inner quality of life. Aesthetics can offer a very, very specific evolutionary agenda in the life and development of the family, both collectively as a unit and individually for all its members.

7.6.3.9.   The Governance or Political Dimension

This dimension of family life can become the domain within which to coordinate, guide, and govern the evolutionary development of all the other dimensions. It can become the learning context and content for (1) genuine participative democracy, (2) the peaceful and negotiated resolution of conflicts, (3) the integration of the family as a social unit, (4) participative management and decision making in the family, and (5) the nurturing and encouragement of the use of these four functions in the community and in larger societal contexts by exercising creative participation in social systems design and public decision making.

7.6.4. Reflection on the Example

Even a cursory exploration of the few core ideas that project the family as an evolutionary guidance system shows the immensity of the task that confronts us in proposing the guidance functions described above. It requires (1) the total reconceptualization of the societal function of the family, (2) a major paradigm shift in the way the family thinks about itself and in what it does, and (3) the development of new arrangements and relationships between the family and other systems of the society.

Extending the "what if" speculation, just imagine what would happen if all of our other societal systems, our communities, our public and private institutions, and our entire governance system from local to global redesigned themselves as evolutionary guidance systems? It would mean a new way of life for us individually and collectively.

In a special issue of *World Futures* a number of authors reported on the design of evolutionary guidance systems (EGS) in a variety of contexts. Included were the design of an EGS for the retirement of a couple (Frantz and Miller, 1993); the design of an EGS for a nonprofit professional/scientific association (Bach, 1993); medical practice for evolutionary learning and the use of EGS (McGee, 1993); EGS for evolutionary systems management of organizations (Wailand, 1993); and a mediation approach that fosters evolutionary consciousness and competence (Pastorino, 1993). In addition, Biach (1995), Blais (1995), and Dils (1995) gave their account of the use of EGS in systems of their interest.

7.6.5. Creating Evolutionary Guidance Systems

In the evolutionary research community, specifically in the general evolutionary research group, we have an ongoing conversation on the "weak" versus "strong" hypothesis of influencing evolution. I have taken, as I do here, the strong position. When Laszlo (1987) proposes that it is possible—in principle—to master the evolutionary process by purposeful action, I say that it is possible not only in principle but in actual practice, provided we become conscious of evolution, attain the will of engaging in conscious evolution, develop evolutionary competence by evolutionary learning, and engage in guiding our evolution by purposeful design.

The activation of evolutionary guidance faces three challenges: (1) how to bring intention, expressed by an evolutionary image, and systems design together; (2) how to transform evolutionary images into functional designs of societal systems; and (3) how to develop societal/organizational arrangements that implement and maintain the design of evolutionary guidance systems.

We all too frequently assume that intention itself leads to action. As Kenneth Boulding said, "intentions are fairly easy to perceive, but frequently do not come about and are not fulfilled. Design is hard to perceive. But it is design and not intention that creates the future" (1985, p. 221). Creating an image of the future, without designing a system that can realize that image, is very much like trying to construct a building from a sketch of the house, without having designed a blueprint for it.

The key proposition advanced here is that societal systems are purposeful systems in which design can guide evolution. With the emergence of the process-oriented, self-organization paradigm in human systems (Jantsch, 1980), evolution became the integral aspect of self-organization, in which the system reaches out beyond its boundaries and design becomes the core process of evolution. Thus, in the evolution of societal systems, design is the central activity and competence in design is a commodity of the highest value.

Design is a creative, decision-oriented, disciplined inquiry that aims to formulate expectations, aspirations, and requirements of the system to be designed; clarify ideas and images of alternative representations of the future system; devise criteria by which to evaluate those alternatives; select and describe or "model" the most promising alternative; and prepare a plan for the development of the selected model. The design and description of alternative evolutionary guidance systems enable their conceptual and empirical testing, and the selection of the most promising "ideal" model.

7.6.6.   The Design of Evolutionary Guidance Systems

Design inquiry that creates an evolutionary guidance system (EGS) operates in several design spaces. Although these spaces were described in earlier sections, here I propose their specific use for the creation of EGS.

7.6.6.1.   The Exploration Image Creation Space (EICS)

In the EICS, a social system, such as a family, creates an evolutionary vision for itself, articulates its collective values and core ideas about its desired future, and develops the first systemic representation of that desired future as the image of its evolutionary guidance system (Frantz, 1995). The design tasks of exploration and image creation are integrated with evolutionary learning aimed at the development of evolutionary competence. This learning is accomplished in the functional context of the exploration/image creation tasks.

7.6.6.2. The Design Solution Space (DSS)

The DSS occupies the center of the design inquiry in which we formulate the systemic representation, or systems model of the particular evolutionary guidance system. It is in this space where we transform the image of the EGS into the design of the EGS. The primary tasks that we undertake in this space include formulation of alternative solutions to (1) the core definition and the purposes of the EGS, (2) the selection and systemic arrangement of functions that enable us to carry out the purposes, and (3) the design of the EGS that will have the human capability and organizational capacity to carry out the functions.

7.6.6.3. The Organized Knowledge Space (OKS)

In the OKS we display the information we generated in the exploration/image creation space as well as collect and organize information and knowledge we generate pertinent to the nine evolutionary dimensions previously described. We also continuously seek information/knowledge as required by carrying out the tasks in the design solution space.

7.6.6.4. The Evaluation/Experimentation Space (EES)

This space is created to enable us to test the emerging solution alternatives both conceptually and in the real world. Such evaluation mitigates against errors in perceiving the real world and ensures the operational/practical implementable quality of the particular evolutionary guidance system.

7.6.6.5. The Space of the Future System (SFS)

It is in this space where we finally display a description or systems model of the evolutionary guidance system we designed. Here we also describe the systemic environment of the EGS and formulate a plan for its development and implementation.

7.6.6.6. The Dynamics of Design

Design is not accomplished in a step-by-step, linear fashion. It is carried out through recurring cycles of several design spirals as we explore, and reexplore, the various spaces of the design inquiry. During this inquiry we integrate aspiration and purpose, information and knowledge, and insight and vision to create a design that best represents and guides the evolution of the societal system we wish to bring to life.

Reflections

It is the power of design that enables us to take part in the continuing process of creation. Our unique gift of creativity enables us to form images of the future, create designs that represent those images, and then develop those designs in our experiential world. It is the unique challenge and responsibility of each of us personally and collectively to engage in the purposeful design of the evolution of our societal systems. There is no more noble and no more important task than meeting this challenge and assuming this responsibility. Creation continues and we can become instruments as coworkers of the Creator in designing and building a better future for all.

Activity #55

First, describe the core ideas of guided evolution. Then, create an image of an evolutionary guidance system for a system of your interest. Note that you are not asked to design an EGS, only to create the first image of it by defining the image in terms of the nine (or more or different) dimensions introduced in this section. In case you apply the dimensions to your family, you should transform my ‚"generic" image into an image specific to your family. In case you elect to work with another system of your interest, your image might be created at the same level of generality as my family example is. Enter your image into your workbook.

7.7.   Contemplating the Contribution of Design to the Creation of a New Societal Way of Life

We now enter the last stretch of the journey toward understanding what design is, how it works, how we can use it, and how it can add value to the life of the society. Having just explored evolutionary consciousness, conscious evolution, and systems design as an instrument to guide the evolution of social systems, we now can move up to the evolution of society as a whole. In this section, I still guide you on the journey, as we contemplate design's contribution to creating a new way of life for society. Then, in the concluding chapter, you commence your own journey and chart your own course to travel the territory of social systems design and discover its new frontiers.

In this section, we listen to scholars projecting comprehensive views of evolution and considering design's role in guiding the ethical evolution of society. We shall consider how the imperatives of social systems design—namely, transcendence, envisioning an image, and bringing the image to life by transforming the system by design—are also applicable at the societal level.

In creating a new societal way of life, we must first transcend the existing state to envision new societal images. Then, we are to engage in design, based on the images, and bring the design to life as we transform society.

7.7.1. Emerging Images of Guided Societal Evolution

In the two preceding sections of this chapter we explored evolutionary consciousness and conscious evolution and the role of systems design as an instrument of the guided evolution of social systems. Now we move up to the overall sphere of societal evolution as we review some comprehensive evolutionary perspectives that enable us to transcend the here and now and envision emerging societal images. These perspectives and images are offered here by several scholars.

7.7.1.1. Societal Evolution Guided by Evolutionary Ethics

In *Life Era,* the astrophysicist Eric Chaisson (1987) projected an arrow of evolutionary time. It started some fifteen billion years ago with the particulate, followed by the galactic, stellar, planetary, biochemical, and the cultural. We are now entering the ethical era. "If our species is to survive to enjoy the future, then we must make synonymous the words 'future' and 'ethical,' thus terming our next grand evolutionary epoch 'ethical evolution'" (p. 201).

The escalating rate and intensity of change at which we evolve is of our own making and it is dominated by technological changes that are often beyond our control as technological intelligence is increasingly outdistancing our sociocultural intelligence. Thus, our challenge is to gain the ability to guide environmental, social, technological, political, and economic changes and become "the agents of change—at least on planet Earth" (p. 177).

Once we understand the permanence of change, we can proceed to guide change in ways that lead to evolution that is beneficial to all humankind.

> Of all the implications for the Life Era concept, the most important in my view is that we, as the dominant species on Earth, develop—evolve if you will, and quickly, too—a global culture. We need to identify and embrace a form of planetary ethics that will guide our attitude and behavior toward what is best for all humankind. (p. 176)

As we have attained evolutionary consciousness and now understand our role in conscious evolution as agents of change, we must take responsibility for guiding evolution. Our actions should be based on a collectively formulated set of ethics and principles.

> We must now develop an integrated worldy culture, including a unified politico-
> economic ideology—which is not just a hackneyed proposal for world government—
> if we as a species are to have a future. . . . Indeed if we act wisely, quite beyond just
> intelligently, then an epoch of something resembling "ethical evolution" should natu-
> rally emerge as the next great evolutionary leap. (p. 8)

Reflections

In the previous section the idea of an evolutionary guidance system (EGS) was developed. An EGS is a systemic arrangement of several key human experiential domains. As designers engage in the creation of an EGS, it will be the ethics of the whole system that shall guide their design decisions and will provide the integrative force that ensures the internal consistency, and the self-creating coherence, of the system. Thus, they will put the system on the path of ethical evolution.

Activity #56

Return to Activity #55. Review the evolutionary guidance system (EGS) for which you have created an image. Examine and evaluate the image from the perspective of evolutionary ethics as an integrating force that gives creative coherence to your system. Describe what is the "whole systems ethic" of your EGS.

7.7.1.2. Directing Evolution toward a Good Society

In *The Evolving Self: A Psychology for the Third Millennium,* Csikszent-mihalyi (1993) examines how the evolutionary principle of complexity can provide meaningful guidance to our efforts of directing evolution, what the role of morality is in societal evolution, and how these two evolutionary perspectives, namely, complexity and morality, can guide us toward the building of a "good society."

7.7.1.2a. The Imperative of Directing Evolution. Csikszentmihalyi commences by suggesting that up to now the societal way of life has been the result of a random chain of changing events. It has not been the result of any planned effort. "And now we suddenly realize that, unless we take things in hand, this process of change will continue under the sway of relentless chance, a chance entirely blind to human dreams and desires" (p. 149). "If there is a central task for humankind in the next millennium, it is to start on the right track in its efforts to control the direction of evolution" (p. 149). In addressing the idea of directed evolution, he sets forth the evolutionary principle of complexity and relates it to morality.

The central theme of evolution is reorganization at ever higher levels of complexity. Such reorganization means increased differentiation and the systemic integration of differentiated parts. There are two opposing tendencies of evolution (Csikszentmihalyi, 1993): (1) changes that lead toward harmony and the ability to obtain energy through cooperation, and (2) changes that lead toward entropy, when energy is attained through exploiting others and causing conflict and disorder. "Harmony is usually achieved by evolutionary changes involving an increase in an organism's complexity, that is, an increase in both differentiation and integration" (p. 156). Differentiation means the degree to which an entity, a system, or a society is composed of parts that differ in function and structure. Integration is measured by the extent to which these parts interact, cooperate, and enhance each other's goals as well as the attainment of their shared goals (e.g., exhibit self-creating coherence).

Today our so-called advanced societies are highly differentiated. The threat to their complexity comes from "an erosion of common values and norms of conduct that may result in a society that disintegrates for lack of integration" (p. 158). "Complexity provides the benchmark for evaluating the direction of evolution. But we have few guidelines to teach us how to enhance complexity in everyday life" (p. 159). Returning to evolutionary learning, we can say that learning how to nurture such enhancement and recognizing and increasing complexity in everyday life become significant features of evolutionary competence. This competence comes into play when we have opportunities to guide the course of evolution.

"In every human group ever known, notions about what is right and what is wrong have been among the central defining concerns of the culture" (p. 159). Moral codes liberate us from dependence on instinct and keep the intergroup harmony that our genes cannot provide. The development of moral systems has been the most successful attempt to guide evolution in a desirable direction.

In the course of the last century, social scientists suggested that the moral systems of different cultures are entirely relative, arbitrary constructions. In fact, says Csikszentmihalyi, "what is so remarkable is how similar the world's moral systems are in considering 'good' to be the achievement of the kind of harmony within consciousness and between people . . . which in turn leads to higher levels of complexity" (p. 159). Ethical systems are efforts to guide us toward the future. They represent the ideals of a life that is freer, more compassionate, more integrated, and which is guided by a vision of what might be. The author notes that the realist easily scoffs at the idealist as being impractical. The realist deals with the concrete, the here and now. Without him we cannot survive. But without the idealist, without investing life energy in pursuing new challenges, we cannot evolve.

7.7.1.2b. The "Good Society." In creating a vision of a good society, Csikszentmihalyi takes us back in time to the French Revolution, which successfully challenged the old world order with the motto of "freedom, equality, brotherhood." These three are key markers of a good society as well as of a complex society. Freedom nurtures differentiation, the formulation of personal goals, and development of unique individuals. But differentiation without integration breaks up society. Here, brotherhood provides the counterweight. Equality connects freedom and brotherhood. "Equality of opportunity and equality before the law are what makes it possible for a group of individuals bent on pursuing their own interests to coexist in peace with one another" (Csikszentmihalyi, 1993, p. 266). Since the time of the French Revolution we have made continuing advancements in nurturing freedom and equality. But what about brotherhood, which should be the integrative force? "Unfortunately, while freedom and equality can be legislated, brotherhood cannot" (p. 267). The framers of the constitution of the United States assumed that Christian morality would moderate unrestrained individualism. The author quotes John Adams, who declared that "our constitution was made for moral and religious people. It is wholly inadequate to the government of any other." The shapers of the Constitution believed that a shared religious morality will provide the force for societal integration. Over the last two centuries, however, while there has been continuing support for freedom, and while, in the course of this century, equality gained strength, we have suffered great losses in the forces of integration. "Freedom without responsibility is destructive, unity without individual initiative is stifling, and equality that does not recognize differences is demoralizing" (p. 269).

The good society nurtures the development of the individual's competence to take part in socially productive activities. It guards against the exploitation of another person, against oppressors and parasites. "Freedom does not apply to doing, but to being." Each person is free to develop a self to the utmost level of its potential complexity, but not to curtail another person's freedom to do so" (p. 269). But we cannot stop at this point. We need to nurture differentiation and integration in our communities and in humanity as a whole. At this point in time, the development of a good society integrates with the attainment of creating greater complexity. "If we are to direct evolution toward greater complexity, we have to find an appropriate moral code to guide our choices. A code that specifies right as being the unfolding of maximum individual potential joined with the achievement of the greatest social and environmental harmony" (p. 162). The development of such a code is no easy task; neither is its application in creating a good society. But how can we accomplish the task of creating a good society? There is no simple step-by-step solution. It would be useless and dangerous to propose "what" is to be done to direct evolution. "We are on a safer grounds in suggesting 'how' we might find out what needs to be done" (p. 270).

Activity #57

First, capture the core ideas of this section. When I contemplate the two basic codes that our society holds dear: the declarations of human rights and the declarations of independence, I suddenly realize what are the shortcomings of these declarations. They are simply one-sided. If we want to build ethical and harmonious societies, we have to develop another set of individual and social and societal declarations that will create a balance. Your task is to develop two such declarations: (1) a declaration of individual and collective responsibilities and (2) a declaration of interdependence. Enter these in your workbook.

Reflections

Reflecting on the "what" and "how" questions raised by Csikszentmihalyi, it should be noted that the whole thrust of this work is to explore approaches for (1) how to design social and societal systems that are based on the shared values and ideals of those who serve the system and are served and affected by it; (2) how to create a moral code that guides the design of those systems? (3) how to make use of the individual and collective initiatives, intelligence, and creativity of the designing community? and (4) how to nurture ethics in both the process and products of design?

7.7.1.3. Destiny to Create

In their book *Creating a New Civilization,* Alvin and Heidi Toffler (1995) set forth a strategy for the third-wave postindustrial knowledge age that is discontinuous and clashing with the second-wave industrial machine age, preceded by the first-wave agricultural society.

"Some generations are born to create, others to maintain a civilization" (p. 104), is the main theme of the Tofflers' new work. This creation is now the task of our generation. An emerging civilization creates and integrates new cultures, technologies, political functions and forms, economies, ethics and morality, scientific orientations, and modes of disciplined inquiry. It reorganizes the society, using the evolutionary concepts discussed above, at a higher level of complexity by increasing and integrating differentiation. This creation does not happen overnight. It is "a consequence of thousands of innovations and collisions at many places over a period of decades" (p. 105). To bring this creation about requires the involvement, the energies, the intelligence, and the commitment of all of us. It is our destiny to create a new way of life, new constitutions, and new institutions by releasing something more powerful than energy: our collective imagination. Thus, we should not think of some massive reorganization from the top, but of hosts of system design efforts and experiments carried out at all levels of society. This creative venture will involve a vast process of social and organizational learning.

The responsibility for creating a new civilization and designing the changes needed to do so lies with us individually and collectively.

> We must begin with ourselves, teaching ourselves not to close our minds to the novel, the surprising, the seemingly radical. This means fighting off the idea-assassins who rush forward to kill any new suggestion on grounds of its impracticality. While defending whatever now exists as practical, no matter how absurd, oppressive, or unworkable it may be. (p. 108)

Obsolescence has its strongest defenders and guardians in our public institutions and, says Toffler, nowhere is "obsolescence more advanced or more dangerous than in our political life. And in no field today do we find less imagination, less experiment, less willingness to contemplate fundamental change" (p. 104). Even people who are most creative in their own field "freeze up" at any notion that our political systems are obsolete and in need of radical change. But we cannot wait any longer. We need a process of reconstruction now so that "we and our children can take part in the exciting reconstitution not merely of our obsolete political structures but of civilization itself. . . . We have a destiny to create" (p. 108).

Reflections

The perspectives articulated by the Tofflers in the context of the main stages of societal evolution are familiar to us. An understanding of the escalating time scale of evolution helps us to appreciate the need for collective competence in social systems design, and a revisiting of societal evolution helps us to value the significance of evolution guided by design. In this section, an examination of the current state of our society and the imperative of creating a new society point directly to the value that social systems design can offer to such a creation.

Activity #58

In the text above, we explored several images that project desired directions that may guide societal evolution. The evolutionary views of the scholars I introduced provide a rich source of information for your task of selecting and defining another set of core ideas about emerging directions of purposeful societal evolution. As you describe these core ideas test for their internal consistency and compatibility. Enter your findings into your workbook.

7.7.2. Emerging Images of Democracy

There is one domain in our societal life that, although mentioned repeatedly in this work, has not been addressed comprehensively. This domain is our relationship to government and its relationship to us. In the text that follows, we discuss the current and possible future states of our political institutions, explore approaches that we might take to organize our public decision-making relationships, and project emerging images of democracy.

7.7.2.1.   The Current State of Democracy and Its Desired Future

"We have not yet tried democracy. . . . We do not even have a conception of what democracy means."

—Mary Parker Follett

This quote is an appropriate introduction to the discussion that follows. First an original definition of democracy is presented. Then we listen to observers of the current state of democracy and consider their projections of desired and much needed changes. I review the work of Philip Slater, visit again with the Tofflers and Csikszentmihalyi, review notions of "teledemocracy," and report on propositions for democracy in the workplace.

7.7.2.1a.   The Original Definition of Democracy.   There are three Greek words we must keep in front of us if we truly want to understand the original meaning of the Greeks' way of life. These are "democracy," *sizitisis,* and *demosophia.* Democracy means "the power of the people." *Sizitisis* stands for "searching together." *Demosophia* is "the wisdom of people." If we integrate these three we get a clear idea of what democracy is. The true meaning of democracy is that people have the power to make decisions about issues affecting their lives. These decisions are made by searching together, by engaging in disciplined and focused conversations. (Socrates said that the only way to arrive at the truth and attain wisdom is by searching together.) To attain wisdom is a prerequisite to exercising the power of people. This type of democracy was practiced in Greece (Christakis, 1993). Citizens of the Athenian Republic would gather at "Agora," an open-air marketplace, and engage in conversation on various issues of shared interest, such as whether to allocate resources to build the Parthenon. Searching together with Pericles, who was at that time President of the Republic, resulted in an agreement in favor of construction.

7.7.2.1b.   Cultural Democracy.   I start with Follett's (1965) observations, quoted earlier as she said that "we have not yet tried democracy. We have not yet learned how to live together. We must find a new principle of association. Crowd philosophy, crowd government, crowd patriotism must go. The herd is no longer sufficient to enfold us. Representative government, party organization, majority rule, with all excrescences, are dead-wood. In their stead must appear the organization of non-partisan groups for the begetting, the bringing into being, of common ideas, a common purpose and a collective will" (pp. 3–4).

Advocating "government by the people," saying that people should do this and that, that people should be in control—today we say "be empowered"—are all useless, Follett says, "unless we provide the procedure within which the people can do this or that" (p. 4). In this work, we have set forth such procedures, namely systems design, the mastery of design by which people can be empowered to take charge of their individual and collective lives and shape their future. People can exercise their power by group organization, which becomes the new method in politics. It "releases us from the domination of mere numbers. Thus, democracy transcends time and place. It can never be understood except as a spiritual force. Majority rule rests on numbers; democracy rests on the well-grounded assumption that society is neither a collection of units nor an organism but a network of human relations. Democracy is not worked out in the polling-booths: it is the bringing forth of a collective will" (p. 7) and the means by which this collective will is manifested is social systems design. "Thus the essence of democracy is creating" (p. 7). Groups organized for design will create the new world at all levels of the society, "the new world we are now blindly feeling after."

"Creative force comes from the group, creative power is evolved through the activity of the group life" (p. 7). Following the boss, a government associa-tion, a political party is all herd life. Democracy says Follett "means a wholly different kind of existence. Democracy depends on the creative power of every man" (p. 6). The potentialities of the individual remains potentialities until they are released by group life. Group organization must be the method of politics because it is the mode by which the individual can be brought forth and made effective as the mode of practical politics. These groups, organized at various levels of the society and operating in a network of relations become designing communities that address issues that affect their lives and create solutions that enable them to take charge of their lives.

Philip Slater (1991) observes that we are experiencing the most radical societal transformation of recorded history. Rejecting the notion that democracy has to do only with forms of government, he sees democracy as an emerging megaculture in a worldwide struggle with the dying megaculture of bureaucratic autocracy. He sees democracy as a system of organizing human relationships under conditions of constant change. Thus, the key is rapid acceptance of new ideas. But those in positions of authority, of the status quo, are not likely to be open to new ideas. "Their minds have a prior engagement" (p. 19). It is people—less invested in the status quo—who can take advantage of change and who usually come up with new ideas anyway.

A cultural democracy is far more comprehensive in organizing human rela-tionships than the token institutions we now have. True democracy is not taking votes periodically and designating others to make decisions for us. It is what Follett called "self-creating coherence," which emerges from those most affected by a particular decision. We are now beginning to grasp that there is order in democracy, not "the order left from our many authoritarian institutions, but the order that comes from emergent democracy itself—the coherence that is self-created, the product of many different individuals interacting" (p. 182). Much of self-creating democracy has to do with its open-endedness. It is to be in a permanent state of reinvention. We don't know where the road to an emergent democracy leads if it is allowed to become its fullest. "Democracy is always on the move. It thrives on uncertainty and unpredictability. There is no formula for democracy, no more than there is formula for any other kind of creativity. Democracy is self-creating and self-governing. Authority and responsibility for making decisions are vested in those who carry out the decision. "Democracy is fueled by the willingness of individuals to involve themselves in directly confronting the issues that concern them. Democracy satisfies one of the strongest needs humans possess—the need to be useful. For democracy is on a permanent talent search: it finds uses in everyone and for everyone" (p. 188).

It is appropriate at this point to differentiate between cultural democracy and a democratic culture. "Cultural democracy" is an idea of a way of life, shared by all members of the culture and put into practice as they engage in collective decision making on issues that affect and interest them. A "democratic culture" provides ongoing learning opportunities and programs, methods, arrangements, resources, and institutions that empower members of the culture to competently practice cultural democracy, participate in collective decision making, and give direction to their evolution.

What Slater describes as democracy is a far cry from our current limited institutions of representative democracy. He projects a new societal way of life, which affects all domains of human experience. We already developed this notion at the level of social systems, which operates as evolutionary guidance systems (EGS). A change in one domain of an EGS brings about change in all others.

Cultural democracy requires a transformation of our individual and collective consciousness. (As we saw it, the same is required in social systems.) Rephrasing Einstein, we cannot bring forth a self-creating democratic culture from the same consciousness that brought forth our current democratic system. We have to think anew. We have to transcend what we have now, envision new images, and, based on those images, transform our society by design. This transformation must be guided by a "unifying purpose" (Csikszentmihalyi, 1990) that like a magnetic field attracts our psychic energy, a greater purpose upon which all lesser purposes depend. This purpose "will define the challenges that a person needs to face in order to transform his or her life into a flow of activity. Without such a purpose, even the best-ordered consciousness lacks meaning" (p. 218).

7.7.2.1c. Twenty-First-Century Democracy. The Tofflers (1995) suggest that "the time has come for us to imagine completely novel alternatives, to discuss, dissent, debate, and design from the ground up the democratic architecture of tomorrow" (p. 90). The apparatus and structures of existing representative governments are increasingly unworkable. They no longer fit the new realities of our radically changed world. The authors set forth three principles of reconceptualization.

The first principle calls for a "configurative society," emerging from the rich diversity and activism of many groups and minorities, providing "a far more varied, colorful, open and diverse society than we have ever known. We can either resist the thrust toward diversity, in a futile last-ditch effort to save the Second Wave political institutions, or we can acknowledge diversity and change those institutions accordingly" (p. 93). We can design imaginative arrangements by which diversity can be legitimized and contribute richly to public life. Fortunately, the technologies of the information age provide pathways toward the design of those arrangements. (This will be discussed later in the section on "teledemocracy.")

The second principle calls for "a shift from depending on representatives to representing ourselves" (p. 96). In worsening practices of our representative institutions, decisions affecting us are still made by a small number of pseudorepresentatives, who are increasingly remote from us and who cannot respond to our needs. Advancements in communication technology open up opportunities for direct citizen participation in political decision making and "democratize a system that is now near break-down and in which few, if any, feel adequately represented" (p. 99).

The third principle, decision division, calls for making decisions at the level where they belong. Some issues cannot be solved at the local level; others cannot be addressed at the national level; some require attention on many levels. "Today's political arrangements violate this principle widely. The problems have been shifted, but the decision power hasn't" (p. 100). Power is still concentrated at the national level. The structures needed to make decisions at the transnational level are highly underdeveloped. Few decision are left for regions, states, and local communities. "The issue is rational allocation of decision-making in a system that has overstressed centralization to the point at which new information flows are swamping the central decisionmakers" (p. 100).

7.7.2.1d. Politics Is the Highest Form of Leisure in a Good Society. Csikszentmihalyi's idea of a good society was already discussed. Commenting on the crudeness and the unresponsiveness of the current representative form, he suggests that

> the amount of information we receive and transmit through an election is woefully meager. If we want our political institutions to represent our goals more clearly, we must find better ways, first, to understand what those goals are, and second, to communicate them to others in a convincing manner. (Csikszentmihalyi, 1993, p. 271)

He finds it incredible that we spend trillions of dollars on programs nobody really benefits from, yet no resources are allocated for programs that would "enhance the match between our dreams and the institutions that are supposed to make them real" (p. 271). He envisions in every community a beautiful place in a park or in a hall where people could meet and discuss their common concerns and make decisions about issues affecting their lives, like the free Athenians, who twenty-five centuries ago instituted a "public sphere" where citizens discussed any issue affecting their city.

> As long as most citizens ignore politics, regarding it as a necessary evil, it will always remain an unsavory practice, controlled by selfish interests. But if we take the shaping of our future as the great challenge it is, we shall discover that the Greeks knew what they were saying when they spoke of politics as the highest form of leisure. The most satisfying way to actualize the self is by building the most complex system—a good society. (p. 272)

7.7.2.1e. Teledemocracy. In the text above, repeated references were made to the power and use of technologies of the information age that enable citizen participation. Teledemocracy (TD) is such a means that can establish direct democracy through the use of communication media. Reporting on thirteen TD projects, Arterton (1987) identifies eleven institutional characteristics of citizen participation:

1. Access tells us the range of participation.
2. Reach accounts for the percentage of citizens available to participate and who actually do get involved.
3. Effectiveness is a measure of participation's direct influence on public policy.
4. Agenda setting reveals the extent to which citizens have an influence on what issues are decided.
5. Diversity of paths accounts for the ways citizens can learn about a TD project.
6. Duration defines the length of time and number of events over which participation lasts.
7. Individual or group-based mode is another dimension of participation.
8. Initiative is the degree to which participants generate the TD opportunity themselves.
9. Cost is the burden that participants carry for their involvement.
10. The educative value reveals the degree to which participants learn about the issues addressed.
11. Participating competence reveals the skills of participation and the confidence attained by becoming politically active.

7.7.2.1f. Democracy at the Workplace. Recent years have brought increased interest in promoting democratic participation at the workplace. Examples of emerging images of democratic participation are described here.

In *The End of Bureaucracy and the Rise of the Intelligent Organization,* Gifford and Elizabeth Pinchot (1993) describe the intelligent organization as one that develops and engages the intelligence and systemwide responsibility of all its members. A key characteristic of an intelligent organization is that it ensures that its members engage their intelligence in making collective decisions, in setting goals and policies, and in having authority to plan and implement their roles in a shared future. "The participative democracy needed by an intelligent organization is any process that involves everyone's intelligence directly in antic-ipating, and acting upon, challenges any organization faces" (p. 193).

In his *Democratic Corporation,* Ackoff (1994) suggests that the essence of democracy is an enterprise, conceptualized as a social system, whose members participate in the selection of both ends and means that are relevant to them. This means that the enterprise "increases the variety of both the means and ends available to its parts, and by so doing, it increases the variety of behavior available to them" (p. 31). In describing his "circular organizational model," Ackoff highlights its three essential characteristics: (1) the circularity of power (the absence of an ultimate authority); (2) the ability of people to participate in all decisions that affect them; and (3) the ability, "individually or collectively, to make and implement decisions that affect no one other than the decision-maker or decision-makers" (p. 115).

In *Putting Democracy to Work,* Adams and Hansen (1992) developed a guide for worker-owned businesses. "Governance is a political function; man-agement is an economic function" (p. 141), they say. In capitalism the two functions are separated. They are united when workers run the enterprise. Some of the markers of this mode of operation include the following (Horvath, 1983). Decisions are made at the work-unit level by face-to-face discussions. Work-units federate in a work community. When a decision at the work-unit level affects other units, decisions are made at the community level. Work-units are expected to make clearly stated decisions and are responsible for their effective implementation. Policies are formulated through a political process, while their implementation is a matter of professional/qualitative judgment.

In *The Fourth Wave: Business in the 21st Century,* Maynard and Mehrtens (1993) trace the evolution of corporate structures. Aligning this evolution with Toffler's (1980) agricultural, industrial, and postindustrial waves, they charac-terize a forth wave for the twenty-first century. They suggest that the forces of the third wave lead to a major societal transformation. They suggest that their projected fourth wave will hasten this transformation. They define the fourth-wave corporation as a community that includes everyone whose life is touched by the enterprise. Its ethos is that people are unified into the collective of the corporate community and they are open to and are supportive of each other. The boundaries between work and personal lives of people are seamless. Diversity, equality, respect for each other, and lifelong learning are valued. Support to all ensures health and the overall well-being of all. The authors' vision for the future corporation is that it will become an exemplar for the entire society. It acts locally while it thinks globally. It practices social and resource accounting. As an organization it is committed to serve others and it is conscious of its moral effects. It sees itself as a community aimed at wellness, it is a model of environmental concerns, and it is a pioneer of appropriate technology. Finally, it considers itself an agent in the service of the greater good of humanity.

Reflections

The notion of "empowering" people to make decisions that affect their lives and their systems is a core idea of true democracy. Much of this power today is delegated to others. In the paragraphs above images have emerged, very much like those developed in this work as we formulated core ideas of empowerment and repeatedly described approaches that enable people to design their own lives, their own communities, and the systems in which they live and work.

But in this section, we took a quantum leap to address society as a whole. At this level we are in a very different ballgame. Up to now in our representative democracy we designated the players of this game. We empowered them to represent us. They are playing the game by their own rules. We became spectators of the game but now we are becoming increasingly dissatisfied and even despise the way the game is played. (A recent bipartisan survey in the United States indicates that three out of four people do not trust their government.) So we want to reclaim empowerment and we want to enter into the arena. Now we have the task to create the methods and the rules of the game. In what follows, building upon what we have attained in working with this book, I set forth the challenge of taking a quantum leap toward designing a new societal way of life.

Activity #59

(1) Review the paragraphs above and identify and describe core ideas of an emerging democratic culture and cultural democracy, explicated by the various authors whose works were introduced above. (2) Bringing forth what you have learned, propose approaches and methods by which these core ideas can be realized in creating a new societal way of life. Enter your findings in your notebook.

7.7.3. Reflections: Looking Back and Looking Ahead

In this closing section, as an overall reflection, I first look back and review the purposes of this work. Then, looking ahead, I propose a challenge that has to be addressed if we aspire to move toward the idea of a self-governing democracy and self-creating society.

7.7.3.1. Looking Back: What Have We Accomplished?

The genesis—and the recurring theme—of this work was the recognition of massive societal changes and transformations that are reflected in the new realities of our postindustrial information/knowledge era. These changes touch the lives of every person, family, community, nation, and the whole of humanity. But we are entering the twenty-first century with organizations, institutions, and political systems designed in the eighteenth and nineteenth centuries. We find that most of our systems are alarmingly out of sync with the new realities. This situation has created a dangerous evolutionary gap, an ever-increasing imbalance that cannot be left unaddressed. We must meet this challenge if there is any chance that our hopes and aspirations for our future and the future of humanity will be realized.

Today, individually and collectively, we are challenged by questions to which we must find answers. Confronted with massive changes that surround us we ask: Is it our fate to be only spectators in the arena of change? Do we have to relegate decisions that affect our lives to others? Are we at the mercy of experts who design systems for us? Or: Is there a role for us in shaping our future and the future of the systems to which we belong? Do we have the right and responsibility to give direction to the evolution of our systems, communities, institutions, and society? If we do have the right and responsibility as we believe we do: What approaches, means, and methods are available to us by which to exercise this right and assume this responsibility? Do we have the knowledge and competence to do so? If not, what do we have to learn to become competent stewards of shaping the future?

These questions became the "triggering questions" of this work. We have explored them, and, I believe, we have answered them. We have understood why we must claim the right to make decisions affecting our lives. We developed approaches and a comprehensive strategy as to how we can go about taking responsibility individually and collectively to give direction to the evolution of our systems. We have certainly attained an understanding of the crucial role of social systems design in answering the questions asked here. We have learned to appreciate the power of social systems design as a means to take responsibility for our lives and the lives of our systems and our communities.

Activity #60

Your very last task is to develop some summary answers—short paragraphs—to the questions I raised in my retrospective review. This should constitute a brief assessment of what we have jointly accomplished in this work.

7.7.3.2.   Designing a Self-Governing, Self-Creating Society

At the close of our journey, ideas and perspectives have emerged that projected images and intentions of directed ethical evolution, of developing a good society, of creating a new societal way of life, and of building a new cultural democracy and a new democratic culture. Now questions arise: how can we bring these images to life? How can we realize these intentions? As I contemplate these questions, I am reminded of what Ken Boulding (1985) said about intention and design. Intentions are easy to come by, but often they are not realized. On the other hand, design is hard work, but it is design that creates the future.

So the "looking ahead" question is: Can social systems design respond to the challenge of creating a new societal way of life, designing a self-governing and self-creating society?

Up to this point in our journey, we worked with design inquiry at the level of social systems embedded in a community and designing communities as social systems. But now, in this closing section, we have reached a level that is much higher, and a system at this higher level is far more complex than the systems we have worked with in this book. The move from the level of social systems and communities to the level of the larger society is a quantum leap of immense escalating dimensions and complexity. It is a challenge the size and weight of which is hard to fathom.

In thinking about meeting this challenge, the challenge of designing a self-governing and self-creating society, we can certainly speculate about how to go about it. But in no way can we project its outcome. The outcome of *what* it is going to be is the product of *how* we shall go about it. Based on what we have learned and what we have accomplished in working with this book, there are certain things we can say with some degree of confidence about how we might approach design inquiry at the societal level.

We have witnessed examples of coercive societal-level designs superimposed on societies form the top down. The recent history of humankind accounts for many of these. But there is no precedence for the employment—at a societal level—of the type of participative, consensus-building, and disciplined design inquiry we have worked with in this book. Acknowledging such lack of precedence and recognizing the limits of transferring design inquiry approaches and models from the (micro) level of working with specific systems to the (macro) level of the overall society, there are still some generalizations that we might make. We can say that designers of a new society:

- Have to transcend what exists now, rather than extrapolate from it. They have to leap out from the boundaries of their existing systems and develop and draw from new consciousness, from a new view of the world, and learn to think anew.

- Must create ideal visions of the future and, based on those visions and their shared values and ideas, create images of a self-creating society.
- Should engage—based on the images they created—in a disciplined inquiry of design, and bring the design to life by commencing with the transformation of the society.

We can further say the following:

- In order for the design to be authentic and sustainable, it has to be genuinely participative. It has to involve people from the various levels of the society and draw upon their individual and collective intelligence, aspirations, and creativity.
- It will be of paramount importance to engage in the design of the design inquiry itself and in the design of all the various designing systems that will have to be established at the various societal levels.
- A prerequisite to engaging in design is the development of a design culture and the creation of evolutionary competence across the society. This will require the development of resources, opportunities, and arrangements for design and evolutionary learning in both the formal and informal systems of learning and human development.
- Design inquiry should pursue the ideal; it should be ethical, it should engage our self-reflective and creating consciousness, draw upon multiple perspectives, and become a never-ending societal venture.
- Existing and quickly emerging information/knowledge technologies and cyber-networks will have to play an essential and crucial role as modes and means of communications in intergroup, intersystem, across-systems, and across-system-levels design inquiry.

Reflection

It will clearly become a major task of those interested in moving from the microlevel of social systems design to the macrolevels of societal design to engage in serious inquiry that develops and tests the parameters of societal-level-systems design. In contemplating moving from the microlevel to several macro-levels, we are confronted with several issues (Hood, personal communication, 1995), discussed next. Although the design principles reviewed above may apply across all levels of the scale, there are some critical qualitative differences that emerge as the scale changes massively. Issues of power, politics, ethics, and morality become increasingly important, if not absolutely critical, for very large scale design, as the substance of the discussion in this section attests. These issues are further burdened by racial, ethnic, and cross-cultural differences and tensions across the world, regions, nations, and within nations. We shall better understand these issues as we learn to make distinctions regarding systems scale, complexity, and dynamics of interactions. We may learn to develop whole families of complementary design approaches and models so that we may eventually be able to select those that appear to be most appropriate to the scale, complexity, and dynamism of a particular system we conceptualize. But until we have the experience of applications, for some of us it may be difficult to see why these scale distinctions may be necessary.

A Closing Thought

Throughout history (Harman, 1988, p. 155), "the really fundamental changes in societies have come about not from dictates of governments and the results of battles but through vast numbers of people changing their minds." If the great venture of designing a self-creating society goes forward and establishes genuine cultural democracy and a democratic culture, then one or two generations from now society will be as different from what it is now as today is from the society of the Middle Ages. "Furthermore, it will be different in ways we can only vaguely intuit, just as the Renaissance futurist had a hard time trying to describe modern society" (p. 168).

Engaging in the type of design venture proposed here will require an immense effort and commitment of individuals, families, groups, systems, institutions, and the society as a whole. I invite you to contemplate what role you might play in this venture, what contributions you might make to the task of designing a society that empowers us individually and collectively not only to govern ourselves but to engage with others in a harmonious way in the creation of a desired and shared future.

Evolutionary creation is an ongoing venture. The core message of this work is that we can take part in it, provided we have the will and the commitment and if we develop the competence needed to do so. And there is no more noble, no more important task than engaging in this creating venture and—by so doing—becoming coworkers of God.

8

The Journey Continues

We have traveled together on the terrain of social systems design, and our journey has been long. As your guide, I introduced to you what I have learned from exploring the work of others, what I have learned from the many design conversations I have had over the years, what I have learned from my students, and what I have learned from my own design experiences. You have done a great deal of work as you have tried to make sense from the material, reflected on it, constructed your own meaning from it, defined core ideas of design, and formulated perspectives that helped you to organize your thinking and prepared you for design actions as you applied what you have learned in the functional context of systems of your interest.

How much you have learned, the depth and the breath of your knowledge and competence in social systems design, depends on the intensity of your interest and the effort you invested in working with the activities. Thus, you may feel that you have now a solid orientation in systems design or that you have gone way beyond it, acquired a firm grounding in it, and are ready to engage in it.

I believe that we have reached the point where you can now guide your continuing learning and application in systems design. Using the journey metaphor, we reached the end of the beginning. We can celebrate your commencement into a journey of learning and doing. We have found out repeatedly that learning never ends and design never ends, so design learning never ends. What follows here is not a prescription for your continuing design work but only an example presented as a design for my own continuing journey.

8.1. The Design of a Journey

Back in Chapter 1, Fig. 1.1 depicted an image of the design of this work. The image displayed seven questions as the seven key strands or vectors of a spiraling exploration of design. As the spirals touched the strands repeatedly, we weaved a tapestry of the many colors of design into patterns that connected into an ever broader and clearer picture and deeper understanding of what social systems design is, why we need it, how it works, who should do it, when to use it, how to use it, and what value it offers to the society.

In the course of my continuing journey, I intend to follow the design of the work discussed above. As I touch a strand, I engage in a repeated pattern of reflection on what I explore as new knowledge and understanding, from which comes the creation and construction of my own knowledge and meaning, based on which I engage in design applications in real-world contexts. Then I reflect upon what I have done and what I have learned from the design experience. This reflection points me to seek new learning and reflection and creation. So the spiral goes on and on as a never-ending process. What follows is a display of my continuing design journey organized in spirals. I offer a description of this journey for you as an idea for your own design journey.

8.1.1. Spiral One: Creative Synthesis of Core Ideas

The first spiral unfolds into a creative synthesis of the core ideas that are embedded in the material of the work. I do here what I have asked you to do as I review the forty-two sections and organize/synthesize the core ideas under the seven questions. This organizing synthesis becomes the platform for moving into the next spiral.

8.1.2. Spiral Two: Reflection

Moving along the second spiral, I reflect on my synthesis as I look at the seven categories and search for internal consistency and compatibility among the seven areas. I expect to find some discrepancies, week connections, and some missing elements. I foresee that this reflection will include others who have reviewed my work. We will engage in an intensive conversation that will enrich reflection and guide us into gaining some new insights.

8.1.3. Spiral Three: Emergence of Creative Insights

I expect that our reflective conversation will lead into the emergence of new creative insights about the seven questions of social systems design. We might raise some additional questions. I predict one such a question might be: Why is there such strong resistance to systems design? And what can be do about it? I know I did not explore this question in depth. The new insights and the new questions will become the most significant fruits of our joint effort. For me the conversations I have been part of have always created unexpected synergy, new discoveries, and many "aha" experiences.

8.1.4. Spiral Four: Search for New Knowledge

The outcome of a creative conversation always points to the need to seek or develop new knowledge that gives substance to the insights gained, the questions raised, and the "aha" experiences. I can now see that the various adding-value-to-design topics presented in Part II will become fertile ground in which to plant new seeds and nurture the development of new design knowledge. Such knowledge will come forth from a continuing exploration of relevant literature, from knowledge that can be created from work with the insights and the new questions that emerged earlier, and from design conversations.

8.1.5. Spiral Five: Another Spiral of Reflection/Creation

During this spiral we reflect on the new knowledge base we developed to create a new system of reorganized and extended core ideas and organizing perspectives that will become the bases for design applications.

8.1.6. Spiral Six: Design Applications

All that was accomplished in the preceding spirals will come to life as we apply what we have learned, what we have created as we engage in systems design in functional contexts of systems of our interest. Such application becomes the source of both design learning and the creation of new design knowledge.

8.1.7. Spiral Seven: New Learning and Creating New Knowledge

Conducting design and learning from it are parallel engagements. In the course of this engagement, we find out how our new formulations, new core ideas, and new perspectives work as they are applied. Their application becomes the source for the emergence of new core ideas, new formulations, and new perspectives. Furthermore, our design work conducted in the decision-oriented disciplined inquiry mode makes it possible for us to test, validate, or question both the philosophical and theoretical bases of our design approach and the methodologies that are grounded in those bases. Here is where we engage in design research as we move into a conclusion-oriented mode and create new design knowledge. We discussed this issue in Chapter 4 and depicted it in Figure 4.2.; now we can apply it.

8.2. A Departing Thought

The example of a design of my continuing journey might be helpful to you as you contemplate your own continuing design journey. I hope that working with the learning resources offered in this book has been a satisfying and rewarding experience for you. My reward and satisfaction will come from the knowledge that I have guided others like you in the development of their design culture, that I helped others in using their design competence for building a higher quality of life for themselves and their families, and that I helped others in working with their systems and their communities and empowered them to give direction to their own evolution and shape their own futures.

References

Ackoff, R. L., 1974, *Redesigning the Future*, Wiley, New York.

Ackoff, R. L., 1981, *Creating the Corporate Future*, Wiley, New York.

Ackoff, R. L., 1994, *The Democratic Corporation*, Oxford University Press, New York.

Ackoff, R. L., 1995, Whole-ing the Parts and Righting the Wrongs, *Systems Research*, 12:43–46.

Ackoff, R. L., and Emery, F. G., 1972, *On Purposeful Systems*, Aldine-Atherton, Chicago.

Adams, F. T., and Hansen, G. B., 1992, *Putting Democracy to Work*, Berrett-Koehler, San Francisco.

Akin, O., 1984, An Exploration of the Design Process, in: *Developments in Design Methodology*, (N. Cross, ed.), Wiley, New York, pp. 189–207.

Alexander, C., 1971, The State of the Arts in Design Methods, *DMG Newsletter*, 5(3):3–7.

Archer, L. B., 1966, *Systematic Method for Designers*, Council of Industrial Design, London.

Archer, L. B., 1984, Whatever Became of Design Methodology? in: *Developments in Design Methodology*, (N. Cross, ed.), Wiley, New York, pp. 347–350.

Argyris, C., 1982, *Reasoning Learning and Action*, Jossey-Bass, San Francisco.

Argyris, C., and Schon, D., 1978, *Organizational Learning*, Addison-Wesley, Reading.

Arterton, C. F., 1987, *Teledemocracy*, Sage Library of Social Research, Sage, Newbury Park, California.

Asimov, I., 1950, *I, Robot*, Doubleday, New York.

Asimow, M., 1962, *Introduction to Design*, Prentice Hall, Englewood Cliffs, New Jersey.

Baburoglu, O., and Garr, M. A., 1992, Search Conference Methodology for Practitioners, in: *Discovering Common Ground* (E. Weisbord, ed.), Berrett-Koehler, San Francisco, pp. 73–82.

Bach, J., 1993, Evolutionary Guidance System in Organizational Design, *World Futures*, 36(2–4):107–128.

Banathy, B. A., 1995, Information Based Design of Social Systems, *IFSR Newsletter*, 38:2–3.

Banathy, B. H., 1973, *Developing a Systems View of Education: A Systems Models Approach*, Fearon Publishers, Palo Alto.

Banathy, B. H., 1979, The Dynamics of Integrative Design, in: *General Systems Research: A Science, A Methodology, A Technology* (R. Erickson, ed.), Society of General Systems Research, Louisville, pp. 191–197.

Banathy, B. H., 1980, The School: An Autonomous or Cooperating Agency, in: *Critical Issues in Educational Policy* (L. Rubin, ed.), Allyn and Bacon, Boston, pp. 112–119.

Banathy, B. H., 1986, A Systems View of Institutionalizing Change in Education, in: *1985–86 Yearbook of the National Association of Academies of Science* (S. Majumdar, ed.), National Association of Academies of Science, Columbus, pp. 12–22.

Banathy, B. H., 1987a, Instructional Systems Design, in: *Instructional Technology Foundations* (R. Gagne, ed.), Earlbaum, Hillsdale, pp. 84–106.

Banathy, B. H., 1987b, The Characteristics and Acquisition of Evolutionary Competence, *World Futures*, 23(1–2):123–144.

Banathy, B. H., 1987c, *The Design of the Far West Laboratory*, Far West Laboratory, San Francisco.

Banathy, B. H., 1988a, Systems Inquiry in Education, *Systems Practice*, 1(4):193–212.

Banathy, B. H., 1988b, The Conceptual Environments of Design Inquiry, in: *Cybernetics and Systems* (R. Trappl, ed.), Kluwer Academic, Vienna, pp. 149–153.

Banathy, B. H., 1988c, Matching Design Methods to System Type, *Systems Research*, 5(1):27–34.

Banathy, B. H., 1989, The Design of Evolutionary Guidance Systems, *Systems Research*, 6(4):289–296.

Banathy, B. H., 1991a, *Systems Design of Education*, Educational Technology Publications, Englewood Cliffs.

Banathy, B. H., 1991b, You Can't Restructure a Horse-and-Buggy into a Spacecraft, *Educational Technology*, XXXI(March):33–35.

Banathy, B. H., 1992a, *A Systems View of Education*, Educational Technology Publications, Englewood Cliffs.

Banathy, B. H., 1992b, Design in the Pursuit of the Ideal, *Educational Technology*, XXXII(January):33–35.

Banathy, B. H., 1992c, Building a Design Culture, *Educational Technology*, XXXII(August):33–35.

Banathy, B. H., 1993a, The Cognitive Mapping of Social Systems, in: *The Evolution of Cognitive Maps* (E. Laszlo, ed.), Gordon and Breach, New York, pp. 205–219.

Banathy, B. H., 1993b, Structuring the Program of the NATO Advanced Workshop: An Architecture of Decision Oriented Disciplined Inquiry, in: *Comprehensive Systems Design* (C. M. Reigeluth, B. H. Banathy, and J. R. Olson, eds.), Springer-Verlag, Berlin, pp. 67–84.

Banathy, B. H., 1994, Creating our Future in an Age of Transformation, *Performance Improvement Quarterly*, 7:87–102.

Banathy, B. H., Armenderiz, J., and Honig, B., 1979, *A Review of Systems Models*, Far West Laboratory, San Francisco.

Banathy, B. H., and Jenks, L., 1990, *The Transformation of Education by Design*, The Far West Laboratory, San Francisco.

Banathy, B. H., and Johnson, D., 1977, Cooperative Group Interaction Curriculum, in: *Curriculum Handbook* (L. Rubin, ed.), Allyn and Bacon, Boston, pp. 570–577.

Barron, F., 1969, *Creative Process and Creative Person*, Holt, Rinehart & Winston, New York.

Barron, F., 1988, Putting Creativity to Work, in: *The Nature of Creativity* (R. Sternberg, ed.), Columbia University Press, New York, pp. 76–98.

Bateson, G., 1972, *Steps to an Ecology of Mind*, Random House, New York.

Beck, M., 1994, *The Concept of Dialogue*, unpublished manuscript.

Beer, S., 1975, *Platform for Change*, Wiley, New York.

Beer, S., 1979, *The Hearth of Enterprise*, Wiley, Chichester, United Kingdom.

Bell, D., 1976, *The Coming of the Post-Industrial Society*, Basic Books, New York.

Bertalanffy, Ludwig von, 1968, *General Systems Theory*, Brazilier, New York.

Bevlin, M. E., 1970, *Design Through Discovery*, Holt, Reinhart & Winston, San Francisco.

Biatch, M., 1995, An Idealized Design of a Design to Guide the Growth and Evolution of a Synagogue Community, *Progress*, 4(Summer):41–58.

Blais, S., 1995, From Work Life to Life Work: Creating the Journey, *Progress*, 4(Summer):141–160.

Blauberg, J. B., Sadovsky, V. N., and Yudin, E. G., 1977, *Systems Theory: Philosophical and Methodological Problems*, Progress Publishers, Moscow.

Block, P., 1993, *Stewardship*, Barrett-Koehler, San Francisco.

Bogdanov, A., 1921–27, *Tektologia*, Proletarskaya Kultura, Moscow.

Bohm, D., 1983, *Wholeness and the Implicate Order*, Routledge and Kegan, London.

Bohm, D., 1985, *Unfolding Meaning*, Ark Paperbacks, London.

Bohm, D., 1990, *On Dialogue*, David Bohm Seminars, Ojai, California.

Bohm, D., and Edwards, M., 1991, *Changing Consciousness*, Harper Collins, San Francisco.

Bohm, D., and Peat, D., 1987, *Science, Order, and Creativity*, Bantam Books, New York.

Booker, P. J., 1964, *Conference on the Teaching of Industrial Design*, Institution of Engineering Designers, London.

Botnik, J. W., and Maltiza, M., 1979, *No Limits to Learning*, Pergamon Press, Oxford.

Boulding, E., 1981, Evolutionary Visions, Sociology and the Human Life Span, in: *Evolutionary Vision* (E. Jantsch, ed.), Westview Press, Boulder, pp. 169–194.

Boulding, K., 1956, *The Image*, The University of Michigan Press, Ann Arbor.

Boulding, K., 1978, *Ecodynamics: A Theory of Social Evolution*, Sage, Beverly Hills.

Boulding, K., 1980, General Systems Theory—The Skeleton of Science, in: *The General Theory of Systems, Applied to Management and Organizations* vol. 1 (D. M. Jamieson, G. Chen, L. L. Schkade, and C. H. Smith, eds.), Intersystems Publications, Salinas, California, pp. 19–26.

Boulding, K., 1985, *Human Betterment*, Sage, Beverly Hills.

Bridges, W., 1991, *Managing Transitions*, Addison-Wesley, Reading.

Briggs, J., and Peat, D., 1984, *Looking Glass Universe*, Simon & Schuster, New York.

Briggs, J., and Peat, D., 1990, *Turbulent Mirror*, Perennial Library, New York.

Broadbent, G., 1984, The Development of Design Methods, in: *Developments in Design Methodology* (N. Cross, ed.), Wiley, New York, pp. 337–346.

Buckley, W., 1968, *Modern Systems Research for the Behavioral Scientist*, Aldine, Chicago.

Cavallo, R., 1979, *Systems Research Movement*, General Systems Bulletin No. 3.

Chaisson, E., 1987, *The Life Era*, The Atlantic Monthly Press, New York.

Checkland, P., 1981, *Systems Thinking, Systems Practice*, Wiley, New York.

Checkland, P., 1995, Model Validation in Soft Systems Practice, *Systems Research*, 12(1):42–54.

Checkland, P., and Scholes, J., 1990, *Soft Systems Methodology*, Wiley, New York.

Christakis, A., 1987, A New Role for Systems Scientists, in: *Design Inquiry*, (B. H. Banathy, ed.), International Systems Institute, Carmel, California, pp. 14–28.

Christakis, A., 1988, National Forum on Non-Industrial Forest Land, *Systems Research*, 5(2):107–114.

Christakis, A., 1993, *The Wisdom of the People*, unpublished manuscript.

Christakis, A., and Conaway, D., 1995, *Building High-Performance Project Teams: Learning about the CogniScope*, Christakis, Whitehouse., LTD, Berwyn, Pennsylvania.

Christakis, A., Christakis, N., Conaway, D., Feudtner, C., and Whitehouse, R., 1995, *Application of the CogniScope System for The Conceptual Design of the AMF Project Plan*, Interactive Management Consultants, Paoli, Pennsylvania.

Churchman, C. W., 1968a, *Challenge to Reason*, McGraw Hill, New York.

Churchman, C. W., 1968b, *The Systems Approach*, Delacorte Press, New York.

Churchman, W. C., 1971, *The Design of Inquiring Systems*, Basic Books, New York.

Churchman, W. C., 1979, *The Systems Approach and Its Enemies*, Basic Books, New York.

Churchman, W. C., 1982, *Though and Wisdom*, Intersystems Publishers, Salinas, California.

Cook, F., 1979, *Planning, Designing, Problem Solving*, paper presented at the annual meeting of the American Educational Research Association, San Francisco.

Cronbach, L. J., and Suppes, P., 1969, *Research for Tomorrow's Schools: Disciplined Inquiry in Education*, Macmillan, New York.

Cross, N., 1984, *Developments in Design Methodology*, Wiley, New York.

Cross, N., 1990, The Nature and Nurture of Design Ability, *Design Studies*, 2(3):127–140.

Csanyi, V., 1982, *General Theory of Evolution*, Akademiai Kiado, Budapest.

Csanyi, V., 1989, *Evolutionary Systems and Society*, Duke University Press, Durham.

Csikszentmihalyi, M., 1991, *Flow: The Psychology of Optimal Experience*, Harper Perennial, New York.

Csikszentmihalyi, M., 1993, *The Evolving Self: A Psychology for the Third Millennium*, Harper Perennial, New York.

Curtis, R. K., 1982, *Evolution or Extinction*, Pergamon Press, New York.

Darke, J., 1984, The Primary Generator and the Design Process, in: *Development in Design Methodology* (N. Cross, ed.), Wiley, New York, pp. 175–188.

Davis, P., 1989, *The Cosmic Blueprint*, Touchstone, New York.

de Zeeuw, G., 1993, The Actor as a Perfect Citizen, in: *Systems Science* (F. Stowell, D. West, and J. Howell, eds.), Plenum, London, pp. 11–18.

Dils, J., 1995, Idealized Design of a Senior Support Group, *Progress*, 4(Summer):91–112.

Ditterle, T., Carr, A., Dolbeck, A., Frantz, T., Jenlink, P., Lieshoff, B., McArthur, I., Nelson, H., Paprock, K., and Rowland, G., 1994, Designing Conversation, in: *Summary Report*, Fifth Annual Asilomar Conversation on Systems Design, International Systems Institute, Carmel, California, pp. 29–34.

Drucker, P., 1989, *The New Realities*, Harper and Row, New York.

Esherick, J., 1963, Problems of the Design of a Design System, in: *Conference on Design Methods* (C. Jones and D. G. Thornley, eds.), Macmillan, New York, pp. 75–82.

Etzioni, A., 1991, *A Responsive Society*, Jossey-Bass, San Francisco.

Etzioni, A., 1993, *The Spirit of Community*, Crown, New York.

Flood, R. L., 1990, *Liberating Systems Theory*, Plenum, New York.

Flood, R. L., and Jackson, M., 1991, *Creative Problem Solving*, Wiley, New York.

Florian, M., 1975, On the Teaching of Model Building, in: *Education in Systems Science*, (B. A. Bayraktar, ed.), Addison-Wesley, New York, pp. 327–332.

Foerster, H., 1984, *Observing Systems*, Intersystems Publications, Salinas, California.

Follett, M. P., 1965, *The New State*, Peter Smith, Gloucester, Massachusetts.

Forrester, J., 1969, *Urban Dynamics*, MIT Press, Cambridge.

Frank, G., and Rehm, B., 1992, Human Resource Future Strategy, in: *Discovering Common Ground*, (M. R. Weisbord, ed.), Berret-Koehler, San Francisco, pp. 143–148.

Franklin, B., and Morley, J., 1992, Contextual Searching, in: *Discovering Common Ground*, (M. R. Weisbord, ed.), Berret-Koehler, San Francisco, pp. 229–246.

Frantz, T. G., and Miller, C., 1993, An Idealized Design Approach, *World Futures*, 36(2–4):83–106.

Gasparski, W., 1984, *Understanding Design: The Praxeological Systems Perspective*, Intersystems Publications, Salinas, California.

Glegg, G., 1971, *The Design of Design*, Columbia University Press, New York.

Gleick, J., 1987, *Chaos: Making a New Science*, Viking, New York.

Golerik, G., 1980, *Essays in Tektology*, Intersystems Publications, Salinas, California.

Gregory, S. A., 1963, *The Design Method*, Plenum, New York.

Hall, A., 1962, *A Methodology of Systems Engineering*, Van Nostrand, Princeton.

Hammer, M., and Champy, J., 1993, *Reengineering the Corporation*, Harper Collins, New York.

Hanneman, G. J., 1975, Models and Model Building in Communication Research, in: *Communication in Behavior* (G. J. Hanneman, ed.), Addison-Wesley, New York, pp. 421–436.

Harman, W., 1984, Offering an Answer to the Question: What Can I Do? *Newsletter, Institute of Noetic Sciences* XII(Spring):3,26.

Harman, W., 1988, *Global Mind Change*, Knowledge Systems, Indianapolis.

Harman, W., and Horman, J., 1990, *Creative Work*, Knowledge Systems, Indianapolis.

Harman, W., and Rheingold, H., 1984, *Higher Creativity*, Tarcher, Los Angeles.

Hartman, N., 1924, Diesseits von Idalismus und Realismus, *Kant-Studien*, 29:160–206.

Haugen, R., 1992, Adapting to Rapid Change, in: *Discovering Common Ground*, (M. R. Weisbord, ed.), Berret-Koehler, San Francisco, pp. 83–94.

Hawley, J., 1993, *Reawakening the Spirit in Work*, Berrett-Koehler, San Francisco.

Hiller, W., Musgrove, J., and O'Sullivan, P., 1984, Knowledge and Design, in: *Developments in Systems Methodology* (N. Cross, ed.), Wiley, New York, pp. 245–264.

Hood, P., and Hutchins, L., 1996, Research Based Development in Education, *Educational Technology*, XXXVI(1):6–13.

Horvath, B., 1983, The Organizational Theory of Workers' Management, in: *International Yearbook of Organizational Democracy* (C. Crouch and F. A. Heller, eds.), Wiley, New York, pp. 279–300.

Huber, G., 1986, The Nature of Design in Post-Industrial Organizations, in: *General Systems, XXIX* (R. Ragade, ed.), Society for General Systems Research, Louisville, pp. 81–104.

Huxley, A., 1945, *The Perennial Philosophy*, Harper Brothers, New York.

Isaacs, W., 1992, Dimensions of Generative Dialogue, in: *The Dialogue Project*, MIT, Cambridge, Massachusetts.

Isaacs, W., 1993, Dialogue, Collective Thinking, and Organizational Learning, *Organizational Dynamics*, Fall:24–39.

Jackson, M., 1992, *Systems Methodology for the Management Sciences*, Plenum, New York.

Jackson, M., 1995, Beyond the Fads: Systems Thinking for Managers, *Systems Research*, 12(1):25–42.

Jackson, M., and Keys, P., 1984, Towards a System of Systems Methodologies, *Journal of Operations Research*, 35:473.

Jantsch, E., 1975, *Design of Evolution*, Braziller, New York.

Jantsch, E., 1980, *The Self-Organizing Universe*, Pergamon Press, Oxford.

Jantsch, E., 1981, *The Evolutionary Vision*, Westview Press, Boulder.

Jantsch, E., and Waddington, C. H., 1976, *Evolution and Consciousness*, Addison-Wesley, Reading.

Jenks, L. C., and Amsler, M., 1991, Assessing the Adequacy of a Social Systems Design, in: *Comprehensive Systems Design*, (C. Reigeluth, B. H. Banathy, and J. Olson, eds.), Springer-Verlag, Berlin, pp. 134–144.

Jones, C. J., 1963, A Method of Systematic Design, in: *Conference on Design Methods* (C. J. Jones and D. Thorney, eds.), Macmillan, New York, pp. 53–74.

Jones, C. J., 1966, Design Methods Revisited, in: *The Design Method* (S. A. Gregory, ed.), Plenum, New York, pp. 295–310.

Jones, C. J., 1977, How My Thoughts about Design Methods Have Changed during the Years, *Design Methods and Theories*, (1):50–62.

Jones, C. J., 1980, *Design Methods: Seeds of Human Futures*, Wiley Interscience, New York.

Jones, C. J., 1984, *Essays in Design*, Wiley, New York.

Kampis, G., 1991, *Self-Modifying Systems*, Pergamon Press, Oxford.

Katzenbach, J. R., and Smith, D. K., 1993, *The Wisdom of Teams*, Harper Business, New York.

Kavanaugh, J., 1989, *Participatory Design of a Citizens Group for Social Empowerment*, unpublished Ph.D. dissertation.

Klir, G., 1981, Systems Methodology: From Youthful to Useful, in: *Applied Systems and Cybernetics* (G. Lasker, ed.), Pergamon Press, New York, pp. 931–938.

Koestler, A., 1964, *The Act of Creation*, Arkana, London.

Laszlo, E., 1972, *The Systems View of the World*, Braziller, New York.

Laszlo, E., 1987, *Evolution: A Grand Synthesis*, New Science Library, Boston.

Laszlo, E., 1995, *Systems Theory at the End of the Millennium: Revolution in Science and Transformation in Society*, presidential address, International Society for the Systems Sciences, Louisville.

Lawson, B. R., 1984, Cognitive Studies in Architectural Design, in: *Developments in Design Methodology* (N. Cross, ed.), Wiley, New York, pp. 209–220.

Lemming, J. S., 1994, Character Education and the Creation of Community, *The Responsive Community,* 4(Fall):49–58.

Lenford, G., and Mohrman, S., 1993, Self-Design for High Involvement: A Large Scale Organizational Change, *Human Relations,* 46(1):144–148.

Levin, P. H., 1966, The Design Process in Planning, *Town Planning Review,* 37:5–20.

Levin, P. H., 1984, Decision Making in Urban Design, in: *Developments in Design Methodology,* (N. Cross, ed.), Wiley, New York, pp. 107–121.

Linstone, H., 1984, *Multiple Perspectives for Decision Making,* North-Holland, New York.

Linstone, H., 1985, Multiple Perspectives, *Interfaces,* 15(4):77–85.

Linstone, H., and Mitroff, I., 1994, *The Challenge of the 21st Century,* State University Press, New York.

Lipnack, J., and Stamps, J., 1993, *The Teamnet Factor,* Wight Publications, Essex Junction, Massachusetts.

Lippitt, G. L., 1973, *Visualizing Change,* University Associates, La Jolla, California.

Lorenz, K., 1977, *Behind the Mirror: A Search for The Natural History of Human Knowledge,* Harcourt Brace Jovanovich, New York.

Luckman, J., 1984, An Approach to the Management of Design, in: *Developments in Design Methodology* (N. Cross, ed.), Wiley, New York, pp. 83–98.

Lynch, K., 1960, *The Image of the City,* MIT Press, Cambridge.

MacCallum, S., 1970, *The Art of Community,* Institute for Human Studies, Menlo Park, California.

MacIntyre, A., 1984, *After Virtue,* Notre Dame University Press, Notre Dame.

Majone, G., and Quade, E., 1980, *Pitfalls of Systems Analysis,* Wiley, New York.

Markley, O. W., 1976, Human Consciousness in Transformation, in: *Evolution and Consciousness* (E. Jantsch and C. H. Waddington, eds.), Addison-Wesley, Reading, pp. 213–229.

Markley, O. W., and Harman, W., 1982, *Changing Images of Man,* Pergamon Press, London.

Mason, R. O., and Mitroff, I., 1981, *Challenging Strategic Planning Assumptions,* Wiley, New York.

Matchett, E., 1968, Control of Thought in Creative Work, *Chartered Mechanical Engineer,* No. 15.

Mathur, K. S., 1978, The Problem of Technology: A Proposed Technology for Design Theories and Methods, *Design Methods and Theories,* 12(2):131–137.

Maynard, H., and Mehrtens, S., 1993, *The Fourth Wave,* Berrett-Koehler, San Francisco.

McGee, D., 1993, A Medical Practice as a Vehicle for Evolutionary Learning, *World Futures,* 36(2–4):129–140.

McWhinney, W., 1992, *Paths of Change,* Sage, Newbury Park.

Miller, J., 1978, *Living Systems,* McGraw-Hill, New York.

Mitroff, I., and Linstone, H., 1993, *The Unbounded Mind,* Oxford University Press, New York.

Montiori, A., 1989, *Toward a Theory Base for A Curriculum in Evolutionary Learning,* unpublished Ph.D. dissertation.

Montiori, A., 1993, *From Power to Partnership: Creating the Future,* Harper, San Francisco.

Moore, C. M., 1987, *Group Techniques for Idea Building,* Sage, Newbury Park.

Morgan, G., 1986, *Images of Organizations,* Sage, Newbury Park.

Morgan, G., 1989, *Creative Organizational Theory,* Sage, Newbury Park.

Morgan, G., 1993, *Imagination,* Sage, Newbury Park.

Nadler, G., 1967, *Work Systems Design: The Ideals Concept,* Richard D. Irwin, Homewood, Indiana.

Nadler, G., 1970, Engineering Research and Design in Socio-Economic Systems, in: *Emerging Methods in Environmental Design and Planning,* MIT Press, Cambridge, pp. 322–330.

Nadler, G., 1981, *The Planning and Design Approach,* Wiley, New York.

Nadler, G., and Hibino, S., 1990, *Breakthrough Thinking,* Prima Publishing, Rocklin, California.

Naisbitt, J., and Aburdene, P., 1990, *Megatrends 2000,* Morrow, New York.

Nelson, H., 1989, Unnatural States of Health in Natural Organizations, in: *The Well-Being of Organizations* (W. Churchman, ed.), Intersystems Publications, Salinas, California, pp. 353–366.

Newell, A., and Simon, H., 1972, *Human Problem Solving,* Prentice Hall, Englewood Cliffs.

Nystrom, C., Helberg, B., and Starbuck, W., 1976, Interactive Process as Organizational Design, in: *Management of Organizational Design* (R. Kilman, L. Pondy, and D. Slevin, eds.), North-Holland, Amsterdam.

Page, J. K., 1966, *Conference Report,* Ministry of Public Building and Works, London.

Papanek, V., 1972, *Design for the Real World,* Bantam Books, New York.

Pastorino, R. E., 1993, A "User"-Designed Mediation Approach, Fostering Evolutionary Consciousness and Competence, *World Futures,* 36(2–4):155–166.

Parness, S. J., 1972, Programming Creative Behavior, in: *Climate for Creativity,* (C. Taylor, ed.), Pergamon Press, New York, pp. 193–228.

Parness, S., Noller, S., and Bondi, A., 1971, *Guide to Creative Action,* Scribner, New York.

Pattakos, A. N., 1995, Searching for the Soul in Government, in: *Rediscovering the Soul of Business,* (B. DeFoore and J. Renesch. eds.), New Leaders Press, San Francisco. pp. 313–326.

Peccei, A., 1977, *The Human Quality,* Pergamon Press, Oxford.

Peck, S. M., 1993, *A World Waiting to be Born,* Bantam Books, New York.

Pinchot, G., and Pinchot, E., 1993, *The End of Bureaucracy and the Rise of the Intelligent Organization,* Berrett-Koehler, San Francisco.

Polak, E., 1973, *The Image of the Future,* Jossey-Bass, San Francisco.

Pollock, S., 1979, Innovation in Education: The Modelling Studio, in: *Education in Systems Science* (B. A. Bayraktar, ed.), Halsted Press, New York, pp. 211–225.

Popper, K. R., 1974, Autobiography of Karl Popper, in: *The Philosophy of Karl Popper* (P. A. Schilpp, ed.), Open Court, LaSalle, Illinois, pp. 3–181.

Prigogine, I., and Stengers, I., 1984, *Order Out of Chaos,* Bantam Books, New York.

Pruzan, P., 1994a, Report of the Ethics Group of the 1992 Asilomar Conversation, *Review of Administration and Informatics,* 6(2):47–56.

Pruzan, P., 1994b, The Renaissance of Ethics and the Ethical Accounting Statement, *Educational Technology,* 34(1):23–28.

Reich, R., 1991, *The Work of Nations,* Knopf, New York.

Reitman, W. R., 1964, Heuristic Decision Procedures, Open Constraints and Ill-Defined Problems, in: *Human Judgements and Optimality* (M. Shelley and G. Bryn, eds.), Wiley, New York, pp. 282–315.

Reswick, J. B., 1965, *Prospectus for Engineering Design Center,* Case Institute of Technology, Cleveland.

Rittel, H., 1984, Second Generation Design Methods, in: *Developments in Design Methodology,* (N. Cross, ed.), Wiley, New York, pp. 317–330.

Rittel, H., and Webber, M., 1984, Planning Problems are Wicked Problems, in: *Developments in Design Methodology* (N. Cross, ed.), Wiley, New York, pp. 135–144.

Rittel, H., 1985, Some Principles for the Design of Educational Systems, *DMG-DRS Journal,* 20(1):359–375.

Rogers, C. R., 1961, *On Becoming a Person,* Houghton Mifflin, Boston.

Rowland, G., 1992, Do You Play Jazz? *Performance Instruction,* November/December:19–25.

Rowland, G., 1994, Designing and Evaluating, *Educational Technology,* January:10–22.

Rowland, G., 1995, Archetypes of Systems Design, *Systems Practice,* 8(3):277–288.

Sage, A., 1977, *Methodology for Large-Scale Systems,* McGraw-Hill, New York.

Salk, J., 1983, *Anatomy of Reality: Merging of Intuition and Reason,* Columbia University Press, New York.

Sallstrom, P., 1992, The Possibility of the Impossible, *Cybernetics and Human Knowing,* 1:49–52.

Schein, E., 1993, On Dialogue, Culture, and Organizational Learning, *Organizational Dynamics,* Fall:40–51.

Schein, E., 1994, The Process of Dialogue: Creating Effective Communication, *The Systems Thinker,* 5(5):1–4.

Senge, P., 1990, *The Fifth Discipline,* Doubleday, New York.

Simon, H., 1969, *The Science of the Artificial,* MIT Press, Cambridge.

Simon, H., 1984, The Structure of Ill-Conceived Problems, in: *Developments in Design Methodology* (N. Cross, ed.), Wiley, New York, pp. 145–166.

Singer, A. A., 1948, *In Search of a Way of Life,* Columbia University Press, New York.

Singer, E. A., 1959, *Experience and Reflection,* University Pennsylvania Press, Philadelphia.

Slater, P., 1991, *A Dream Deferred,* Beacon Press, Boston.

Sless, D., 1978, A Definition of Design: Origination of Useful Systems, *Design Methods and Theories,* 12(2):123–130.

Smith, W. E., 1992, Planning for the Electricity Sector in Colombia, in: *Discovering Common Ground,* (M. R. Weisbord, ed.), Berrett-Koehler, San Francisco, pp. 171–186.

Sutherland, J., 1973, *A General Systems Philosophy for the Behavioral Sciences,* Braziller, New York.

Teilhard de Chardin, P., 1959, *The Phenomenon of Man,* Harper & Row, New York.

Theobold, R., 1987, *The Rapids of Change,* Knowledge Systems, Indianapolis.

Thomas, J. C., and Carroll, J. M., 1984, The Psychological Study of Design, in: *Developments in Design Methodology* (N. Cross, ed.), Wiley, New York, pp. 221–236.

Toffler, A., 1970, *Future Shock,* Bantam Books, New York.

Toffler, A., 1980, *The Third Wave,* Bantam Books, New York.

Toffler, A., and Toffler, H., 1995, *Creating a New Civilization,* Turner, Atlanta.

Toynbee, A., 1964, Is America Neglecting Her Creative Minority? in: *Widening Horizons in Creativity* (C. W. Taylor, ed.), Wiley, New York, pp. 3–9.

Tsivacou, I., 1990, An Evolution of Design Methodology, *Systems Practice,* 3(6):545–560.

Ulrich, W., 1983, *Critical Heuristics in Social Planning,* Haupt, Bern.

Ulrich, W., 1993, Some Difficulties with Ecological Thinking, Considered from a Critical Systems Perspective: A Plea for Critical Holism, *Systems Behavior,* (6):583–612.

Van der Ryn, S., 1966, Problem and Puzzles: Searching for a Science of Design, *AIA Journal,* 45(1):37–42.

Vickers, G., 1970, *Freedom in a Rocking Boat,* Basic Books, New York.

Vickers, G., 1981, Some Implications of Systems Thinking, in: *Systems Behavior* (Open University Systems Group, ed.), Harper & Row, London, pp. 19–25.

Vickers, G., 1982, Social Ethics, in: *Changing Images of Man* (O. W. Markley and W. Harman, eds.), Pergamon Press, London.

Vickers, G., 1983, *Human Systems Are Different,* Harper & Row, London.

Vitruvius, M. P., 1955, *D'Vitruvius on Architecture,* Harvard University Press, Boston.

Wailand, C., 1993, Evolutionary/Systemic Management of Organizations, *World Futures,* 36(2–4):141–154.

Wallace, G., 1926, *The Art of Thought,* Harcourt, New York.

Warfield, J., 1973, *An Assault on Complexity,* Battelle Memorial Institute, Columbus.

Warfield, J., 1976, *Societal Systems,* Wiley, New York.

Warfield, J., 1982, Organizations and Systems Learning, in: *General Systems XXVII* (R. Ragade, ed.), Society for General Systems Research, Louisville.

Warfield, J., 1987, Developing Design Culture in Higher Education, in: *Design Inquiry,* Intersystems Publishers, Salinas, California, pp. 173–177.

Warfield, J., 1990, *A Science of General Design,* Intersystems Publishers, Salinas, California.

Warfield, J., and Cardenas, R., 1994, *A Handbook of Interactive Management,* Iowa State University Press, Ames.

Webster, 1979, *New Collegiate Dictionary,* Merriam, Springfield.

Weick, K., 1979, *The Social Psychology of Organizing,* Addison-Wesley, Reading.

Weisbord, M., 1989, *Productive Workplaces: Organizing and Managing for Dignity, Meaning, and Community,* Berrett-Koehler, San Francisco.

Weisbord, M., 1992, Applied Common Sense, in: *Discovering Common Ground* (M. Weisbord, ed.), Berrett-Koehler, San Francisco.

Wheatley, M., 1992, *Leadership and the New Science,* Berrett-Koehler, San Francisco.

Whitehead, A. N., 1968, *Modes of Thought,* Free Press, New York.

Wilson, J. Q., 1993, *The Moral Sense,* The Free Press, New York.

Zeldin, T., 1994, *Intimate History of Humanity,* Harper Collins, New York.

Index

363